Birds
of the
Wadden Sea

*Final report of the section 'Birds'
of the Wadden Sea Working Group*

edited by
C.J.Smit
W.J.Wolff

A.A.Balkema / Rotterdam / 1981

Cover photograph: flock of waders - Jan van de Kam, Griendtsveen
Drawings: Michel Binsbergen, Texel; Rick Nichols, Texel; Hans Schoon-
          heden, Texel; birds: Jos Zwarts, Utrecht
Typing: Riny Wielinga, Texel

Report 6 of the Wadden Sea Working Group, ISBN 90 6191 056 0
Clothbound edition of all eleven reports, ISBN 90 6191 062 5

Printed in the Netherlands

TABLE OF CONTENTS

LIST OF CONTRIBUTORS AND MEMBERS OF THE SECTION "BIRDS"

OF THE WADDEN SEA WORKING GROUP

Dr. P.H. Becker	– Institut für Vogelforschung "Vogelwarte Helgoland", Wilhelmshaven, Federal Republic of Germany
Dr. G.C. Boere	– Staatsbosbeheer, Inspectie Natuurbehoud, Utrecht, The Netherlands
G. Busche	– Ornithologische Arbeitsgemeinschaft Schleswig-Holstein und Hamburg e.V., Heide, Federal Republic of Germany
Drs. B.S. Ebbinge	– Rijksinstituut voor Natuurbeheer, Leersum, The Netherlands
Cand. Mag. M. Fog	– Vildtbiologisk Station Kalø, Rønde, Denmark
Dr. F. Goethe	– Institut für Vogelforschung "Vogelwarte Helgoland", Wilhelmshaven, Federal Republic of Germany
Dipl.-Ing. G. Grosskopf	– Stade-Schölisch, Federal Republic of Germany
Dr. G.W. Harmsen	– Hilversum, The Netherlands
Drs. J.B. Hulscher	– Zoölogisch Laboratorium, Rijksuniversiteit Groningen, Haren, The Netherlands
Dr. D. König	– Kronshagen bei Kiel, Federal Republic of Germany
H. Meltofte	– Dansk Ornithologisk Forening, København, Denmark
Dipl.-Biol. P. Prokosch	– Institut für Haustierkunde, Universität Kiel, Federal Republic of Germany
Drs. J. Rooth	– Rijksinstituut voor Natuurbeheer, Leersum, The Netherlands
Drs. C.J. Smit	– Rijksinstituut voor Natuurbeheer, Texel, The Netherlands
C. Swennen	– Nederlands Instituut voor Onderzoek der Zee, Texel, The Netherlands
Dr. J. Verwey	– Nederlands Instituut voor Onderzoek der Zee, Texel, The Netherlands
Dr. W.J. Wolff	– Rijksinstituut voor Natuurbeheer, Texel, The Netherlands
Drs. L. Zwarts	– Rijksdienst voor de IJsselmeerpolders, Leeuwarden, The Netherlands

1 CONCLUSIONS AND RECOMMENDATIONS

For about 50 species of ducks, geese, waders, gulls and terns the Wadden Sea is of vital importance because the total population, or at least considerable parts are dependent on the area for at least a part of the year.

The Wadden Sea is highly important for birds breeding in an area that reaches from Ellesmere Island in Canada in the west to the Taymyr peninsula in Central Siberia in the east.

The few existing simultaneous bird counts show that in late summer when maximum numbers are present, over 3 million birds may be counted in the Wadden Sea area. The number of birds actually using the Wadden Sea will probably be 2-3 times higher.

Although large numbers of local bird counts are already available, the picture for the Wadden Sea as a whole still needs improvement. Simultaneous counts of the whole area are strongly recommended. Coverage of especially the Niedersachsen area has to be improved. Further research into the accuracy of counts of very large flocks (over 10,000 birds) is needed.

For the 32 most important estuarine bird species of the Wadden Sea data on distribution, annual cycle, numbers, and food relationships are given in separate chapters.

Of the very small NW European Spoonbill *(Platalea leucorodia)* population 20-30% breeds and feeds in the Wadden Sea area, whereas another 20% probably feeds during part of the year in the Wadden Sea.

Of the world population of the Barnacle goose *(Branta leucopsis)* about 60-70% is dependent on the Wadden Sea.

The whole world population of the Dark-bellied Brent goose *(Branta b. bernicla)* visits and is largely dependent on the Wadden Sea. The very small Spitsbergen population of the White-bellied Brent goose *(B. b. hrota)* also visits the Wadden Sea.

Nearly the whole NW European Shelduck *(Tadorna tadorna)* population uses the Wadden Sea as a moulting area and for a large part also as a wintering area.

Over 30% of the NW European population of Wigeon *(Anas penelope)* visits the Wadden Sea area.

In autumn about half of the NW European population of Teal *(Anas crecca)* may be counted in the Wadden Sea area.

10-20% of the NW European population of Mallard *(Anas platyrhynchos)* visits the Wadden Sea area.

About 40% of the NW European population of Pintail *(Anas acuta)* may be present in the Wadden Sea area at the same time.

Although only a few per cent of the NW European population of the Eider *(Somateria mollissima)* breeds in the Wadden Sea area, this bird

is nevertheless the main avian predator of bottom fauna. This is due
to its heavy weight and to the influx of birds from the Baltic in winter. About 20-40% of all Baltic Eiders stay in the Wadden Sea in winter.

15-20% of the NW European population of the Red-breasted merganser
(Mergus serrator) winters in the Wadden Sea.

Possibly about 30-40% of the Goosander *(Mergus merganser)* wintering
in NW Europe does so in the Wadden Sea.

At least 80-90% of the Oystercatchers *(Haematopus ostralegus)* breeding in NW Europe visits the Wadden Sea area. Next to the Dunlin *(Calidris alpina)* the Oystercatcher is the most numerous bird species of
the Wadden Sea.

Of the isolated NW European population of the Avocet *(Recurvirostra
avosetta)* about 50% breeds in the Wadden Sea area. In autumn probably
nearly all NW European Avocets stay in the Wadden Sea.

It is not well-known which percentages of the NW European populations of Ringed plover *(Charadrius hiaticula)*, Kentish plover *(Charadrius alexandrinus)* and Grey plover *(Pluvialis squaterola)* visit the
Wadden Sea. Tens of thousands of birds, a substantial part of the Eurasian population, are involved, however.

The very large numbers of Knots *(Calidris canutus)* in the Wadden
Sea belong to the breeding populations of NE Canada and Greenland and
probably also of N Asia. Probably at least 50% of the birds belonging
to the nominate subspecies and maybe nearly 100% visit the Wadden Sea.

Many thousands of Sanderlings *(Calidris alba)* and Curlew sandpipers
(Calidris ferruginea) visit the Wadden Sea area, but it is unknown
what percentages of the arctic breeding populations they constitute.

At least 50% of the NW European population of Dunlin *(Calidris alpina)* visits the Wadden Sea, whilst also being the most numerous bird
species there.

At least 40% but probably much more of the NW European and NW Siberian population of Bar-tailed godwit *(Limosa lapponica)* visits the
Wadden Sea.

At least 50% of the population of Curlews *(Numenius arquata)* in NW-
Europe visit the Wadden Sea.

Many thousands of Spotted redshanks *(Tringa erythropus)* and Greenshanks *(Tringa nebularia)* migrate through the Wadden Sea area, but it
is unknown what percentage of the total populations this constitutes.

About 10-20% of the Icelandic and NW European population of Redshanks *(Tringa totanus)* may be counted at one day in the Wadden Sea
area.

Many thousands of Turnstones *(Arenaria interpres)* occur in the Wadden Sea area, but it is unknown what percentage of the total population they constitute.

Nearly 100,000 pairs of Black-headed gulls *(Larus ridibundus)* breed
in the Wadden Sea area. Comparable numbers feed in the area outside
the breeding season. It is not exactly known which part of the European breeding population visits the area.

About 4000 pairs of Common gulls *(Larus canus)* breed in the Wadden
Sea area but over 100,000 may occur in the area outside the breeding
season. This constitutes about 10-20% of the total European population.

Nearly 70,000 pairs of Herring gulls *(Larus argentatus)* breed in the Wadden Sea area, constituting about 10% of the NW European breeding population. A large percentage of the breeding population remains in the Wadden Sea all year round, although many feed on refuse dumps and fishery discards.

Before 1958 over 50% of the NW European population of the Sandwich tern *(Sterna sandvicensis)* bred in the Wadden Sea area. This percentage declined strongly due to mortality induced by temporary pollution around 1965, but since then it has increased to about 30% again. Nearly 10,000 pairs of Sandwich terns breed in the Wadden Sea nowadays.

About 10,000 pairs or 25% of the NW European population of the Common tern *(Sterna hirundo)* breed in the Wadden Sea nowadays. Before 1954 in the Dutch part of the area alone 30,000-35,000 pairs bred, the decline most probably being mainly due to chemical pollution.

About 400-700 pairs of Little terns *(Sterna albifrons)* breed in the Wadden Sea area, constituting about 15% of the total NW European population.

The distribution of feeding waders of a particular species over the tidal flats depends on food density and the distance from the feeding area to the roost, high densities and short distances being preferred. The total number of birds also seems important, since at higher densities the birds tend to spread out over the tidal flats more evenly. It is likely that bird densities are limited by competition for space and thus for food. It has not yet been proven, however, that this competition also limits the size of the populations. Further research into this problem is strongly recommended.

The available data suggest resource partitioning between the different bird species in the Wadden Sea with regard to space, time and food. On the other hand it has been found that different bird species influence each others distribution over the tidal flats and therefore may be in competition. Further research into this problem is recommended.

In the entire Wadden Sea probably about 2.5-3.4 million birds occur at the same time maximally. Waders comprise 1.7-2.3 million specimens. The maximum numbers of birds in the Wadden Sea area in Denmark, Schleswig-Holstein, Niedersachsen and the Netherlands are estimated at 0.3-0.6, 1.1-1.3, 0.3-0.5, and 0.8-1.0 million, respectively. These figures need support from further counts. No other estuarine area along the E Atlantic coasts of Europe and N Africa shows bird numbers as high as those in the Wadden Sea. The peak occurrences in the Wadden Sea occur in autumn and spring; in winter bird numbers are lower, although the Wadden Sea still is the most important wintering area for estuarine birds in Europe. Only the Banc d'Arguin in Mauritania, NW Africa, has higher numbers of birds in mid-winter.

For different species the Wadden Sea fulfills different functions. The birds may use the Wadden Sea area as a breeding ground, a moulting area, a wintering ground and/or as an area where prior to breeding a large reserve of energy may be built up, for example.

Bird are estimated to take on average about 4 g dry organic matter $.m^{-2}.year^{-1}$ as food from the Dutch Wadden Sea, nearly completely in the form of larger benthic invertebrates. Together with other predators about 13-14 g dry organic matter $.m^{-2}.year^{-1}$ are taken from the benthic fauna. Since about 15-20 g dry organic matter $.m^{-2}$ become annually available as potential food, this forms an indication that food might be limiting bird numbers. Therefore a profound study of the carrying capacity of the Wadden Sea for birds is strongly recommended.

Any factor (potentially) negatively influencing the size of the bird populations dependent on the Wadden Sea, is defined as a threat. As major threats to the birds of the Wadden Sea are considered reclamations for agriculture, embankment for other purposes, pollution (especially by chlorinated hydrocarbons), introduction of terrestrial predators, like rats, disturbance through several causes, hunting and oil pollution. As possible threats may be listed introductions of exotic plant and animal species, fisheries and changing agricultural management of breeding places. Opinions differ on the effect of predation by Herring gulls: in the Netherlands no effects on the population sizes of other bird species are observed, but in Germany it is claimed that Herring gulls cause considerable damage.

Quantitative research of the effects of the threats and possible threats listed is strongly recommended.

2 INTRODUCTION

C.J. Smit

One of the most obvious phenomena in the Wadden Sea is its large numbers of birds. Even people unfamiliar with ornithology are soon impres-
sed by the immense flocks of birds that can be observed. For about 50
species of ducks, geese, waders, gulls and terns the Wadden Sea is of
vital importance because total or at least considerable proportions
of populations are dependent on the area for at least a part of the
year. The Wadden Sea is highly important for birds from an area that
reaches from Ellesmere Island in Canada in the west to the Taymyr peninsula in Central Siberia in the east (fig. 1).

This immense area where the birds originate from is about 100-125
times larger than the Danish, German and Dutch Wadden Sea area (fig.
2: page 6/10 and 6/11) as a whole.

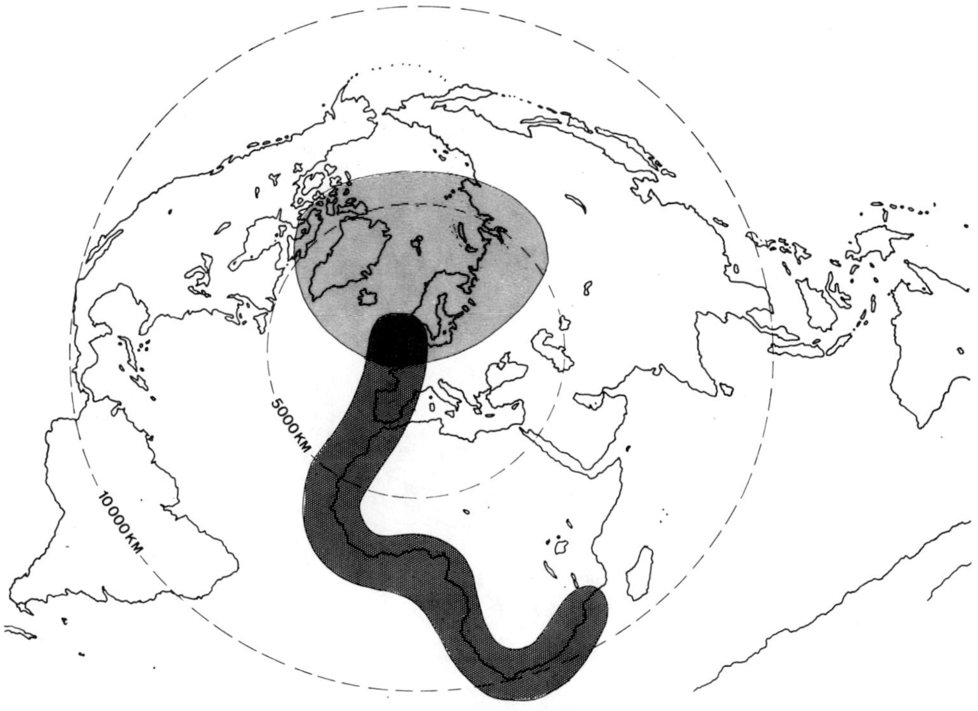

Fig. 1. Breeding- (light shaded) and wintering- (dark shaded) areas
of birds passing through the Wadden Sea.

Simultaneous bird counts show that in late summer when maximum numbers are present more than 3 million birds may be counted in the Wadden Sea in these three countries. The number that actually uses the Wadden Sea will probably be 2-3 times higher because there is a continuous flow of birds through the area (chapter 5).

For some species the Wadden Sea is important as a wintering area, for others as a moulting area, again for others as an area to stay for some time to build up energy resources after a long and energy consuming flight, sometimes for several thousands of kilometers. The area has the possibility to feed all these birds because of a relatively high production of benthic prey animals and fish (Dankers, Wolff & Zijlstra, 1979; Dankers, Kühl & Wolff, 1981).

This report firstly tries to quantify the occurrence of birds: to show the importance of the Wadden Sea in their annual cycle, and to compare the numbers in the Wadden Sea with those of other estuarine areas. Second goal of the report is to give some information on the interrelationships between species, and the relations between the occurrence of birds and the biotic and abiotic environment. Finally the report tries to give an account of threats to the bird populations of the Wadden Sea.

Fig. 2. Survey of the Wadden Sea area.

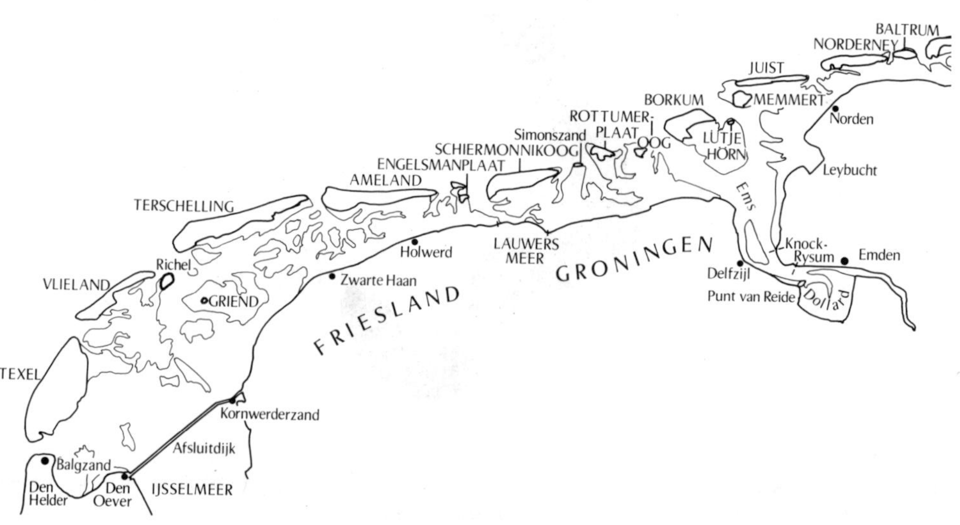

Varde
Blåvands Huk
Ho Bugt
Skallingen
LANGLI
Esbjerg
FANØ
Ribe
Kilsand
MANDØ
Koresand
RØMØ
Ballum
JORDSAND
SYLT
Højer
Rodenäs
SCHLESWIG-
FÖHR
OLAND
Hauke Haienkoog
AMRUM
GRÖDE
HABEL
LANGENESS
HAMBURGER HALLIG
Japsand
HOOGE
NORDSTRANDISCHMOOR
Norderoogsand
PELLWORM
Dithmarscher Bucht
NORDEROOG
NORDSTRAND
SÜDFALL
Husum
Süderoogsand
SÜDEROOG
EIDERSTEDT
St.Peter
Eider
HOLSTEIN
BLAUORT
TERTIUS
Meldorfer Bucht
HELMSAND
TRISCHEN
SCHARHÖRN
Elbe
NEUWERK
KNECHTSAND
Cuxhaven
Hullen
SPIEKEROOG
WANGEROOGE
LANGEOOG
OLDEOOG
BALTRUM
MELLUM
NORDERNEY
Neuharlingersiel
JUIST
Westeraccumersiel
Jade
Weser
BORKUM
MEMMERT
Norden
Hooksiel
LÜTJE
HÖRN
Leybucht
OSTFRIESLAND
Wilhelmshaven
Bremerhaven
Ems
Jadebusen
Knock-
Rysum
Emden
NIEDERSACHSEN
Delfzijl
Punt van Reide
Dollard

Birds living in estuaries are generally counted at the high tide roosts, places where they gather into groups when the tidal flats are flooded and foraging becomes impossible. The size of these groups varies from only a few birds to several thousands, sometimes even tens of thousands. By counting birds on the roosts generally a rather complete impression can be obtained of the numbers foraging on the nearby tidal flats, at least if all census areas are covered. This holds especially for waders, but less so for gulls, ducks and geese as sometimes large numbers of these birds stay on the water during high tide and remain out of sight from the shore. The only possibility to determine their number is counting them from plane and/or boat.

Observers carrying out bird counts may have to face many problems (Boere & Zegers, 1976). Investigations on the accuracy of bird counts however showed that the results of observers counting flocks of birds were rather consistent. Because random fluctuations in counting results neutralize one another, the results of large scale counts, in which the results of many observers and areas are combined, become even more consistent. The number of birds counted in a sitting flock however usually tends to be an underestimation of the number present (Kersten et al., in press).

In the Danish Wadden Sea, bird counts were being carried out regularly already as early as 1930-1932 when the birds were counted daily on the island of Jordsand. Counts of geese in the Danish Wadden Sea have taken place almost annually since 1961 and of ducks especially during 1965-1973. Large scale counts of waders in the Danish Wadden Sea were carried out for the first time in 1969. From 1973-1977 they frequency increased and in 1978 counts were carried out in all months, both by plane as well as from the ground.

From 1965 to 1970 in Schleswig-Holstein counts have been carried out in all months of the year, although a complete coverage of the whole area could not always be achieved. After 1970 however the frequency of counts decreased considerably. In 1979 a revival of activities took place.

In Niedersachsen where regular counts have been carried out in several coastal areas, only few complete counts of the whole area exist. In nature reserves on some of the Wadden Sea islands daily counts have been carried out by the wardens in spring and summer months.

In the Dutch Wadden Sea area the first attempt to count the birds in the whole area already has been made in 1931 (Van Oordt, 1932). The first attempts to count the birds in parts of the area became more or less regular in the fifties of this century. The counting areas were generally restricted to a certain island or part of the mainland coast. In 1963 the Dutch Wadden Sea as a whole was counted again and for the first time an overall picture of the numbers in the whole area could be developed. From 1972-1979 about 15 more or less complete counts followed.

The data from all counts, published as well as unpublished, have been collected as far as possible for 32 species (Smit, 1977). This compilation forms the basis for most of the information on numbers presented in this report.

For the selection of the species mentioned in this report the following criteria have been used:
1) a substantial part of the NW European flyway population of a species depends on the Wadden Sea;
2) such densities of a species occur locally that the birds may influence prey densities, at least during a part of the year.

References

Boere, G.C. & P.M. Zegers, 1976. Het hoe en waarom van wadvogeltellingen. Waddenbulletin 11: p. 20-31.
Dankers, N., H. Kühl & W.J. Wolff, 1981. Invertebrates of the Wadden Sea. Final report of the section "Marine Zoology" of the Wadden Sea Working Group. Balkema, Rotterdam.
Dankers, N., W.J. Wolff & J.J. Zijlstra, 1980. Fishes and fisheries of the Wadden Sea. Final report of the section "Fishes and fisheries" of the Wadden Sea Working Group. Balkema, Rotterdam: 157 pp.
Kersten, M., K. Rappoldt & C. Smit. Over de nauwkeurigheid van wadvogeltellingen. Limosa, in press.
Oordt, G.J. van, 1932. Verslag van een tocht met Hr. Ms. Watervliegtuig L 7 boven de Nederlandse Waddenzee op 5 October 1931. Eerste jaarverslag der Stichting "Vogeltrekstation Texel".
Smit, C.J., 1977. On the occurrence of 32 bird species in the Danish, German and Dutch Wadden Sea. Unpublished report Wadden Sea Working Group (3 parts): 445 pp.

3 ECOLOGICAL DATA ON BIRD SPECIES OF THE WADDEN SEA

3.1 INTRODUCTION
C.J. Smit & W.J. Wolff

In the following chapters an ecological portrait will be sketched
for the 32 most important estuarine bird species occurring in the
Wadden Sea area. These portraits are based on four areas of knowledge, viz. on distribution, annual cycle, numbers and food relationships.

The distribution of each species is described in terms of breeding area, migration routes, wintering areas and moulting areas.

Data on migration patterns and periods as well as on moult and
weight changes are given to describe the annual cycle.

Under the heading "Numbers" the population size is estimated and
the numbers of individuals in different parts of the Wadden Sea area
are given.

The chapter on food gives information on the composition and the
amount of food taken as well as on the methods employed to obtain
the food.

Finally the most important literature references are given.

3.2 SPOONBILL *(PLATALEA LEUCORODIA L.)*
J. Rooth

Da: Skestorken; G: Löffler; Du: Lepelaar

3.2.1 Distribution

3.2.1.1 Breeding area
The distribution is discontinuous palaearctic, oriental and ethiopian. The Spoonbill is a colonial breeding bird from boreal to tropical climate zones (Voous, 1960 and fig. 3). Formerly its occurrence
was scattered over Europe and eastwards. Especially in Western Europe
it is nowadays much reduced due to drainage of nesting and feeding
areas. In earlier years human exploitation of eggs and nestlings for
food also reduced numbers (Brouwer, 1964). In Western Europe the species is only breeding in Southern Spain (Guadalquivir - Cota Doñana)
and in the Netherlands. Until recently it was breeding in Denmark.
During the breeding season it is frequently seen in East Anglia, England.

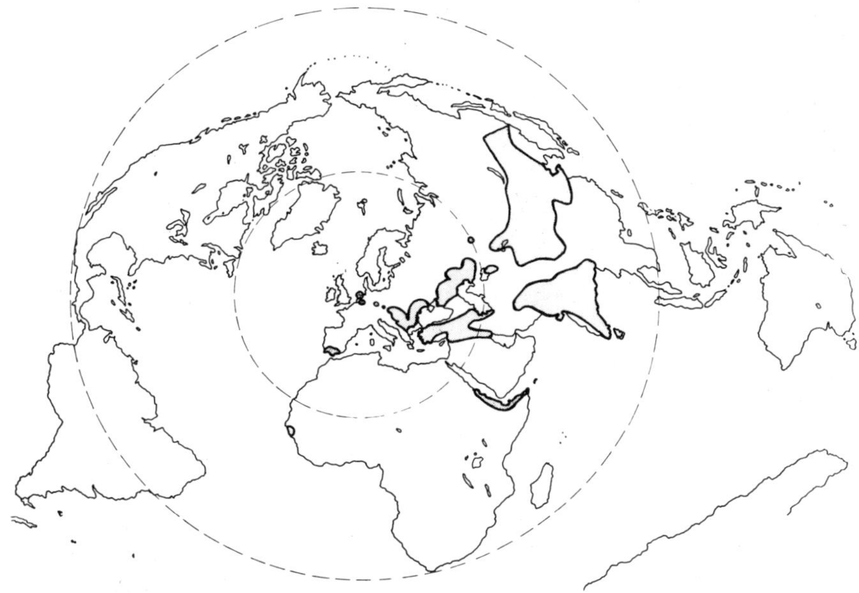

Fig. 3. Breeding area of the Spoonbill (after Voous, 1960).

3.2.1.2 Migration routes

Dutch breeding birds migrate along the Atlantic coast and more inland in France and Spain. They stay in Morocco for some time and are wintering in West Africa. Some more westward recoveries originate from the Azores, Madeira, Canary islands and Cape Verde islands (fig. 4).

3.2.1.3 Wintering areas

A small part of the population is wintering in Southern Spain and Morocco. The main part winters in Mauritania and Senegal, where respectively the Banc d'Arguin and the Delta of the Senegal river belong to the main wintering areas.

3.2.1.4 Moulting areas

The moult of wing feathers starts in August and September when breeding birds from the Wadden Sea area wander around. Probably moult is interrupted during migration and completed in the winter quarters.

3.2.2 Annual cycle

3.2.2.1 Migration

After the breeding season in July and August Spoonbills disperse. They may be seen in the whole Wadden Sea area then (see 3.2.3.2), but also wander to the Central and Southwestern parts of the Netherlands, some even to SE England. The main part of the population leaves the

Fig. 4. Recoveries of Spoonbills banded as pullus on the Dutch Wadden Sea islands. (After unpublished data Vogeltrekstation, Arnhem).

Netherlands in September, some in October. Winter observations are exceptional. During migration Spoonbills visit several coastal and inland marshes in France, Spain and Morocco, where they occur together with the Spanish breeding population. During winter months they stay in Mauritania and Senegal, from where migration to the north already starts in the second half of January, possibly initiated by low water levels in the Senegal delta in this time of the year (Poorter, 1978). Spoonbills arrive in the Wadden Sea by the end of February but later arrivals do occur as well. Egg-laying and breeding has been observed in March, April, May, June, July and August. Second and third calendar-year birds stay in NW Africa and Spain but fourth and fifth year birds are summering in the Netherlands, some of them in England (Poorter, 1978).

3.2.2.2. Moult

A complete post-breeding moult occurs. The wings start with inner primaries in August-September. Probably moult is interrupted during migration and finished in winter quarters. A partial pre-breeding moult takes place in January-March (Cramp & Simmons, 1977).

3.2.2.3 Weight chances

No information available.

3.2.3 Numbers

3.2.3.1 Population size

Nowadays breeding colonies occur in "de Geul" and "de Muy" on the island of Texel and on the "Boschplaat" on the island of Terschelling. There is another breeding colony in the "Zwanenwater" on the Dutch mainland, but close to the Wadden Sea. These birds feed in the Wadden Sea as well.

In this century the "Muy" colony on Texel was the first settlement of Spoonbills in the Wadden Sea area. After several unsuccessful attempts in 1904, and from 1921-1931 in 1933 5 pairs succeeded in rearing young. From 1934 their number increased rapidly. In 1938 there were at least 110 nests and this was about the average for the next ten years. In 1953 and 1954 about 150 pairs were breeding. This number decreased to about 80 pairs in the early sixties, which was followed by a more serious decline caused by chlorinated hydrocarbons (Koeman, 1971; Rooth & Jonkers, 1972). The minimum was 16 pairs in 1968. This number increased slowly during the next decade to 25-30 pairs and reached a maximum of 34 pairs in 1979. Another colony on the island of Texel started in "de Geul" in 1953 with one pair. During the last 25 years the number of breeding pairs fluctuated between 5 and 15 pairs.

In 1962 one pair bred on the "Boschplaat" on the island of Terschelling in a dry and sandy dune area, which harbours a very large Herring gull colony. In 1963 8 pairs were breeding in this area and since then the number fluctuates between 7 and 15 pairs.

The number of breeding pairs in the Wadden Sea area generally amounts to about 20-30% of the total breeding population in the Netherlands (250 pairs in 1979). The development of the breeding population on the Dutch Wadden Sea islands is shown in fig. 5.

An occasional breeding attempt took place in Germany on the island of Memmert in 1962 (Pundt & Ringleben, 1963).

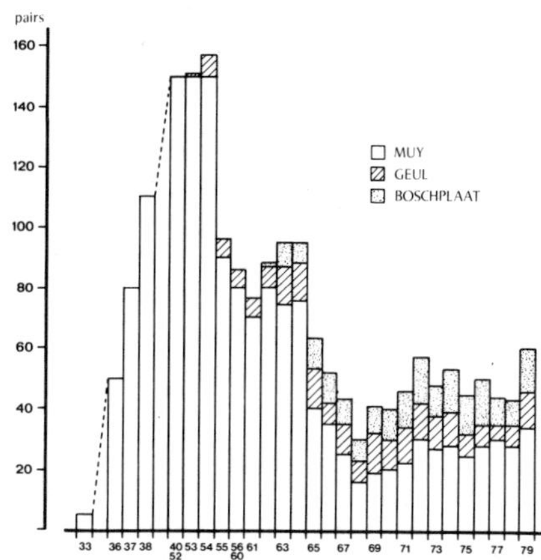

Fig. 5. Development of the Spoonbill population breeding on the Dutch Wadden Sea islands Texel (Muy and Geul nature reserves) and Terschelling (Boschplaat nature reserve) (after Dijksen & Dijksen (1977) and Rooth, unpubl. data)).

3.2.3.2 Numbers per area

During the breeding season Spoonbills frequently visit ditches in the polder areas of the islands of Texel and Terschelling. On some occasions the distance between colonies and feeding areas may be up to 15 km. There are observations that birds from the Terschelling colony migrate to the polder of the island of Ameland, which is as near as the polder of Terschelling itself. Later in the breeding season and in summer Spoonbills mostly feed in the Wadden Sea in shallow water. In July and August flocks of some tens to a hundred birds can be observed in the Dutch Wadden Sea area, especially in the Lauwersmeer, near the islands of Terschelling, Vlieland and Texel and at the Balgzand. Smaller numbers may occur in the German and Danish part of the Wadden Sea (Kortegaard, 1973; Berndt & Drenckhahn, 1974; Goethe et al., 1978).

Between 1953 and 1963 flocks of Spoonbills were after the breeding season regularly present on Memmert. The maximum number that has been recorded here in 1956 was 34 individuals. Sometimes 30% of these flocks consisted of first calender-year birds (Goethe, in litt.). Probably these birds originated from Dutch colonies. The dessication of the brackish foraging pool at Memmert and perhaps the decrease of the Dutch population, resulted in the absence of regular presence after 1963.

3.2.4 Food

3.2.4.1 Food composition

The main food in spring consists of Sticklebacks (*Gasterosteus aculeatus*) which migrate in that period from the sea to brackish or fresh inland water. The great amount of shallow ditches in the Netherlands plays an important role in providing Sticklebacks and other small fish as well as aquatic insects and their larvae. During spring the Spoonbills in the Wadden Sea area frequently visit areas with ditches on Texel and Terschelling. Later in the breeding season and in summer Spoonbills mostly feed in the Wadden Sea in shallow water where they catch crustaceans like *Crangon crangon* and small fish.

3.2.4.3 Feeding activities

Spoonbills feed by day and night. They feed solitary as well as socially. The prey is mainly found tactily during the characteristic sweeping of the bill from side to side.

3.2.4.4 Total food consumption

Poorter (pers. comm.) estimates the daily intake per adult as well as per growing juvenile at 500 g wet weight.

References

Berndt, R.K. & D. Drenckhahn (ed.), 1974. Vogelwelt Schleswig-Holsteins, Vol. 1. Ornithol. Arbeitsgemeinschaft Schleswig-Holstein und Hamburg e.V., Kiel: p. 188.

Brouwer, G.A., 1964. Some data on the status of the Spoonbill, Platalea leucorodia L., in Europe, especially in the Netherlands. Zoölogische Mededelingen 39: p. 481-521.
Cramp, S. & K.E.L. Simmons (eds.), 1977. Handbook of the birds of Europe, the Middle East and North Africa, Vol. 1. Oxford University Press, Oxford: p. 352-357.
Dijksen, A.J. & L.J. Dijksen, 1977. Texel, Vogeleiland. Thieme, Zutphen: p. 55-57.
Goethe, F., H. Heckenroth & H. Schumann, 1978. Die Vögel Niedersachsens und des Landes Bremen. Naturschutz und Landschaftspflege Niedersachsen, Hannover. Sonderreihe B, Vol. 2.1: p. 92-93.
Koeman, J.H., 1971. Het voorkomen en de toxicologische betekenis van enkele chloorkoolwaterstoffen aan de Nederlandse kust in de periode van 1965-1970. Thesis, University Utrecht: p. p. 103-110.
Kortegaard, L., 1973. Skestorken Platalea leucorodia i Danmark 1900-1971. Dansk Orn. Foren. Tidsskrift 67: p. 3-14.
Poorter, E.P.R., 1978. Population ecology and conservation of West European Platalea leucorodia. IWRB-symposium Conservation of colonially nesting waterfowl,Carthage, Tunisia, November 1978.
Pundt, G. & H. Ringleben, 1963. Der Löffler (Platalea leucorodia) 1962 erstmals deutscher Brutvogel auf der Insel Memmert. J. Orn. 104: p. 97-100.
Rooth, J. & D.A. Jonkers, 1972. The status of some piscivorous birds in the Netherlands. TNO-nieuws 27: p. 552-555.
Voous, K.H., 1960. Atlas of European birds. Nelson, London: p. 20-21.

3.3 BARNACLE GOOSE *(BRANTA LEUCOPSIS* (Bechstein)*)*
Barwolt Ebbinge

Da: Bramgaas; G: Nonnengans, Weisswangengans; Du: Brandgans

3.3.1 Distribution

3.3.1.1 Breeding area

The Barnacle goose is a West palaearctic species breeding only in the tundra climatic zone. It is one of the very few species of birds endemic to the Northeast Atlantic region (Voous, 1960). It breeds on the steep rocky tundra, on cliffs along fjords and rather often on small islands off the coast hardly accessible for Polar foxes.

There are no subspecies, but ringing results have shown the existence of three separate populations: one breeding along the East coast of Greenland, another breeding on Spitsbergen, and a third one, the so-called Barents Sea population, breeding on Novaya Zemlya and Vaygach (Boyd, 1961, 1968).

3.3.1.2 Migration routes

The migration routes follow the coasts as much as possible, but when necessary short-cuts are made over land, e.g. from the White Sea

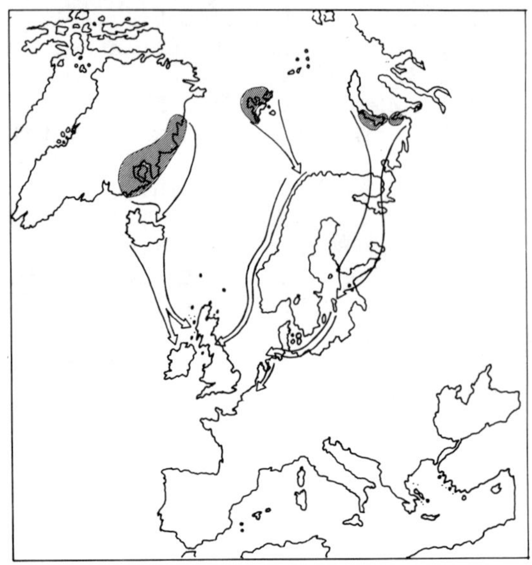

Fig. 6. Breeding areas and migration routes of the three Barnacle goose populations. (after Owen, 1977).

to the Gulf of Finland and across Schleswig-Holstein (fig. 6).

3.3.1.3 Wintering areas

The Greenland population winters along the west coasts of Ireland and Scotland (mean temperature December-February + 4.4°C). The Spitsbergen population winters along the Solway Firth, SW Scotland (mean temperature December-February + 4.5°C). The Barents Sea population winters from the West coast of Schleswig-Holstein (mean temperature December-February + 1.4°C) till the Delta area, SW Netherlands (mean temperature December-February + 3.1°C). In extremely cold winters many geese from this latter population may spend part of the winter even in Northern France (Roux, 1963; Timmerman et al., 1976).

3.3.1.4 Moulting areas

Moult partly takes place in the breeding areas and partly in the wintering areas (for details see section 3.3.2.2).

3.3.2 Annual cycle

3.3.2.1 Migration

In autumn the birds belonging to the Barents Sea population fly almost directly from the Barents Sea to the Wadden Sea, passing the East coast of Schleswig-Holstein between the 8th of October and the 2nd of November (data over the period 1952-1972; after Schmidt, 1973).

Till the end of November this population is concentrated entirely in the Wadden Sea area: along the West coast of Schleswig-Holstein and in the Lauwersmeer-area (Busche, 1977; Ebbinge et al., 1975) (fig. 7).

Fig. 7. Haunts of Barnacle geese
in the Wadden Sea area
(data: Busche, Ebbinge
& Hummel, unpubl.).

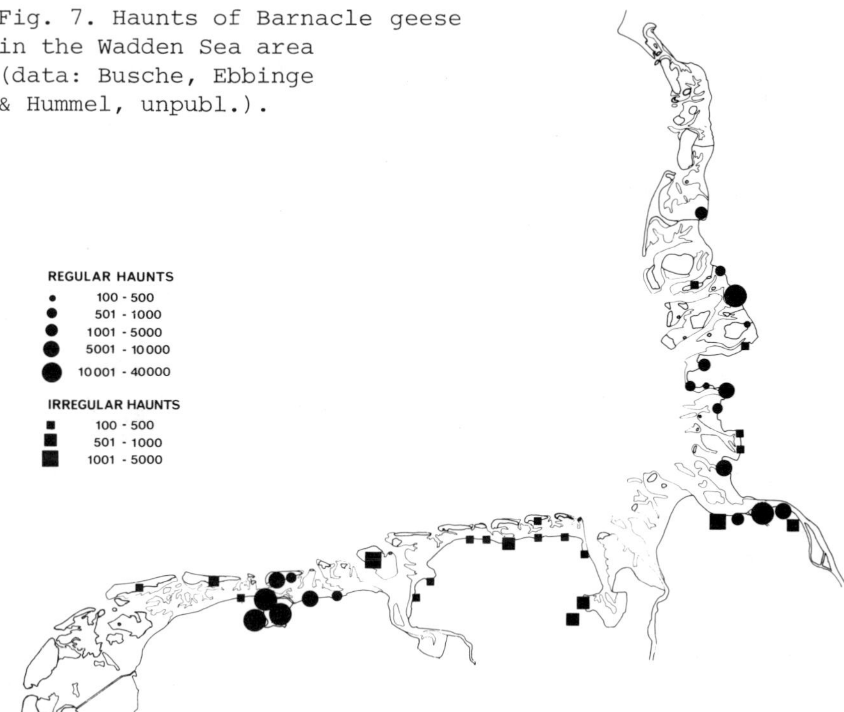

During the winter months (December-February) part of the population remains in the Wadden Sea area, while others move further south
to the Central and Southern parts of the Dutch province of Friesland
and to the Delta area.

In March the majority of the population is present along the west
coast of Schleswig-Holstein, but even until mid-April several thousands of birds can be seen as far south as the Delta area.

From the end of March onwards large numbers leave for the spring
staging areas in the Baltic (Gotland and Estonia). By the end of
April almost the entire population is concentrated in this area, where
it stays for several weeks before leaving for their breeding grounds
in the end of May (Kumari, 1971).

3.3.2.2 Moult

A complete moult of flight feathers occurs in summer in the breeding areas. From mid-August until October-November moult of head-, bo-
dy- and tail feathers takes place (Cramp & Simmons, 1977). Of birds
caught in late November in the Dutch Wadden Sea area 20% were still
moulting their tail feathers (Ebbinge, unpubl.).

3.3.2.3 Weight changes

Weights from the Barents Sea population are presented in fig. 8.
Especially in January there is a marked loss of weight in both adult
males and females.

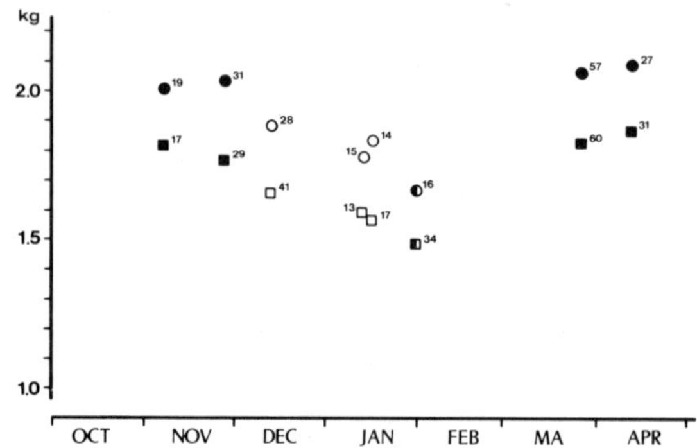

Fig. 8. Mean weights of the Barents Sea population (Ebbinge, unpubl.)
● adult males (n=134);■ adult females (n=137), cannon-netted and
immediately weighed after capture; O adult males (n=57);□ adult females (n=71), clapnetted and weighed after one day in captivity;
◐ adult males (n=16);◨ adult females (n=34), clapnetted during a
cold spell and weighed after one day in captivity. Figures indicate
sample size.

3.3.3 Numbers

3.3.3.1 Population size
   In 1959-1960 the world population of the Barnacle goose was estimated at 30,000 birds (Boyd, 1961), two-thirds of this number (19,700)
belonging to the Barents Sea population. In 1972-1973 another complete
census revealed a considerable increase in numbers with a world total
of 70,000 birds (Ebbinge et al., 1975), the Barents Sea population
again accounting for two-thirds (41,000) of the world population.

Table 1. Population dynamic parameters of the three Barnacle
geese populations. The mean annual mortality rate is calculated as
(j-i(1-j)); i is taken from the slopes of the lines presented in
fig. 9. 1) data from Boyd (1961, 1968), Ogilvie (1978) and Ogilvie
& Boyd (1975); 2) data from Ogilvie (1978) and Owen & Norderhaug
(1977); 3) data from Ebbinge (unpubl.).

Population	Greenland [1]	Spitsbergen [2]	Barents Sea [3]
mean percentage of first-year birds (j)	14% (n=11)	22% (n=8)	24% (n=9)
mean increase of the population per year (i)	5.3%	11%	6.0%
calculated mean annual mortality rate	9%	13%	19%
period	1958-1977	1970-1978	1959-1978

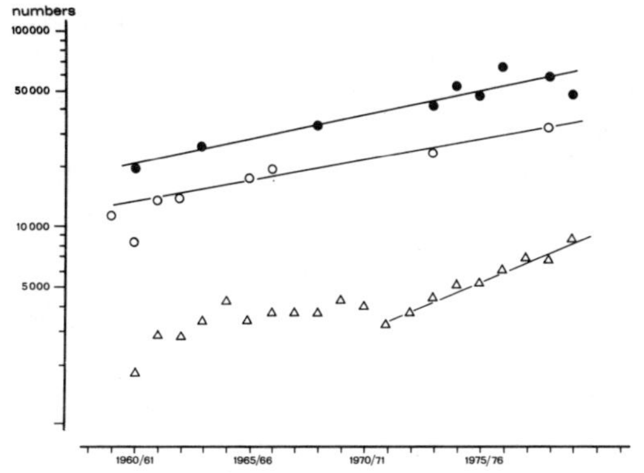

Fig. 9. Development of the three Barnacle geese populations.
• Barents Sea population (data: Boyd, 1961; Timmerman, 1976; Ebbinge et al., 1975; Ebbinge, unpubl.; Lok, in prep.);
o Greenland population (data: Boyd, 1961, 1968; Ogilvie, 1978);
△ Spitsbergen population (data: Boyd, 1961; Ogilvie, 1978).

Fig. 9 shows for the two larger populations that this increase in numbers follows a more or less straight line, when plotted on a logarithmic scale. This indicates that over a period of almost 20 years the rate of increase is fairly constant, and very similar for both populations (5-6% per year). The reproductive rates of these populations show, rather surprisingly, considerable differences (Table 1). This means that also the average annual rate of mortality over the last twenty years must diverge from 9% for the Greenland population to 19% for the Barents Sea population. Analyses of 230 ring recoveries from a total of 3133 ringed birds from this latter population yields a fairly similar figure for the yearly mortality rate: 17% on average over the last 20 years (Ebbinge, in prep.). The higher mortality rate in the Barents Sea population when compared to the two others (Table 1) is most likely caused by a heavier hunting pressure. Fig. 10 shows that within the Barents Sea population hunting as a mortality factor gradually decreased during the last 20 years, as more and more countries banned hunting. In the sixties still nearly half of the ringed birds reported shot, were killed in spring (April-May), when almost the entire population stayed in the Baltic. But in the seventies this spring shooting decreased considerably. Autumn shooting (mainly in Schleswig-Holstein, see f.ig. 11) continued in the seventies. From the ring recoveries can be calculated that each autumn

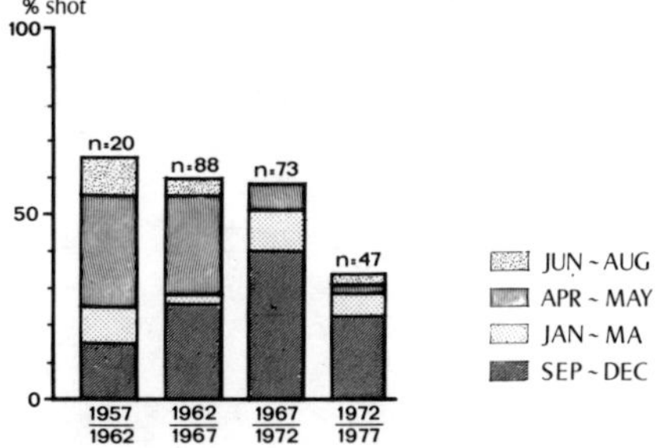

Fig. 10. Hunting as a mortality factor in the Barents Sea population of the Barnacle goose (Ebbinge, unpubl.). For each period the percentage of shot birds in all recoveries is given.

on average 200-300 Barnacle geese were shot, assuming all hunters report the rings of shot birds. Now that hunting is also closed in Germany, since April 2, 1977, the mean yearly mortality rate is very likely to be much lower than the overall mean over the last twenty years as given in Table 1. However, this does not necessarily imply a higher rate of increase for the Barents Sea population, because of a possible decrease in the rate of reproduction.

Though not all counts are available yet, the world population in the 1977-1978 season will have totalled approximately 100,000 birds.

3.3.3.2 Numbers per area

Because of the gradual increase in numbers during the past 20 years the mean numbers given in Fig. 12 will generally be too low to describe the present situation, but they are still useful to illustrate the fluctuation in numbers within the seasons.

Up to 1970 Schleswig-Holstein was an important autumn staging area with 20,000 Barnacle geese in November. In the seventies, and especially since 1973, the importance of both Schleswig-Holstein and Niedersachsen as autumn staging areas was greatly reduced, because most Barnacle geese were spending the autumn in the recently reclaimed Lauwersmeer-area in the Netherlands. It is not clear whether this westward shift is due to avoidance by the geese of the comparatively heavily hunted areas in Germany (compare fig. 11), or

Fig. 11. Recoveries in the Wadden Sea area of Dutch-ringed Barnacle geese (Ebbinge, unpubl.).

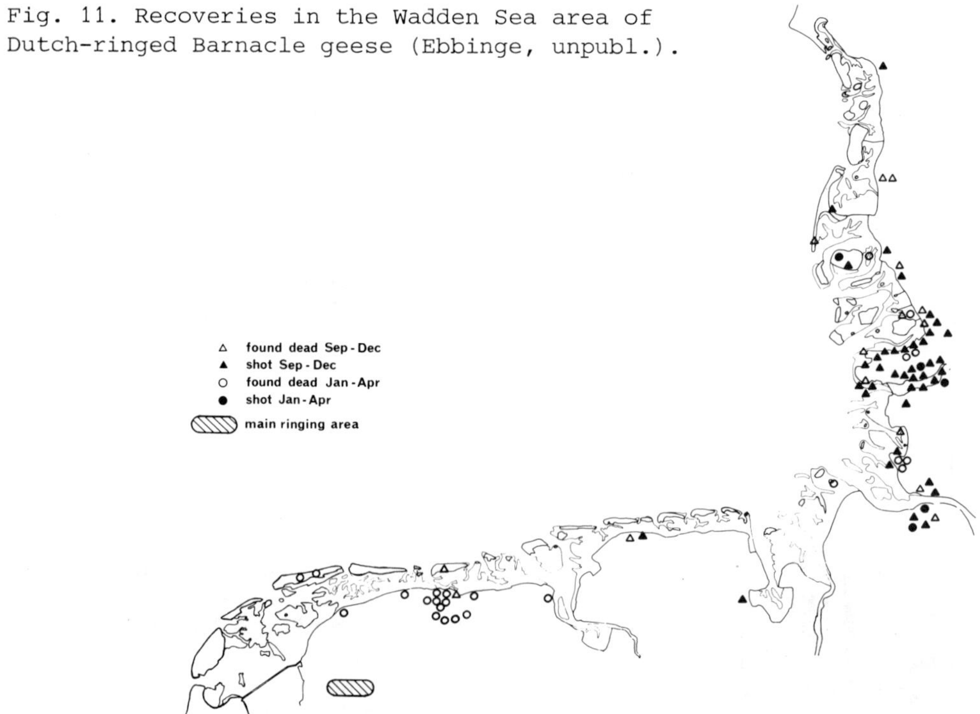

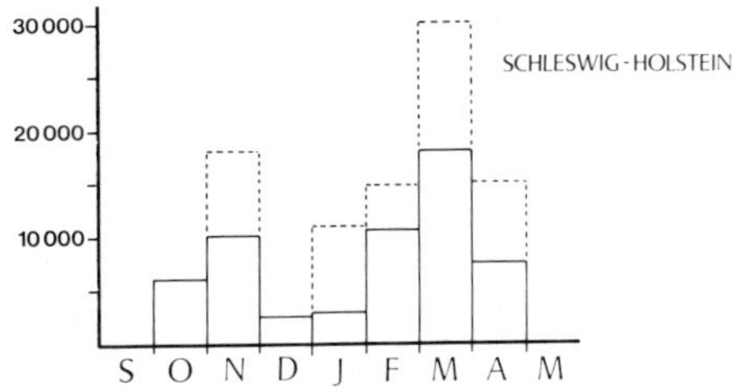

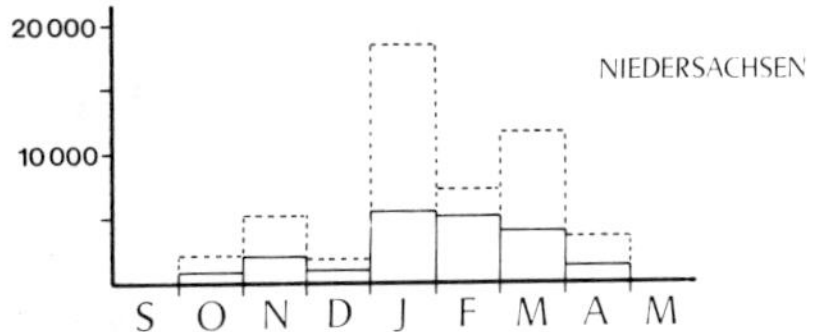

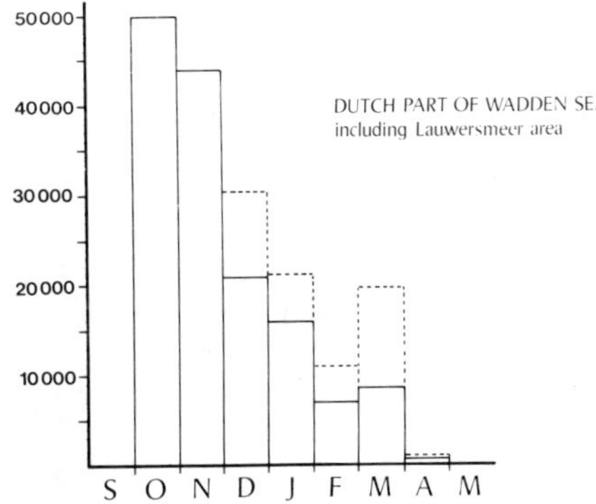

Fig. 12. Mean numbers (dotted lines:peak numbers) of Barnacle geese per month in the Wadden Sea area in Schleswig-Holstein (1965-1976; after Busche, 1977), Niedersachsen (1969-1976; after Dircksen & Hummel, unpubl.) and the Netherlands (1972-1977, including the Lauwersmeer-area; Ebbinge, unpubl.).

to the preference of the geese for the food situation in the newly reclaimed Lauwersmeer-area.

If hunting was the main factor causing this westward shift of their autumn staging area, the geese should start again to use their former haunts in Schleswig-Holstein in autumn now that they are fully protected since 1977. Indeed, in the autumn of 1978 16,000 Barnacle geese stayed in Germany again. In most winters a small part of the population stays in Schleswig-Holstein all the time. Only complete snow cover forces these geese to move westward.

From January onwards there is a marked eastward shift from the Netherlands to Niedersachsen and Schleswig-Holstein, but it is not until March that the majority of the birds is concentrated in Schleswig-Holstein (Busche, 1977).

Peak numbers for each month from 1959 till 1977 are shown in fig. 12.

3.3.4 Food

3.3.4.1 Food composition

In the Wadden Sea area the main feeding areas are saltings with a predominant vegetation of *Festuca rubra* on the higher terraces and *Puccinellia maritima* on the lower parts. Within the Wadden Sea area these salt marshes can only be found in their original "wild" state on a few islands (e.g. the "Boschplaat" on Terschelling and the "Oosterkwelder" on Schiermonnikoog), whereas along the mainland coast almost all salt marshes are characterized by an elaborate system of man-made ditches to enhance the growth of new land. This latter type of salt marsh is generally intensively grazed with sheep or cattle. Locally however, for instance in the province of Groningen, the grazing pattern has a greater diversity with lower average rates.

Apart from the massive *Salicornia* fields that have developed temporarily in the Lauwersmeer-area, intensively cultivated inland pastures, with a predominant vegetation of *Lolium perenne*, *Poa pratensis* and *Poa annua*, are important feeding areas in the Dutch part of the Wadden Sea area. To some extent grazing on winter cereals and winter rapes may occur.

In March and April Barnacle geese preferably graze the new growth of *Festuca rubra* on Schiermonnikoog and of *Puccinellia maritima* on the salt marshes in Schleswig-Holstein. Recently feeding on clover stolons in the autumn has been observed in the Dutch part of the Wadden Sea area on the "Oosterkwelder" on Schiermonnikoog (Prins, pers. comm.) as well as in inland pastures in the Bandpolder area in the province of Friesland (Ebbinge, unpubl.).

Both quantity and quality of the various food plants taken by Barnacle geese vary in the course of the season, but very little is known about how this influences the choice between different food items and to what extent this choice again influences the condition of the geese.

3.3.4.2 Feeding activities

Usually Barnacle geese withdraw to tidal flats or permanent sandbanks to rest. During the dark phase of the moon, feeding is restricted to the daylight period, and the nights are spent on these roosts. During bright moonlight, however, a complete reversal of this 24-hour rhythm occurs, the geese spending the daytime at rest on tidal flats (Ebbinge et al., 1975).

Though feeding on salt marsh vegetation, Barnacle geese have never been observed to drink salt water. When feeding on the salt marsh vegetation on Schiermonnikoog regular flights of 5-10 kilometers are carried out to drink from the freshwater outlets of the Banckspolder. Similar drinking flights have been recorded in the Eider estuary in Schleswig-Holstein (Busche, pers. comm.). In the recently reclaimed Lauwersmeer-area (1969) Barnacle geese showed a strong preference for the halophyte *Salicornia*, but also spent much more time drinking fresh water as compared to when feeding on inland pastures (Ebbinge et al., 1975). The fact that the majority of the Barents Sea population fed on the vast *Salicornia* fields (3300 ha in 1972-73) in the Lauwersmeer-area, but largely ignored *Salicornia* fields of similar size along the coasts of Groningen and on Schiermonnikoog must be due to the unique situation that in this Lauwersmeer-area adjacent to *Salicornia* on the still saline plains fresh water was readily available.

3.3.4.3 Total food consumption

The daily food consumption varies in the course of the season. For December Ebbinge et al. (1975) estimated the daily consumption of grass *(Poa pratensis* and *Lolium perenne)* at 135-160 grams dry weight, of which 1/3 was retained by the geese. As Barnacle geese loose weight in this period (section 3.3.2.3) their daily food intake in the autumn and in spring will have to be higher.

References

Boyd, H., 1961. The number of Barnacle Geese in Europe in 1959-60. Wildfowl Trust Ann. Rep. 12: p. 116-124.

Boyd, H., 1968. Barnacle Geese in the west of Scotland, 1957-1967. Wildfowl 19: p. 96-107.

Busche, G., 1977. Gänse im Westen Schleswig-Holsteins. Die Heimat 84: p. 340-349.

Cramp, S. & K.E.L. Simmons (eds.), 1977. The handbook of the birds of Europe the Middle East and North Africa, Vol. 1. Oxford University Press, Oxford: 722 pp.

Ebbinge, B., K. Canters & R. Drent, 1975. Foraging routines and estimated daily food intake in Barnacle Geese wintering in the northern Netherlands. Wildfowl 26: p. 5-19.

Kumari, E., 1971. Passage of the Barnacle Goose through the Baltic area. Wildfowl 22: p. 35-43.

Ogilvie, M.A., 1978. Wild geese. Poyser, Berkhamsted: 350 pp.

Ogilvie, M.A. & H. Boyd, 1975. Greenland Barnacle Geese in the British Isles. Wildfowl 26: p. 139-147.

Owen, M., 1977. Wildfowl of Europe. Mc Millan, London & Wildfowl Trust: 256 pp.

Owen, M. & M. Norderhaug, 1977. Population dynamics of Barnacle geese (Branta leucopsis) breeding in Svalbard 1948-1976. Ornis Scand. 8: p. 161-174.

Roux, F., 1963. Les stationnements d'anatides en France pendant la vague de froid de 1962-63. Bureau Int. Rech. Sauv. Publ. 6.

Schmidt, G.A.J., 1973. Das Winterhalbjahr der Nonnengänse aus Schleswig-holsteinischer Sicht. Die Heimat 80: p. 289-295.

Timmerman, A., 1976. De brandgans. In: Lebret, T., Th. Mulder, J. Philippona & A. Timmerman, Wilde ganzen in Nederland. Thieme, Zutphen: p. 113-124.

Timmerman, A., M.F. Mörzer Bruyns & J. Philippona, 1976. Survey of the winter distribution of Palaearctic geese in Europe, Western Asia and North Africa. Limosa 49: p. 230-292.

Voous, K.H., 1960. Atlas of European birds. Nelson, London: 284 pp.

3.4 BRENT GOOSE *(BRANTA BERNICLA L.)*
Barwolt Ebbinge, Mette Fog & Peter Prokosch

Da: Knortegaas, G: Ringelgans; Du: Rotgans

3.4.1 Distribution

3.4.1.1 Breeding area
   The breeding range of the Brent goose covers the whole circumpolar high arctic zone (Voous, 1960). Within this range three subspecies occur. All of these have been recorded in the Wadden Sea area (fig. 13a):
- *Branta b. bernicla*, the Dark-bellied Brent goose, breeding in Western Siberia from the Yamal peninsula in the east to the West coast of the Taymyr peninsula (Owen, 1977). The whole world-population of this subspecies visits the Wadden Sea area.
- *Branta b. hrota*, the Light-bellied Brent goose. Of this subspecies only a part of the small population breeding on the Spitsbergen archipelago visits the Wadden Sea area (Owen, 1977).
- *Branta b. nigricans*, a very dark subspecies, breeding in Eastern Siberia and the Western part of the Nearctic, occurs as a straggler in the Wadden Sea (Lambeck, 1977; Beintema et al., 1976).

3.4.1.2 Migration routes
   B. b. bernicla migrates through the Baltic and the Gulf of Finland (fig. 13b). First stop in autumn is the Southern part of the Danish

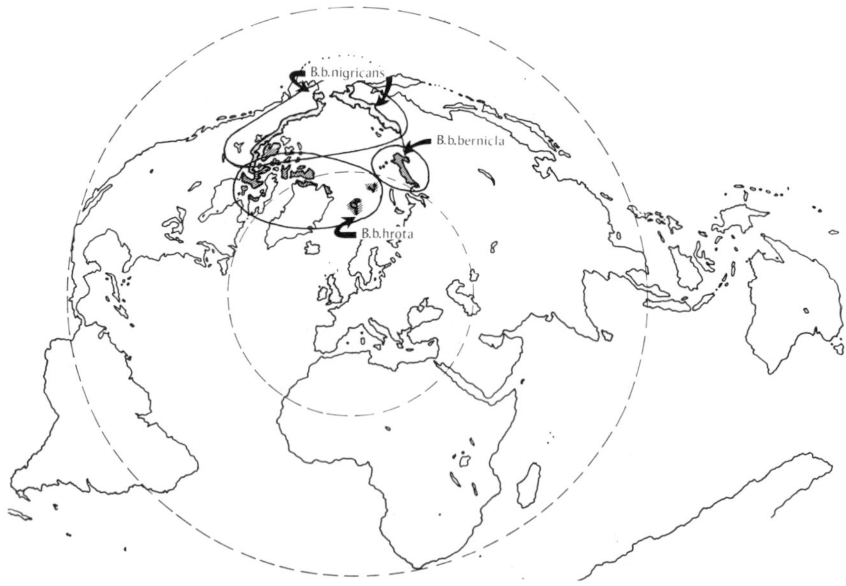

Fig. 13a. Breeding areas of the Brent goose. After Voous (1960), Owen (1977) and Timmerman et al. (1976).

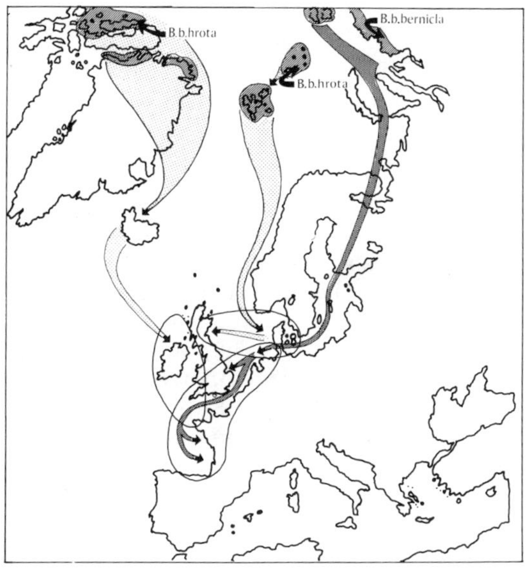

Fig. 13b. Migration routes of those Brent goose populations wintering in Western Europe. Approximate boundaries of wintering areas are encircled. After Voous (1960), Owen (1977) and Timmerman et al. (1976).

waters in the Baltic and the Wadden Sea. The Wadden Sea is also the last major stop in spring before migration towards the breeding grounds. Migration routes of *B. b. hrota* are unknown, but their first stop in autumn is the Danish Wadden Sea.

3.4.1.3 Wintering areas

B. b. bernicla: 80-90% of the world population of this subspecies winters south of the Wadden Sea area along the East and South coast of England, in the Delta area in the Netherlands, and along the West coast of France. In mild winters up to 20% of the world population may stay in the Wadden Sea area throughout winter (fig. 14).

B. b. hrota: winters mainly along the East coast of England in Northumberland. Smaller numbers winter in Denmark (Timmerman et al., 1976).

3.4.1.4 Moulting areas

Primaries are moulted in the arctic in summer. In the Wadden Sea area only some body feathers are moulted.

3.4.2 Annual cycle

3.4.2.1 Migration

B. b. bernicla: First arrivals in the Wadden Sea area take place from mid-September till early October. Peak numbers are reached in October/November. From December until March the birds are present in the wintering areas. From the end of March onwards numbers in the Wadden Sea start to rise again sharply to reach a peak between the end of April and early May. By this time almost the whole world population of this subspecies is concentrated in the Wadden Sea area

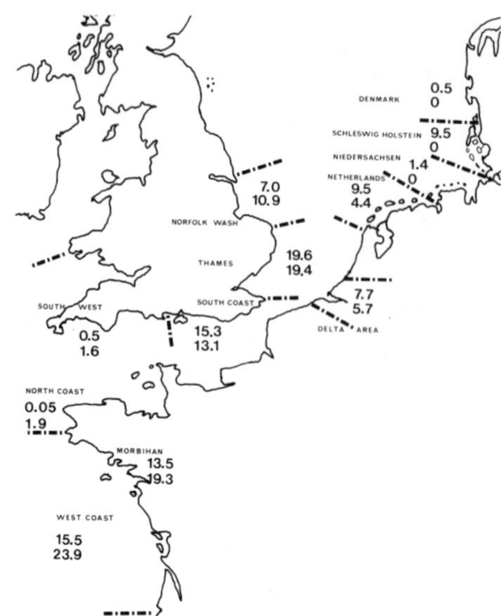

Fig. 14. January-distribution of the Dark-bellied Brent goose during counts in 1975 (mild winter, upper figure) and 1979 (cold winter, lower figure) (after St. Joseph, 1979). The figures indicate the percentage of the total number counted, e.g. 74,000 in January, 1975 and 140,000 in January, 1979.

(see fig. 15). By the end of May virtually all Brents have left the area for their arctic breeding grounds though a few hundreds are staying till mid-June.

   B. b. hrota: Occurs regularly in the Northern part of the Wadden Sea area in September and October mixed in flocks with *B. b. bernicla*, irregularly in other parts of the Wadden Sea.

3.4.2.3 Weight changes

   In autumn Brent geese gain weight while feeding mainly on Eelgrass *Zostera noltii* and *Enteromorpha* (Fog, 1967). Fig. 16 shows that during the early part of winter they loose weight to reach a fairly stable weight of about 1330 g in adult birds throughout the winter (St. Joseph et al., in prep.). Only during severe cold spells weights drop below this level. Lowest weights were recorded in early April when many birds had just arrived from their French and British wintering areas in the Wadden area. During the first three weeks of May weights increase drastically in the Wadden Sea area. By the end of May adult birds weigh 1600 g on average.

3.4.3 Numbers

3.4.3.1 Population size

   B. b. bernicla: In the past wintering Brent geese have always been numerous on the extensive *Zostera* beds along the coasts of Western Europe. In the thirties, however, simultaneously with a wide-spread disease affecting all *Zostera marina* beds along the Atlantic coasts, numbers of Brent geese dropped very seriously. Only 16,500 Dark-bellied Brent geese were left in the early fifties. Because hunting pressure was thought to be the main factor preventing the geese to recover

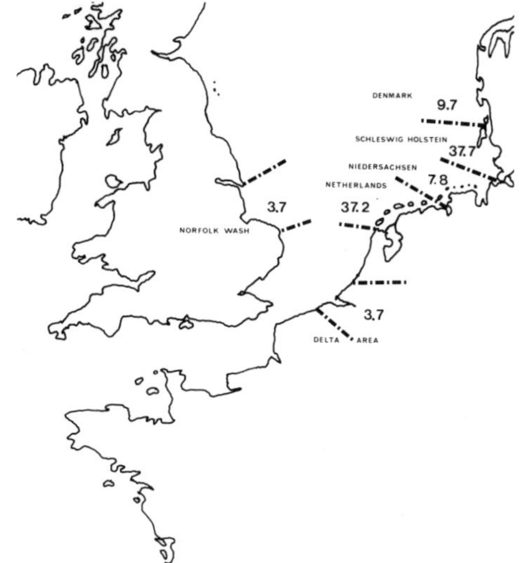

Fig. 15. Spring-distribution of
the Dark-bellied Brent goose
(after St. Joseph, 1979). The
figures indicate the percentage
of the total number counted,
viz. 134,000 in April 1979.

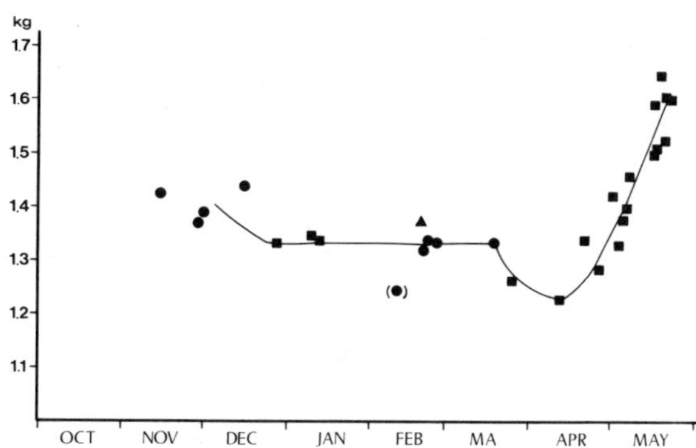

Fig. 16. Mean weights of
Dark-bellied Brent geese
during their stay in western Europe. After St. Joseph et al., in prep.
(●) = after cold spell;
● = England; ▲ = France;
■ = Wadden Sea area;
n = 918.

again (Salomonsen, 1958), the species was given full protection in the
Netherlands, England and France. Despite this protection in their wintering areas numbers increased very little (fig. 17). It was not until 1972, when also Denmark granted full protection for the Brent
goose, that numbers started to increase considerably. At present there
is still an open season in Schleswig-Holstein in November and December.
In January 1979 the whole world population consisted of 140,000 birds.
It is not clear how long the present rate of increase will continue,
because nothing is known about the carrying capacity of their breeding grounds.

   B. b. hrota: The very small population from Spitsbergen consisted
in 1977-1978 of only 2200 birds (St. Joseph, in litt.). This population has also been much more numerous earlier this century. Mulder
(1976) estimates that in the beginning of this century *B. b. hrota*
contributed to about 10% in the annual catches of Brent goose hunters
in the Western part of the Dutch Wadden Sea.

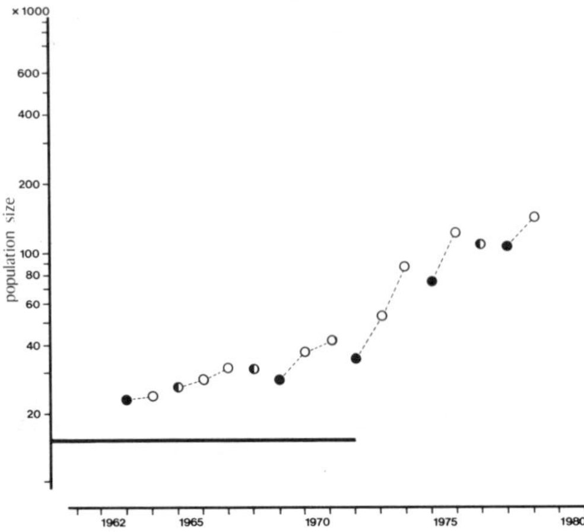

Fig. 17. Population size of the Dark-bellied Brent goose in 1962-1979 (after St.Joseph,1979). The symbols denote the percentage of juvenile birds: o > 35%, ◐ 5-18% and ● < 1%. ─────── denotes autumn hunting in Denmark.

3.4.3.2 Number per area

Because of the recent increase in numbers it is hardly meaningful to calculate monthly means over the past twenty years. Therefore only counts collected during the last five years were examined to describe the present situation. Monthly variations in numbers are shown in fig. 18 for two well documented areas in Schleswig-Holstein and the Dutch part of the Wadden Sea area as a whole. Langeness is a typical example for the Wadden Sea area in Denmark and Schleswig-Holstein, because both in the autumn and in spring very large numbers are present. In the Dutch part of the Wadden Sea area the phenology of the Brent geese is quite different. Numbers in the autumn are much lower than in spring and the birds also occur in winter. This phenomenon is further illustrated in fig. 19 giving the main Brent haunts in the Wadden Sea area for three seasons separately.

Table 2 shows the maximum numbers of *B. b. bernicla* counted during the seventies in the Danish, German and the Dutch parts of the Wadden Sea area.

Table 2. Maximum number of *Branta b. bernicla* per month in the four main areas in the Wadden Sea. Data from the Netherlands are from Ebbinge (unpubl.); Niedersachsen and Schleswig-Holstein: Prokosch (1978-1980); Denmark: Fog (unpubl.).

	The Netherlands		Niedersachsen		Schleswig-Holstein		Denmark	
October	13,000	(1976)	5,000	(1978)	57,000	(1978)	?	
November	29,000	(1978)	?		?		?	
December	?		?		?		?	
January	11,000	(1976)	2,600	(1978)	8,000	(1976)	1,000	(1978)
February	?		?		?		?	
March	38,000	(1979)	?		30,000	(1979)	?	
April	50,000	(1979)	10,000	(1979)	50,000	(1979)	13,000	(1979)
May	54,000	(1977)	9,000	(1979)	51,000	(1979)	?	

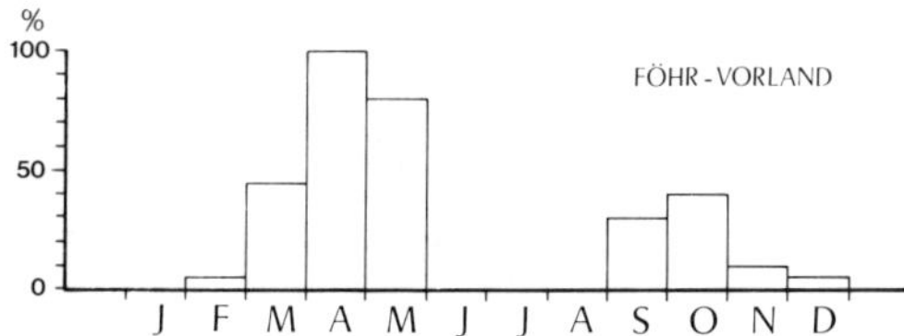

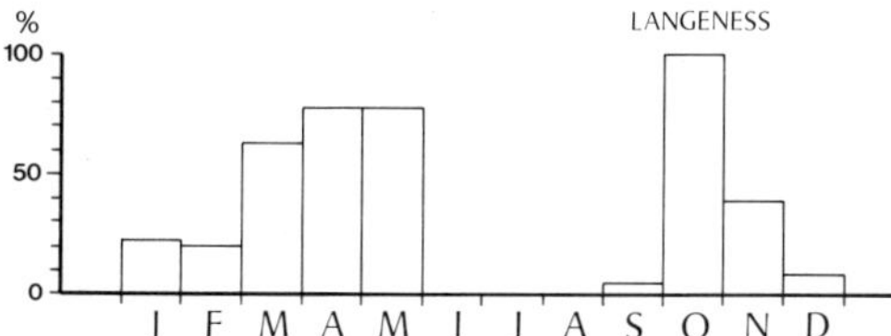

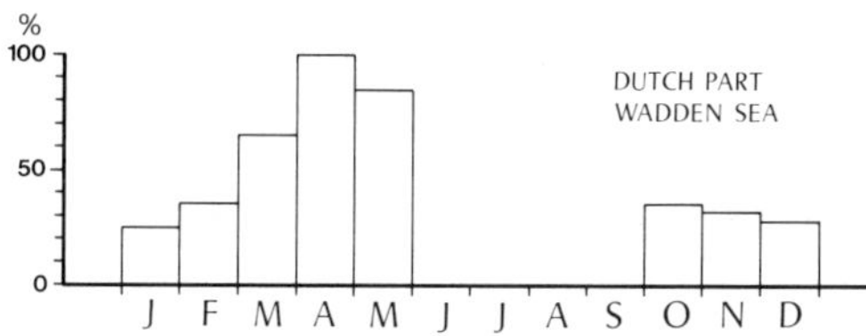

Fig. 18. Brent goose phenology in the Wadden Sea area. Monthly means are expressed as a percentage of the mean seasonal peak count for the area concerned. Data from the Föhr-Vorland (1975-1978) and Langeness (1974-1978) (both in Schleswig-Holstein) are taken from Prokosch (1979), from the Dutch part part of the Wadden Sea (1974-1978) from Ebbinge (in press).

Fig. 19a. Autumn distribution of Dark-bellied Brent geese in the Wadden Sea area in 1975-1979. After Ebbinge, Fog and Prokosch (unpubl.).

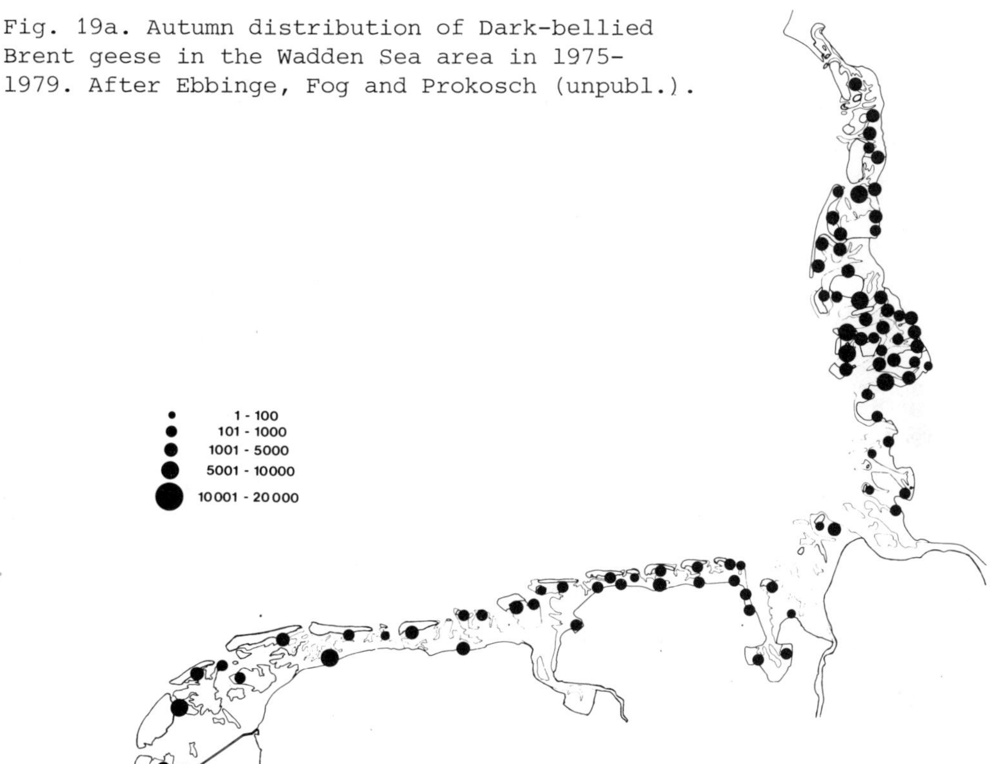

Fig. 19b. Winter distribution of Dark-bellied
Brent geese in the Wadden Sea area in 1975-
1979. After Ebbinge, Fog and Prokosch (unpubl.).

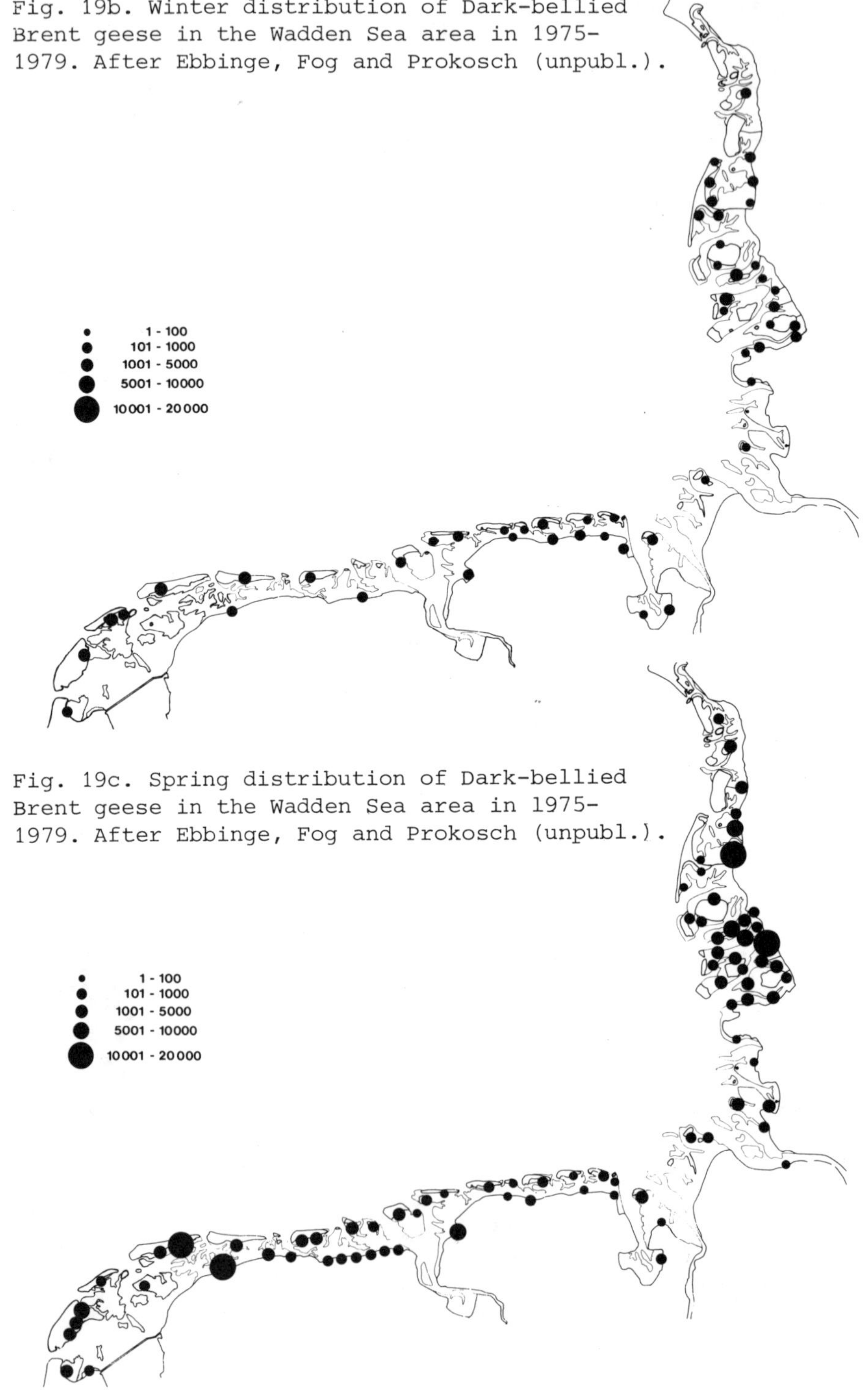

Fig. 19c. Spring distribution of Dark-bellied
Brent geese in the Wadden Sea area in 1975-
1979. After Ebbinge, Fog and Prokosch (unpubl.).

3.4.4 Food

3.4.4.1 Food composition

During their stay along the coasts of Western Europe the composition of the diet gradually changes from strictly marine plants like green algae and *Zostera* to terrestrial plants growing on the salt marsh in spring (Mörzer Bruyns & Tanis, 1955; Ranwell & Downing, 1959).

Just after their arrival in Western Europe in autumn the major concentrations of Brent geese are found on *Zostera*. At present the most important *Zostera*-beds (predominantly *Zostera noltii*) lie in the Northern part of the Wadden Sea area (Denmark and Schleswig-Holstein), along the East coast of England and the West coast of France. When these *Zostera*-beds are largely eaten out, or have lost their leaves in the annual cycle, Brents either depart to other areas, or stay in the same area and change diet. If they stay they start to feed on other plant species on the mudflats, *Enteromorpha* and *Ulva* spec. being the most important. From December to February however Brents in the Dutch part of the Wadden Sea area and part of those wintering in England feed entirely inland on pastures or on winter wheat. From March to May feeding on the newly emerging salt marsh vegetation becomes more and more important, but in this period both young green algae (mainly *Enteromorpha*) and grass in meadows and pastures are also taken.

Within the Wadden Sea area the most important feeding areas are:
- *Zostera* fields (mainly *Zostera noltii*) in the intertidal zone in the Wadden Sea in Denmark and Schleswig-Holstein.
- *Enteromorpha* fields in the intertidal zone on sheltered places all over the Wadden Sea.
- Natural salt marshes on the island of Terschelling (Boschplaat), Schiermonnikoog (Oosterkwelder) and Rottumerplaat.
- Cattle and/or sheep grazed natural salt marshes, like the "Groede" on the island of Terschelling, "Nieuwlandsrijd" on the island of Ameland and on various Halligen in Schleswig-Holstein.
- Cattle and/or sheep grazed salt marshes created by land reclamation works. Along the mainland coast all over the Wadden Sea area the development of new salt marshes has been greatly stimulated by a system of simple man-made ditches, resulting in very clayey salt marshes with a predominant vegetation of *Puccinellia maritima*. Most important specimens of this type are the area from Zwarte Haan to Holwerd in the Dutch province of Friesland, the Leybucht in Niedersachsen, the salt marshes in the Nordstrander Bucht in Schleswig-Holstein, and those of Rodenäs and Højer on the border of Schleswig-Holstein and Denmark.

3.4.4.2 Feeding activities

When feeding on the tidal flats Brent geese generally feed by day but during periods of full moon when there is sufficient moonlight also by night (Charman, pers. comm.). When foraging on land however they are strictly diurnal. During the night Brent geese generally withdraw to the Wadden Sea or shallow bays to roost floating on the water. During the day, when Brents are feeding ashore they may withdraw to the same roosts, often after disturbance.

Brent geese have well-developed salt glands which enable them to drink salt water. In their first autumn Brents (accompanied by their parents) feeding on *Enteromorpha* on the island of Vlieland, have been observed to carry out drinking flights to inland freshwater ponds. Adult birds without young stayed continuously on the *Enteromorpha*-beds and drank salt water only (Ebbinge, unpubl.). This might suggest that the salt glands in birds in their first calandar-year have not yet been fully developed.

3.4.4.3 Total food consumption

Provisional estimates (Ebbinge, unpubl.) for Brent geese feeding entirely in meadows in midwinter yield a daily intake of grass leaves of 100-120 g dry weight which is approximately 500 g fresh weight. In May however, while fattening up for their spring migration towards the breeding grounds their daily intake amounts to approximately 200-250 g dry weight (Ebbinge, unpubl.).

References

Beintema, A.J., E. van der Bilt, B. Helming & T. Thuyls, 1976. Een nieuwe ondersoort van de Rotgans, Branta bernicla nigricans, voor Nederland. Limosa 49: p. 131-134.

Ebbinge, B.S., (in press). Ganzen. In: Rijksinstituut voor Natuurbeheer, Natuurbeheer in Nederland, Vol. 2. Pudoc, Wageningen.

Fog, M., 1967. An investigation on the Brent goose in Denmark. Dan. Rev. Game Biol. 5: p. 1-40.

Lambeck, R.H.D., 1977. Het voorkomen van Rotgans Branta bernicla ondersoorten in het Nederlandse Waddengebied tijdens het voorjaar van 1976. Limosa 50: p. 92-97.

Mulder, T., 1976. Bijgeloof, domesticatie en vangst. In: T. Lebret, Th. Mulder, J. Philippona & A. Timmerman, Wilde ganzen in Nederland. Thieme, Zutphen: p. 141-147.

Mörzer Bruyns, M.F. & J. Tanis, 1955. De rotganzen, Branta bernicla (L.), op Terschelling. Ardea 43: p. 261-271.

Owen, M., 1977. Wildfowl of Europe. McMillan, London & Wildfowl Trust: 256 pp.

Prokosch, P., 1978-1980. Ringelgans Rundbriefe 1-10.

Prokosch, P., 1979. Zur Bedeutung der Salzwiesen der Halligen und Vorländer im Nordfriesischen Wattenmeer für die Ringelgans (Branta bernicla). Bericht Landesamt für Naturschutz und Landespflege Schleswig-Holstein, Kiel: 90 pp.

Ranwell, D.S. & B.M. Downing, 1959. Brent goose (Branta bernicla) winter feeding pattern and Zostera resources at Scolt Head Island, Norfolk. Anim. Beh. 7: p. 42-56.

Salomonsen, F., 1958. The present status of the Brent goose (Branta bernicla, L.) in Western Europe. Vidensk. Medd. dansk naturh. Foren. 120: p. 43-80.

St. Joseph, A.K.M., 1979. A review of the status of Branta bernicla

bernicla. Proceedings Second technical meeting on western Pale-
arctic migratory bird Management, Paris, 1979.
St. Joseph, A.K.M., B. Ebbinge, O. Fournier & P. Prokosch (in prep.).
Weight changes in the Dark-bellied Brent goose on their wintering
and spring staging areas.
Timmerman, A., M.F. Mörzer Bruyns & J. Philippona, 1976. Survey of
the winter distribution of palaearctic geese in Europe, Western
Asia and North Africa. Limosa 49: p. 230-281.
Voous, K.H., 1960. Atlas of European birds. Nelson, London: 284 pp.

3.5 SHELDUCK *(TADORNA TADORNA* (L.)*)*
F. Goethe

Da: Gravand; G. Brandgans; Du: Bergeend

3.5.1 Distribution

3.5.1.1 Breeding area
This characteristic Wadden Sea species has a discontinuous distribution in the Southern and Western palaearctic (fig. 20). It is breeding in the temperate, Mediterranean, steppe and desert climate zones.
In NW Europe it breeds along the coasts of NW France, Ireland, Norway, S Sweden and along the Baltic. In Britain, Ireland, the Netherlands, Germany and Denmark it breeds along the coast as well as inland. Nests are found in open areas in dense vegetation, outlet-pipes
or old and partly ruined buildings, along the North Sea coast rather
often in rabbit-holes in the dunes. They often breed in colony-like
communities.

3.5.1.2 Migration routes
The migration of the NW European population is dominated by its
moult migration. A noticable southward migration takes place along
the West coast of Jutland at Blaavandshuk (Petersen, 1974). The Baltic population crosses Jutland and Schleswig-Holstein. Shelducks
from Britain and Ireland cross the British mainland at a great alti-
tude and the North Sea (Coombes, 1950; Doornbos 1977). Especially
young birds disperse in southward directions, generally following
coast lines. Some go as far south as the Bay of Biscay (Bauer & Glutz,
1968; Cramp & Simmons, 1977).

3.5.1.3 Wintering areas
When the Wadden Sea is free of ice, Shelducks stay over winter
in the whole area. After moult breeding birds from France and the British Isles generally return to their breeding areas to winter. Populations breeding along the Baltic, winter in the Wadden Sea, along
the French coast and along the British Isles (Cramp & Simmons, 1977).
At least a part of the population breeding on the continent winters
along the British Isles (Lohse, 1977; unpubl. data Institut für Vogelforschung, Wilhelmshaven). Some birds go as far south as Spain
(Bauer & Glutz, 1968).

Fig. 20. Breeding distribution of Shelduck (based on Voous, 1960;
Haftorn, 1971; Bauer & Glutz, 1968 and Cramp & Simmons, 1977).
The dark shaded area indicates the region from which birds visiting
the Wadden Sea originate.

3.5.1.4 Moulting areas

The moult of the Shelduck is one of the most fascinating phenomena of bird life in the Wadden Sea (fig. 21). The area where wing
moult takes place is situated between the rivers Weser and Eider,
particularly in the nature- and game reserve Knechtsand-Eversand,
in the area around Trischen and on the sandflats in the Dithmarscher
Bucht, north to the Westcoast of the Eiderstedt peninsula (see fig.
27). Occasionally moult takes place in the area around Norderoog,
Scharhörn and Mellum and in the Danish Wadden Sea around Fanø and
Rømø (Goethe, 1957, 1961a, 1961b; Oelke, 1969a; Drenckhahn et al.,
1971). Smaller moulting places are the area south of the island of
Vlieland and the Dutch Delta area. In the Danish part of the Wadden
Sea especially, groups of one-year old non-breeding Shelducks and
adults from NW European breeding areas occur (Joensen, 1974). The
moulting centre in the German Wadden Sea is of vital importance in the
annual cycle of most of the NW European Shelduck population. Moult of
body feathers has been observed on a large scale in the Wadden Sea
area in Schleswig-Holstein (Drenckhahn et al., 1971).

3.5.2 Annual cycle

3.5.2.1 Migration

The breeding birds and immatures from Britain, Ireland, Belgium, the Netherlands, coastal Germany, Denmark, Scandinavia, Poland and Baltic S.S.R. (fig. 22) gather in the breeding areas in June. At Blaavand migration of Shelducks has a maximum from mid-June till mid-July (Petersen, 1974). They arrive in the moulting area in June and July. The first flocks with completed moult of flight feathers leave the Knechtsand area from mid-August on. Peak numbers of migrating Shelducks can be observed by the end of this month. In Sep-

Fig. 21. Moulting Shelducks, Grosser Knechtsand, August 1956 (Photo Dr. F. Goethe). Released by R.P. Düsseldorf, Nr. BN 14.

tember and October there are still birds present in the area because the timing of migration of the individual birds varies and
depends on the breeding period. Shelducks breeding on the continent
which have completed moult of flight feathers to a certain extent
migrate tardily towards the breeding areas. Some even go in northern
directions before leaving towards the wintering quarters. Scandinavian and Baltic populations concentrate in the Wadden Sea north of
the Elbe. British and Irish Shelducks return to the breeding areas
that can also function as a wintering area as a result of the relatively warm climate. They pass the Western part of the Wadden Sea
(Bauer & Glutz, 1968; Cramp & Simmons, 1977).

First calendar-year Shelducks of the North Sea population remain
in the breeding areas until October and November. They may join the
passing adults or migrate in SW directions towards the Atlantic
coast. They often go farther than the adults (Bauer & Glutz, 1968).
The birds arrive in the breeding areas from the beginning of February to mid-April. Later in the season from the end of March until
April, non-breeding Shelducks pass by (Bauer & Glutz, 1968).

3.5.2.2 Moult

About 3 weeks after fledging, first calendar-year birds moult body feathers, tail, some inner secondaries and perhaps wing coverts
and particular coverts. This moult is completed in December. Moult
of the immatures and the post-breeding moult of adults begins in
June (non-breeding birds) or July (breeding birds). By this time
also the change of body feathers begins. It generally ends by the
time wing and tail feather moult begins. The latter are moulted
from early July until mid-October, generally however in August.
In this period the birds are unable to fly (Bauer & Glutz, 1968;
Cramp & Simmons, 1977). The period of flightlessness lasts 25-31
days (Hoogerheide, 1958; De Groot, 1975; Goethe, unpubl.). The time
at which moult and flightlessness begins may vary considerably between specimens. Birds which are still flightless in October probably
are breeding birds with late broods (Bauer & Glutz, 1968; Cramp & Simmons, 1977).

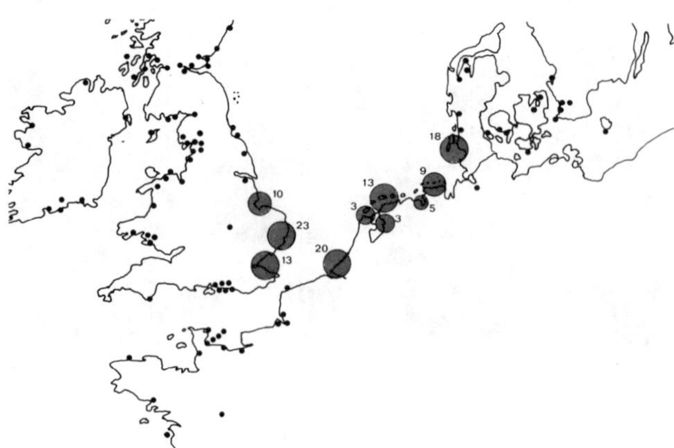

Fig. 22. Recoveries of
Shelducks ringed on Gros-
ser Knechtsand. Figures
denote numbers of recov-
eries. Two recoveries
in N Scotland have not
been included. Modified
after Bauer & Glutz
(1968).

3.5.2.3 Weight changes

Bauer & Glutz (1968) give some weights of birds from Denmark, the German Baltic coast and Belgium. Only very few weights of Shelducks from the Wadden Sea area have been published. An adult female from Grosser Knechtsand weighed 1000 g (Goethe, unpublished).

3.5.3 Numbers

3.5.3.1 Population size

Nearly the whole NW European Shelduck population potentially uses the Wadden Sea as a moulting and partly as a wintering area. The total size of this population is:

Denmark: 2000-4000, an increase has been noted in the past years (Dybbro, 1976). In the Danish Wadden Sea area about 70-90 pairs breed (data: Dansk Ornithologisk Forening).

Germany (Fed. Rep.). The whole German population amounts to about 2500-3500 pairs and increases. Along the Wadden Sea coast in Niedersachsen 1400-1900 and in Schleswig-Holstein 400-500 pairs breed (data: Szijj, 1973; Institut für Vogelforschung, Wilhelmshaven.

The Netherlands: 3500-4500 pairs, the inland population increases (Swennen, 1979) but in the Wadden Sea area, where 1300-1450 pairs are breeding, only slightly (Smit, in litt.).

Belgium: 115, increase (Lippens & Wille, 1972).

France (Atlantic and Channel coasts populations only) 312, increase (Duncombe, in litt.).

Britain and Ireland: 12,000, increase (Yarker & Atkinson-Willes, 1971).

Sweden: 10,000, perhaps not all of them breeding, increase Curry-Lindahl, in litt.).

Norway: unknown but there are indications for an increase: Holgerson, in litt.).

Germany (Dem. Rep.): 200, increase (Klafs & Stübs, 1977).

Estonian S.S.R.: 150, stable numbers (Onno, 1970).

The whole NW European breeding population therefore comprises of at least 60,000 birds. This number excludes Norwegian Shelducks, first calendar-year birds and immatures.

A high percentage of paired birds, often about 50% do not breed. The population wintering in NW Europe is estimated at 130,000 birds (Atkinson-Willes, 1976). The distribution of the population breeding in the Wadden Sea area is shown in fig. 23.

3.5.3.2 Numbers per area

Aerial surveys in the Danish part of the Wadden Sea gave numbers listed in Table 3. The species is most numerous in September and October when more than 20,000 may be present. Later in autumn the birds spread northward and become increasingly numerous in coastal waters throughout the country. Nevertheless the Wadden Sea still remains one of the most important wintering areas in Denmark though numbers may vary considerably from year to year. Numbers decrease in March and April and in late April and May only a few thousands or less are left (Joensen, 1974).

Fig. 23. Distribution of the breeding
population of the Shelduck in the Wadden Sea area (after data from the Institut für Vogelforschung Wilhelmshaven
and Smit, in litt.).

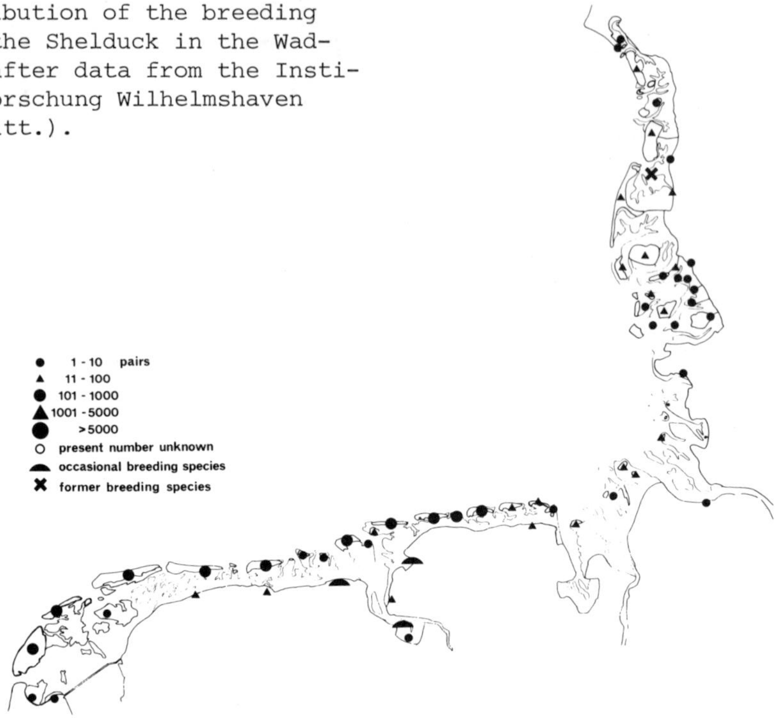

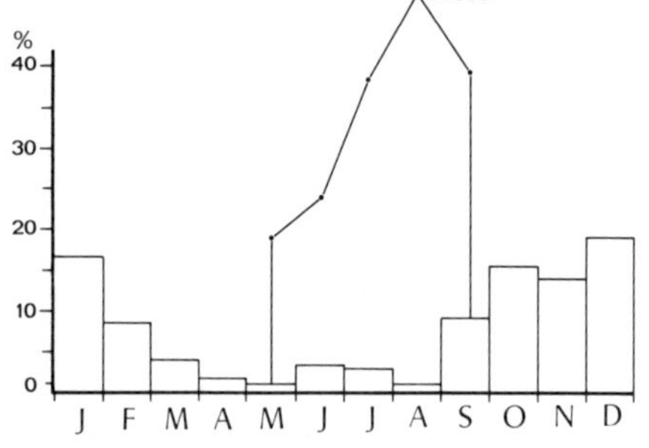

Fig. 24. Occurrence of the Shelduck in the Wadden Sea area in
Schleswig-Holstein. Numbers for
each month are expressed as a
percentage of the total number
of birds observed in all months.
The slender line shows the results of aerial surveys of moulting birds (after Busche, 1980).

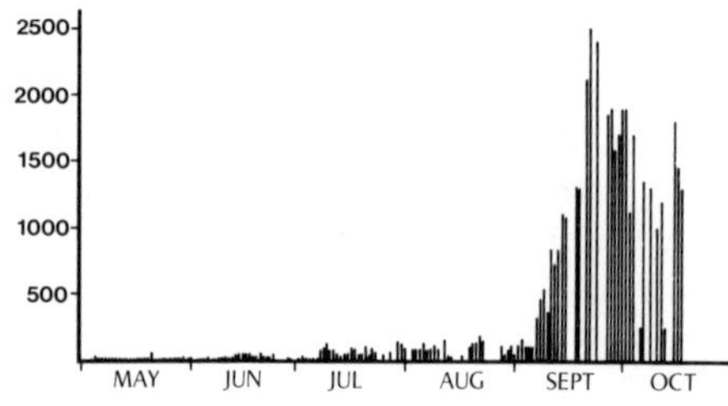

Fig. 25. Numbers of Shelducks in
the Southern part of the Jadebusen in 1955 (after Goethe,
1961b).

Table 3. Number of Shelducks in the Danish part of the Wadden Sea (after: Joensen, 1974).

November	1968	16,400
November	1969	6,400
January	1969	15,300
January	1970	400
January	1971	14,100
January	1972	7,200
March	1969	15,700

Around Trischen about 30,000 Shelducks gather to moult (Dircksen, 1968). With the aid of aerial surveys in the moulting areas in the Dithmarscher Bucht and around Trischen on August 12th 1972 55,000 and on August 25th 1971 59,000 Shelducks have been counted (Drenck-hahn in Busche, 1980). In mild winters 20,000-40,000 birds are present in the Wadden Sea area in Schleswig-Holstein. In cold winters numbers may drop to 3400 (January, 1970). In February departure towards the breeding places already begins and numbers decrease. In May generally about 3000 to 4000 are present which approximizes the size of the breeding population along the West coast of Schleswig-Holstein (Busche, 1980). Fig. 24 shows the fluctuation in numbers present in the area at the Knechtsand moulting area. Aerial surveys from August 8 until August 14, 1955 yielded a maximum of approximately 100,000 birds. This was the maximum number that was recorded in 1955 (Goethe, 1961a). In the first week of August 1965 the number was estimated at 70,000-80,000 (Oelke, 1969b). There seems to be a certain "exchange" between the moulting areas of Knechtsand and Trischen. At the end of August 1971 about 59,000 birds were recorded here while the number of moulting Shelducks on Knechtsand in that year was relatively small (Drenckhahn et al., 1971).

Numbers recorded in the Jadebusen are shown in fig. 25. Numbers recorded in the Wadden Sea area between Juist and the Jadebusen are shown in Table 4. Additional figures from aerial midwinter surveys of the German Wadden Sea area between Borkum and Sylt had the fol-

Table 4. Number of Shelducks in the Wadden Sea between the island of Juist and the Jadebusen (source: Wissenschaftliche Arbeitsgemeinschaft für Natur- und Umweltschutz e.V., Jever and Arbeitskreis Natur- und Umweltschutz Harlingerland, Esens, unpubl.).

September	1976	5,355
October	1976	16,115
November	1976	18,435
December	1976	17,615
January	1977	9,560
February	1977	10,030
March	1977	7,180
April	1977	3,045

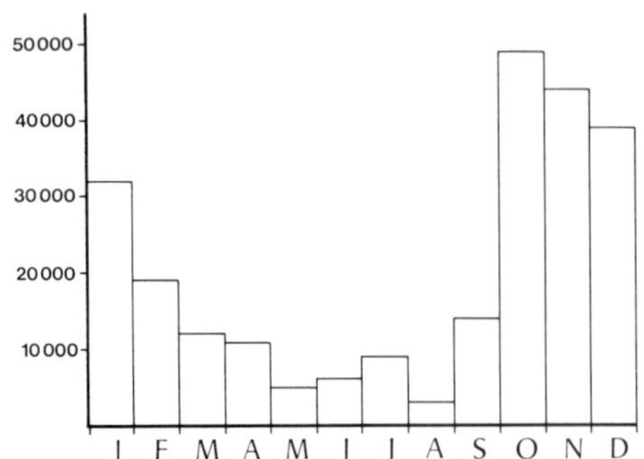

Fig. 26. Mean number of Shelducks per month in the Dutch part of the Wadden Sea (after Smit, in litt.).

Table 5. Number of Shelducks counted in the Dutch part of the Wadden Sea (after various publications in Limosa and Watervogels and Zegers, in litt.).

January		1973	33,820
January	12	1974	35,180
January	18	1975	30,150
January	17	1976	21,750
January	8	1977	44,230
January	14	1978	31,520
March	12	1977	12,630
March	13	1976	7,600
April	6	1973	15,800
April	19	1975	8,660
May	11	1974	5,800
May	13	1976	7,210
July	29	1972	5,670
August	22	1963	5,900
August	30	1975	7,855
September	1	1973	11,850
October	19	1974	35,000
November	16	1974	19,200
November	17	1973	44,655
November	25	1972	15,130
November	14	1976	42,390
December	29	1966	48,000

lowing results: January 11, 1964 7,581; January 19, 1965 20,960; January 16, 1968 1,040; February 4, 1969 11,391.

On January 16, 1968 there was pack ice on the coast of Schleswig-Holstein (source: Institut für Vogelforschung, Wilhelmshaven, unpubl.).

Fig. 26 gives mean numbers per month and Table 5 actually counted numbers in the Dutch part of the Wadden Sea showing low numbers until August and a sudden increase in September. This increase corresponds with the end of the moulting season in the German part of the Wadden Sea. The Dutch part is an important wintering area for the

Fig. 27. Regular peak occurrences of Shelducks in autumn per census area (after: Joensen, 1974; Smit, 1977; Busche, 1980). The dotted line indicates the major moulting areas around Knechtsand and Trischen.

Shelduck. Already in February birds leave for the breeding quarters (Doornbos, 1977; Smit 1977).

Fig. 27 shows regular peak occurrences in autumn in the Wadden Sea area. Data from all parts of the Wadden Sea show evidently that from October to February the whole area is a wintering region for a considerable part of the NW European Shelduck population.

3.5.4 Food

3.5.4.1 Food composition

The most important food organisms are molluscs, especially *Cerastoderma edule*, *Hydrobia ulvae*, *Macoma balthica*, small *Mya arenaria*, crustaceans and diatoms (Goethe, 1961a, 1961b; Bauer & Glutz, 1968).

Droppings collected south of Schiermonnikoog contained rests of filiform algae, *Enteromorpha*, *Laomedea*, *Nereis* and larvae of chironomids, in small quantities seed of phanerogames, Foraminifera, Coleoptera, Cyanophyta, parts of *Sphagnum*, Gramineae, *Anaitides*, *Pectinaria*, Oligochaeta, Nematoda, *Littorina*, *Gammarus* and Ostracoda were found. Chicks are chiefly eating crustaceans at first (Beintema, 1969).

The feeding grounds in the Wadden Sea generally are muddy sandflats (Mischwatt). According to Comes & Goethe (1978) the feeding areas in the Elbe estuary near Scharhörn largely coincide with the main areas of *Macoma balthica*, *Cerastoderma edule*, *Hydrobia ulvae* and *Zostera* spec. On Schiermonnikoog they often feed in the *Hydrobia* zone.

3.5.4.2 Feeding activities

Feeding techniques depend on mud- and water conditions. Shelducks use trampling, scything or dabbling, head dipping, upending and diving, the latter obviously in young birds (Goethe, 1961a; Swennen & Van der Baan, 1959). These types of food acquisition behaviour are very well adapted to the feeding situations in the Wadden Sea.Shelducks probably prefer fresh water as drinking water but many also drink sea water. If fresh water on the salt marshes is available, they make regular drinking flights to these areas (Swennen, in litt.). During their flight feather moult on Knechtsand no freshwater is available for several weeks. On the island of Schiermonnikoog it has been determined that Shelducks prefer feeding areas that can be used during low tide as well as during high tide. When numbers increase however these areas become fully occupied and areas that can only be used during low tide are visited as well. In such a case Shelducks carry out flights towards high tide roosts on the land (Zwarts, 1969). During moult the birds rest along beaches in shallow water. In case of bad weather they generally rest on high sand ridges. Dircksen (1968) stated, and Oelke (1969a) confirmed that flightless Shelducks use tidal drift in the gullies to reach the feeding grounds or resting places.

3.5.4.3 Total food consumption

No data available.

References

Atkinson-Willes, G.L., 1976. The numerical distribution of ducks, swans and coots as a guide in assessing the importance of wetlands in midwinter. Proc. Intern. Conf. Conservation Wetlands and Waterfowl, Heiligenhafen, 1974. IWRB, Slimbridge: p. 199-254.

Bauer, K.M. & U.N. Glutz von Blotzheim, 1968. Handbuch der Vögel Mitteleuropas, Vol. 2. Akademische Verlagsgesellschaft, Frankfurt/Main: 535 pp.

Beintema, A.J., 1969. Biologie van de bergeend op Schiermonnikoog. Doctoraalverslag Zool. Lab. University of Groningen.

Busche, G., 1980. Vogelbestände des Wattenmeeres von Schleswig-Holstein. Kilda, Greven (in press).

Comes, P. & F. Goethe, 1978. Die ornitho-ökologischen Verhältnisse im Seevogelschutzgebiet Scharhörn und im Scharhörn-Neuwerk-Watt. Hamb. Küstenforschung 38: 110 pp.

Coombes, R.A.H., 1950. The moult migration of the Shel-Duck. Ibis
    92: p. 405-418.

Cramp, S. & K.E.L. Simmons (eds.), 1977. Handbook of the Birds of
    Europe, the Middle East and North Africa, Vol. 1. Oxford Univer-
    sity Press, Oxford: 722 pp.

Dircksen, J., 1968. Brandgansmauserzug und tidenbedingte Bewegungen
    von Brandgans Tadorna tadorna und Eiderente Somateria mollissima
    im Raum von Trischen. Vogelwarte 24: p. 179-184.

Doornbos, G., 1977. Het belang van Groningse- en Friese Waddenkust
    voor de Bergeend (Tadorna tadorna L.) in de herfst en winter.
    Watervogels 2: p. 25-35.

Drenckhahn, D., R. Heldt jun. & R. Heldt sen., 1971. Die Bedeutung
    der Nordseeküste Schleswig-Holsteins für einige eurasische Wat-
    und Wasservögel mit besonderer Berücksichtigung des nordfriesischen
    Wattenmeeres. Natur und Landschaft 46: p. 338-346.

Dybbro, T., 1976. De danske ynglefugles udbredelse. Dansk Ornitholo-
    gisk Forening, København: 293 pp.

Goethe, F., 1957. Ueber den Mauserzug der Brandenten zum Grossen
    Knechtsand. In: Fünfzig Jahre Seevogelschutz. Festschrift Verein
    Jordsand. Meise, Hamburg: p. 96-106.

Goethe, F., 1961a. A survey of moulting Shelduck on Knechtsand. Bri-
    tish Birds 54: p. 106-115.

Goethe, F., 1961b. The moult gatherings and moult migration of the
    Shelduck in north-west Germany. British Birds 54: p. 145-161.

Groot, R.A. de, 1975. Slagpengroei en aanvullende gegevens over de
    energie balans van de bergeend (Tadorna tadorna L.). Doctoraal-
    verslag, University of Groningen.

Haftorn, S., 1971. Norges fugler. Universitetsforlaget, Oslo: 862 pp.

Hoogerheide, C. & J., 1958. Slagpenrui van de Bergeend in Artis. Ar-
    dea 46: p. 149-158.

Joensen, A.H., 1974. Waterfowl Populations in Denmark 1965-1973. Dan-
    ish Rev. Game Biology 9 (1): 206 pp.

Klafs, G. & J. Stübs, 1977. Die Vögel Mecklenburgs. Jena, 358 pp.

Lippens, L. & H. Wille, 1972. Atlas des oiseaux de Belgique et de
    l'Europe occidentale. Lannoo, Tielt: 833 pp.

Lohse, C., 1977. Ringfunde bei Brandgänsen (Tadorna tadorna). Aus-
    picium 6: p. 257-282.

Oelke, H., 1969a. Die Bedeutung des Grossen Knechtsandes als Mauser-
    gebiet der Brandgans. Landschaft und Stadt 3: p. 104-115.

Oelke, H., 1969b. Die Brandgans im Mausergebiet Grosser Knechtsand.
    J. Orn. 110: p. 170-175.

Onno, S., 1964. The number of waterfowl in Estonia. Wildfowl Trust
    Ann. Report 16: p. 110-114.

Petersen, F.D., 1974. Traekket af aender Anatidae ved Blaavand 1963-
    1971. Dansk Orn. Foren. Tidsskrift 68: p. 25-37.

Smit, C.J., 1977. On the occurrence of 32 bird species in the Danish,
    German and Dutch Wadden Sea. Unpubl. report International Wadden
    Sea Working Group, Part 2: 171 pp.

Swennen, C., 1979. Bergeend. In: R.M. Teixeira (ed.), Atlas van de
    Nederlandse broedvogels. Natuurmonumenten, 's-Graveland: p. 56-
    57.

Swennen, C. & G. van der Baan, 1959. Tracking birds on tidal flats
    and beaches. British Birds 52: p. 15-18.
Szijj, J., 1973. Breeding populations of Anatidae in the Federal Re-
    public Germany. Bull. Int. Waterfowl Res. Bureau 35: p. 14-15.
Voous, K.H., 1960. Atlas of European birds. Nelson, London: 284 pp.
Yarker, B. & G.L. Atkinson-Willes, 1971. The numeral distribution of
    some British breeding ducks. Wildfowl 22: p. 63-70.
Zwarts, L., 1969. Bergeend. Schierboek 3: p. 51-55.

3.6 WIGEON *(ANAS PENELOPE L.)*
   M. Fog

Da: Pibeand; G: Pfeifente; Du: Smient

3.6.1 Distribution

3.6.1.1 Breeding area
   The species has a trans-palaearctic breeding distribution, mainly in the boreal zone, to a lesser extent in the temperate and tundra
climatic zones (Voous, 1960). No subspecies are distinguished (fig.
28). It breeds along fresh-water pools, lakes and riverbanks in the
coniferous belt and scrub tundra.

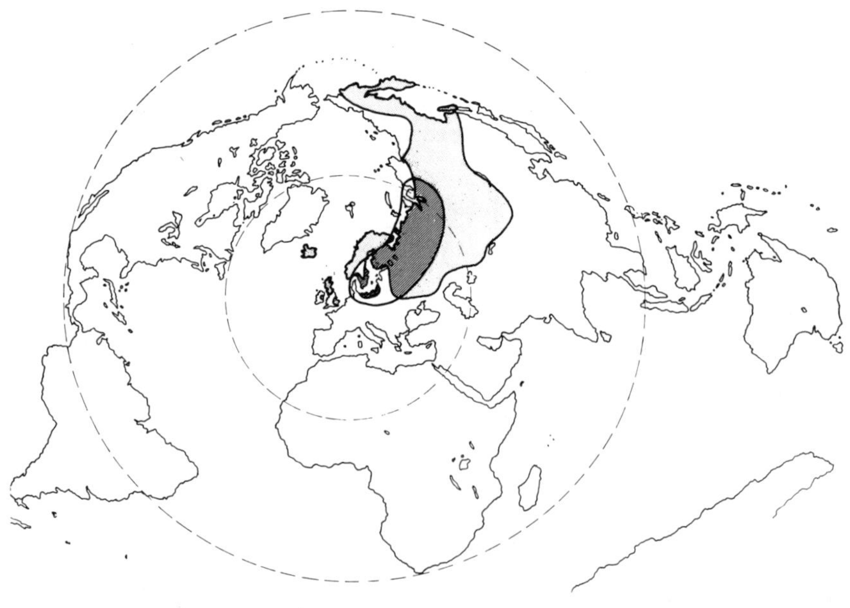

Fig. 28. Breeding area of Wigeon (after Voous, 1960). No subspecies
are distinguished. The dark shaded area approximately indicates the
region where birds visiting the Wadden Sea originate from.

3.6.1.2 Migration routes

According to ringing data the population passing or wintering in the Wadden Sea originates from Fennoscandinavia, European Russia and a huge area throughout West- and Central Siberia (fig. 28). There are indications that Wigeon carry out "Schleifenzug" (see also 3.6.2.1).

3.6.1.3 Wintering areas

The wintering area of the NW European flyway population ranges from the Wadden Sea (especially the Dutch part (Table 6)), inland grassland and wetland areas throughout the Netherlands (especially in the provinces Friesland and Noord-Holland (Table 7)), the British Isles (Prater, 1976), southward along the Atlantic coasts of France and the Iberian peninsula (Donker, 1959). Norwegian Wigeon probably winter on the British Isles (Haftorn, 1971). There are no indications that these birds pass through the Wadden Sea.

3.6.1.4 Moulting areas

Early arriving Wigeon moult in suitable places in the Wadden Sea area. Numbers however are small. At least a part of the few hundred males that are present every year in the Hauke Haienkoog in June-July moult flight feathers (Drenckhahn et al., 1971). In autumn the Wadden Sea, like other areas, acts as a moulting area when body feathers are renewed.

Table 6. Number of Wigeon determined by counts from the shore in the Dutch Wadden Sea. Results of recent January counts are given in Table 7.

Date	Year	Number	Source	Remarks
March 12	1977	42,259	v.d. Bergh et al., 1978	Lauwersmeer incl.
March 13	1976	24,479	"	"
April 4	1973	1,040	Boere & Zegers, 1975	
April 19	1975	10,500	Boere & Zegers, 1977	
May 1	1976	130	Zegers, in litt.	
May 11	1974	7	Boere & Zegers, 1977	
July 29	1972	13	Boere & Zegers, 1974	
August 22	1963	50	Rooth, 1966	
August 30	1975	850	Boere & Zegers, 1977	
September 1	1973	1,130	Boere & Zegers, 1975	
October 19	1974	89,470	Boere & Zegers, 1977	
November 16	1973	38,946	v.d. Bergh & Schäffner, 1977	
November 17	1974	153,101	"	
November 25	1972	80,324	"	Area covered partly, Lauwersmeer incl.
November	1976	163,655	Zegers, in litt.	
December 29	1966	16,500	Spaans, 1967	

Table 7. Number of Wigeon in January counts in the Netherlands.
Source 1972-1974: Van der Bergh & Schäffner (1977); 1975: Van der
Bergh et al. (1978); 1978: Van der Bergh (1979). These Wadden Sea
data all include Lauwersmeer except 1976 and 1977 (data for these
latter two years: Zegers, in litt.).

Area	1973	1974	1975	1976	1977	1978	Mean 1973-78	%
Wadden Sea	16,910	64,544	76,890	101,400	30,040	111,083	66,811	28
Friesland	4,702	13,721	8,776	56,474	71,184	103,205	43,010	18
Delta area	30,920	30,305	72,491	23,074	57,316	36,219	41,721	17
Noord-Holland	28,115	27,335	28,110	26,322	57,090	67,255	39,038	16
IJsselmeer	2,176	12,838	38,495	28,553	35,238	30,245	24,591	10
Rest of country	7,763	19,640	21,409	18,248	37,723	44,816	24,933	10
Total	90,586	168 383	246 171	254 071	288 591	392 823	240 104	100

3.6.2 Annual cycle

3.6.2.1 Migration

Recoveries of Wigeon ringed in the Netherlands and Great Britain
show that in July and August the population is present in the breeding area and along the coasts of the Baltic Sea (Donker, 1959). In
September and October numbers increase in all parts of the Wadden
Sea. Seasonal fluctuations in Schleswig-Holstein and the Netherlands
are shown in figs. 29 and 30. At least half of the NW European flyway population winters in Britain, the rest in the Netherlands, Ireland and along the Atlantic coasts of France (Atkinson-Willes, 1976;
Prater, 1976) and the Iberian peninsula (Donker, 1959). Ringing recoveries indicate that birds visiting the Wadden Sea mainly winter
in the Netherlands, the British Isles and along the Atlantic coast
of France. Low numbers in spring in Denmark (Joensen, 1974), Schleswig-Holstein (fig. 29 and the Netherlands (fig. 30), and recoveries
from Wigeon ringed in the Netherlands indicate that during spring
migration the birds take a more direct route over Central Europe towards the breeding grounds (Donker, 1959).

3.6.2.2 Moult

After the breeding season males (end July-January) and females
October-April) moult. Flight feathers of males are renewed from end
July-August, females follow later on (until October) (Bauer & Glutz,
1968).

3.6.2.3 Weight changes

Young male Wigeon (23) from Denmark weighed in their first winter
555-888 g (mean 706 g), 20 young females 576-811 g (mean 633 g), adult
males (42) 610-1073 g (mean 819 g), adult females (24) 552-962 g
(mean 724 g) (Schiøler in Bauer & Glutz, 1968). Unfortunately no detailed information from the Wadden Sea area was available.

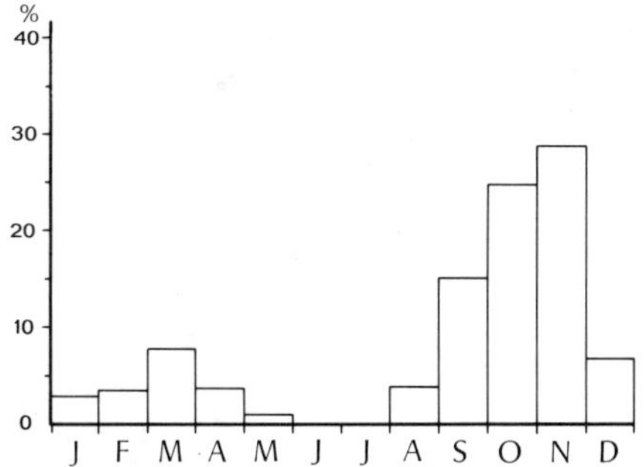

Fig. 29. Occurrence of the Wigeon in the Wadden Sea area in Schleswig-Holstein. Numbers for each month are expressed as a percentage of the total number of birds observed in all months (after Busche, 1980).

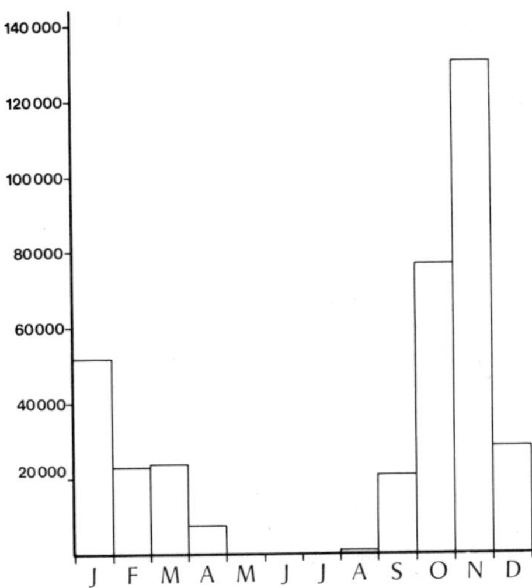

Fig. 30. Mean number of Wigeon per month in the Dutch part of the Wadden Sea (excluding Lauwersmeer) (after Smit, 1977).

3.6.3 Numbers

3.6.3.1 Population size

The size of the NW European, Mediterranean, Caspian Sea/Persian Gulf and Turkestan/Pakistan populations is estimated to amount to about 1,500,000 birds, the NW European population to about 400,000-500,000 (Atkinson-Willes, 1976).

Wigeon occasionally breed in the Wadden Sea area. This has been ascertained on Texel, Terschelling, Ameland, Borkum, Baltrum, Oland, Föhr and Rømø and in the Hullen area at the mouth of the Elbe estuary (Smit, pers. comm.). Partly these breeding birds originate from semi-domesticated Wigeon from duck decoys.

3.6.3.2 Numbers per area

In the Danish part of the Wadden Sea numbers reach maxima in October. During this month 20,000-55,000 Wigeon are present. In November numbers drop to 10,000-20,000, in winter and spring numbers are

Fig. 31. Regular peak occurrences of Wigeon
per census area in autumn. Based on Joensen
(1974), Smit (1977) and Busche (1980).

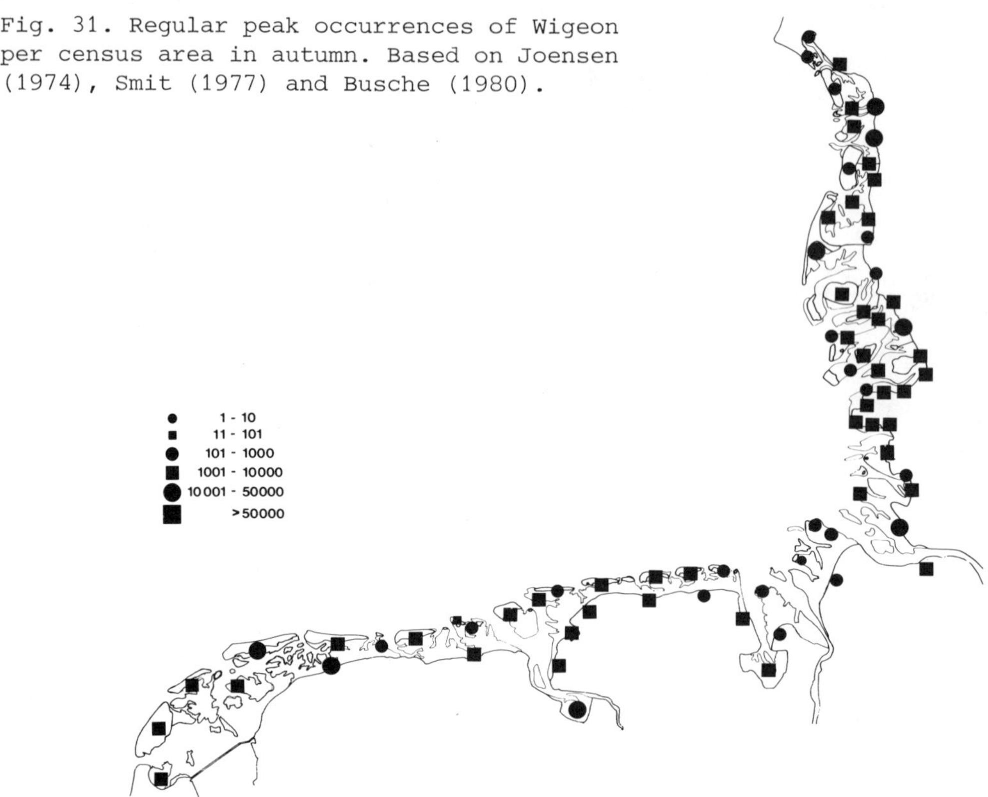

less than 10,000. Habitat preferences and behaviour during periods
with westerly storms resemble those of Mallards. It has also been
observed that movements to more sheltered waters take place, for example to the shallow coastal coves along the East coast of Jutland.
Flocks gathering here disappear again as soon as conditions in the
Wadden Sea improve (Joensen, 1974).

In Schleswig-Holstein a maximum number of 70,000 may be expected
in October-November. Occasional concentrations of 50,000 or more
(1964: 70,000 in Rantum Becken (Sylt)) have not been recorded any
more in the past years. Usually midwinter numbers vary from 2,000 to
6,000 (Busche, 1980).

According to Rettig (1979) Wigeon is most numerous in the NW part
of Ostfriesland in March, especially in the Leybucht. The same counts
for the Dollard where peak numbers in February and March are higher
than those in autumn (Dantuma & Glas, 1974). Fluctuations in numbers
in the Dutch Wadden Sea area are shown in fig. 30 and Tables 6 and 7.
Fig. 30 excludes numbers occurring in the Lauwerszee after its en-
closure. In October-November up to 60,000 Wigeon are found here regularly (Ebbinge, in litt.).

Distribution of Wigeon in the Wadden Sea area is shown in fig. 31.

3.6.4 Food

3.6.4.1 Food composition
The Wigeon feeds mostly on vegetable matter, such as leaves, stems, seeds as well as germules. On the tidal flats they may eat *Zostera*, *Enteromorpha* and *Chaetomorpha*, on the salt marshes *Salicornia*, *Ruppia* and grasses. Inland they take a great variety of plants, mono- as well as dicotyledonous (Bauer & Glutz, 1968). British studies show that on salt marshes they prefer *Puccinellia maritima* and *Agrostis stolinifera* over *Festuca rubra* (Cadwalladr et al., 1972).

Analysis from Wigeon killed on Terschelling in winter showed that stomachs contained especially seeds from *Heleocharis palustris*, several *Carex* and *Polygonum* species, grasses and *Empetrum nigrum* (De Vries, 1939). In the Wadden Sea no further investigations on the food of Wigeon have been carried out.

3.6.4.2 Drinking water
Like many other dabbling ducks Wigeon visit freshwater ponds, ditches etc. to drink.

3.6.4.3 Feeding activities
In undisturbed areas Wigeon feed by day in salt marshes and inland polders. After frequent disturbance however feeding only occurs in the dark while the day is spent by roosting on land as well as on the water. A thorough study of feeding and resting activities on salt marshes and tidal flats in the Wadden Sea has not been carried out.

3.6.4.4 Total food consumption
No information available.

References

Atkinson - Willes, G.L., 1976. The numerical distribution of ducks, swans and coots as a guide in assessing the importance of wetlands in midwinter. In: M. Smart (ed.), Proc. Int. Conf. Conserv. Wetlands and Waterfowl, Heiligenhafen 1974. IWRB, Slimbridge: p. 199-254.

Bauer, K.M. & U.N. Glutz von Blotzheim, 1968. Handbuch der Vögel Mitteleuropas, Vol. 2. Akademische Verlagsgesellschaft, Frankfurt/Main: 535 pp.

Bergh, L.M.J. van den, 1979. Verslag van de watervogeltellingen in januari en maart 1978. Watervogels 4: p. 48-72.

Bergh, L.M.J. van den & B.E. Schäffner, 1977. Verslag van de watervogeltellingen in de jaren 1972 t/m 1974. Watervogels 2 (extra nummer): p. 119-143.

Bergh, L.M.J. van den, B.E. Schäffner & J.J. Smit, 1978. Verslag van de watervogeltellingen in de jaren 1975-1977. Watervogels 3 (extra nummer): p. 43-71.

Boere, G.C. & P.M. Zegers, 1974. Wadvogeltelling in het nederlandse

Waddengebied in juli 1972. Limosa 47: p. 23-28.

Boere, G.C. & P.M. Zegers, 1975. Wadvogeltellingen in het Nederlandse Waddengebied in april en september 1973. Limosa 48: p. 74-81.

Boere, G.C. & P.M. Zegers, 1977. Watervogeltellingen in het Nederlandse Waddenzeegebied in 1974 en 1975. Watervogels 2: p. 161-173.

Busche, G., 1980. Vogelbestände des Wattenmeeres von Schleswig-Holstein. Kilda, Greven (in press).

Cadwalladr, D.A., M. Owen, J.V. Morley & R.S. Cook, 1972. Wigeon (Anas penelope L.) conservation and salting pasture management at Bridgwater Bay National reserve, Somerset. J. Appl. Ecol. 9: p. 417-425.

Dantuma, R. & P. Glas, 1974. Aantalsverloop van de trekvogels in de Dollard. In: Werkgroep Dollard, Dollard, portret van een landschap: p. 35-45.

Donker, J.K., 1959. Migration and distribution of the Wigeon, Anas penelope L., in Europe, based on ringing results. Ardea 47: p. 1-27.

Drenckhahn, D., R. Heldt jun. & R. Heldt sen., 1971. Die Bedeutung der Nordseeküste Schleswig-Holsteins für einige eurasische Wat- und Wasservögel mit besonderer Berücksichtigung des Nordfriesischen Wattenmeeres. Natur und Landschaft 46: p. 338-346.

Haftorn, S., 1971. Norges fugler. Universitetsforlaget, Oslo: 862 pp.

Joensen, A.H., 1974. Waterfowl populations in Denmark 1965-1973. A survey of the non-breeding populations of ducks, swans and coot and their shooting utilization. Dan. Rev. Game Biol. 9(1): p. 1-206.

Prater, A.J., 1976. Birds of Estuaries Enquiry 1973-1974. British Trust for Ornithology, Royal Society Protection of Birds, Wildfowl Trust: 47 pp.

Rettig, K., 1979. Erweiterte Artenliste der Vogelwelt im nordwestlichen Ostfriesland nebst Literaturübersicht, Teil I. Emden, 30 pp.

Rooth, J., 1966. Vogeltelling in het hele Nederlandse Waddengebied in augustus 1963. Limosa 39: p. 175-181

Smit, C.J., 1977. On the occurrence of 32 bird species in the Danish, German and Dutch Wadden Sea. Unpubl. report Intern. Wadden Sea Working Group, part 2: 171 pp.

Spaans, A.L., 1967. Wadvogeltelling in het gehele Nederlandse Waddengebied in december 1966. Limosa 40: p. 206-215.

Voous, K.H., 1960. Atlas of European birds. Nelson, London: 284 pp.

Vries, V. de, 1939. Bijdrage tot de voedselbiologie van een viertal eendensoorten, naar aanleiding van materiaal, afkomstig van Vlieland en Terschelling. Limosa 12: p. 87-98.

3.7 TEAL *(ANAS CRECCA* L.*)*
M. Fog

Da: Krikand; G: Krickente; Du: Wintertaling

3.7.1 Distribution

3.7.1.1 Breeding area
The species breeds holarctic in the boreal, temperate, steppe, tundra and the Mediterranean climatic zones (fig. 32) (Voous, 1960). It breeds along freshwater ponds, ditches and lakes with a well developed shore vegetation, in moorlands and locally in coastal marshland.

3.7.1.2 Migration routes
Teal passing and wintering in the Wadden Sea originate mainly from Fennoscandinavia and the Western part of the USSR (fig. 32). After the breeding season part of the population interrupts autumn migration and gathers to moult. Probably autumn migration takes place on a broad front. Because of good feeding- and roosting possibilities the Wadden Sea area acts as a bottle neck in which migrating Teal concentrate (Lebret, 1947; Wolff, 1966). As numbers in spring in all parts of the Wadden Sea are relatively small Busche (1980) suggests that Teal carry out "Schleifenzug". Maps 11-18 in Wolff (1966) indicate that Teal after being ringed in autumn in the Netherlands in

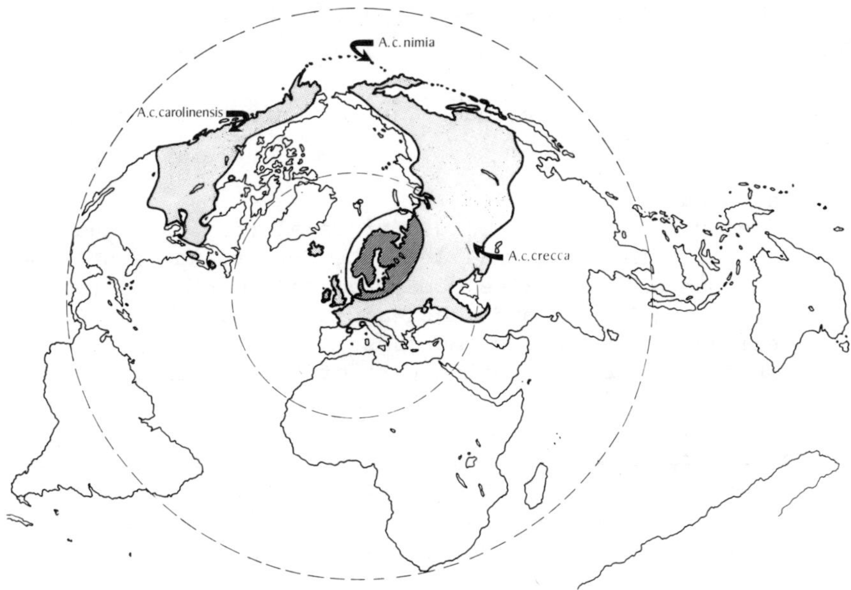

Fig. 32. Breeding area of Teal (after Voous, 1960). The dark shaded area approximately indicates the region where birds visiting the Wadden Sea originate from.

Fig. 33. Recoveries of Teal ringed on the island of Texel from 1965-1973 (after unpublished data Vogeltrekstation, Arnhem).

spring migrate on a broad front through Central Europe towards the breeding places in the north.

3.7.1.3 Wintering areas

According to recoveries of Teal ringed at Texel (fig. 33) and Fanø, wintering areas are the British Isles, the Netherlands and France (Wolff, 1966). Incidental recoveries come from the Iberian peninsula, Italy and Greece. In the Netherlands the Wadden Sea acts as a wintering area, as well as the Delta and the IJsselmeer (Table 8). Small numbers winter in the German and Danish part of the Wadden Sea.

3.7.1.4 Moulting areas

Male Teal moult throughout the breeding area and to the south and southwest of it where ever they find a suitable place. Females which succeed in raising their brood seem to moult mainly in the breeding area (Wolff, 1966). On some places groups of Teal, mainly males, gather to moult. Moulting concentrations are found on lakes in the USSR, Finland and Sweden and along the coasts of the Baltic. Parts of the Wadden Sea act as a moulting area too, probably especially for those birds breeding in its surroundings. Moulting concentrations are known from the Rantum Becken at Sylt, Hauke Haienkoog (Busche, 1980) and possibly the Dollard. For moulting areas, Teal prefer shallow waters, rich in food with a dense vegetation along the edges which offers possibilities to hide.

3.7.2 Annual cycle

3.7.2.1 Migration

Shortly after the beginning of the breeding season males already leave the breeding area and wander around for some time until moult begins. Part of these Teal gather in moulting groups. In the Wadden Sea in Schleswig-Holstein arrival already starts in July, in Denmark, Niedersachsen and the Netherlands in August. Teal arriving first are mostly males, females follow later on in the season after termination of the breeding season. Wolff (1966) showed that Teal present in the Netherlands in September and October interrupt their autumn migration for some time. Whether this also holds for the population present in the whole Wadden Sea area is uncertain, fact is that numbers in the Schleswig-Holstein and Dutch part are rather constant from September to November. In November and December Teal are passing rapidly through the Netherlands, but in mild winters a number of them stays here (Wolff, 1966; Tables 8 and 9), as well as along the Schleswig-Holstein West coast (Busche, 1980). Teal return from their winter quarters from February to April (figs. 34 and 35), in Denmark some do not leave until May (Joensen, 1974).

Table 8. Number of Teal in January counts in the Netherlands. Source 1972-1974: Van der Bergh & Schäffner (1977); 1975-1977: Van der Bergh et al. (1978); 1978: Van der Bergh (1979). All data for the Wadden Sea include the Lauwersmeer.

Area	1973	1974	1975	1976	1977	1978	Mean 1973-1978	%
Delta	11,543	9,011	22,374	9,684	11,587	8,505	12,117	36
Wadden Sea	333	3,383	13,175	25,994	3,794	5,583	8,710	26
Duck decoys	1,209	3,006	6,280	8,075	3,176	3,864	4,268	13
Rivers Rhine, Meuse, Waal, IJssel	1,382	1,624	3,333	3,012	1,615	2,220	2,198	6
IJsselmeer	827	990	6,262	1,771	628	2,058	2,089	6
Rest of country	836	1,394	5,206	4,621	1,350	14,408	4,639	14
Total	16,130	19,408	56,630	53,157	22,150	36,638	34,019	100

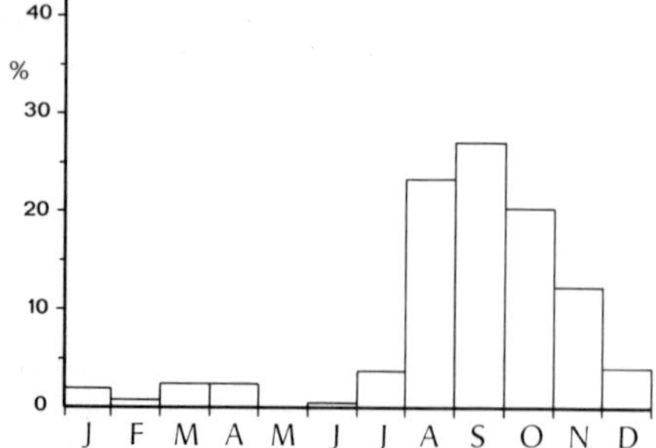

Fig. 34. Occurrence of the Teal in the Wadden Sea area in Schleswig-Holstein. Numbers for each month are expressed as a percentage of the total number of birds observed in all months (after Busche, 1980).

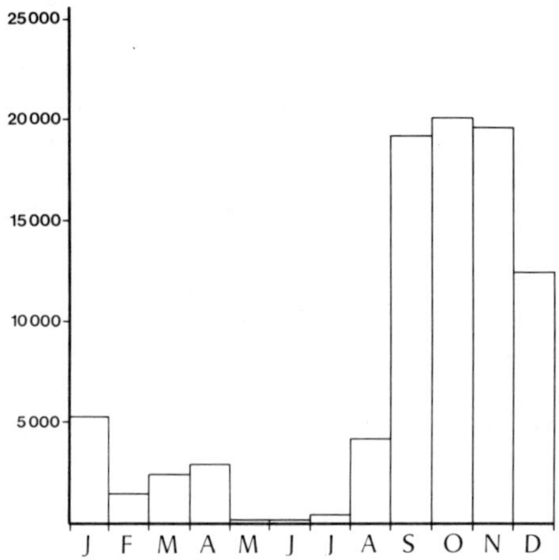

Fig. 35. Mean number of Teal per month in the Dutch part of the Wadden Sea (excluding Lauwersmeer) (after Smit, 1977).

Table 9. Number of Teal determined by counts from the shore in the Dutch Wadden Sea. Results of recent January counts are given in Table 8.

Date	Year	Number	Source	Remarks
March 12	1977	6,675	Van der Bergh et al., 1978	Lauwersmeer incl.
March 13	1976	1,957	"	"
April 6	1973	6,900	Boere & Zegers, 1975	
April 19	1975	3,350	Boere & Zegers, 1977	
May 1	1976	510	Zegers, in litt.	
May 11	1974	67	Boere & Zegers, 1977	
July 29	1972	780	Boere & Zegers, 1974	
August 22	1963	30,000	Rooth, 1966	
August 30	1975	2,300	Boere & Zegers, 1977	
September 1	1973	5,950	Boere & Zegers, 1975	
October 19	1974	10,670	Boere & Zegers, 1977	
November 16	1973	41,589	Van der Bergh & Schäffner, 1977	"
November 17	1974	80,754	"	"
November 25	1972	7,840	"	"
November	1976	13,145	Zegers, in litt.	
December 29	1966	13,500	Spaans, 1967	Area only partly covered. Lauwersmeer incl.

3.7.2.2 Moult

Males and those females of which the brood has been disturbed moult wing feathers from mid-June until mid-August. Ducks which succeeded in raising young, moult later. Body feathers are renewed from June-July until September-October (males) and October-November (females). Moult of Teal however has not been studied intensively (Bauer & Glutz, 1968).

3.7.2.3 Weight changes

Extensive information on weight changes in the Camargue is given by Bauer & Glutz (1968). The only information on weights from the Wadden Sea area (including Teal from Vejlerne (W·Jutland)) are some autumn weights. For adult males (n=70) the range was 220-405 g (mean 341 g), for adult females (n=45) 245-355 g (mean 311 g) (unpubl. data Game Biology Station, Kalø). During the moulting season weights drop.

3.7.3 Numbers

3.7.3.1 Population size

The size of the Western palaearctic population is estimated to amount to at least 2.4 million Teal. Of these 150,000 belong to the

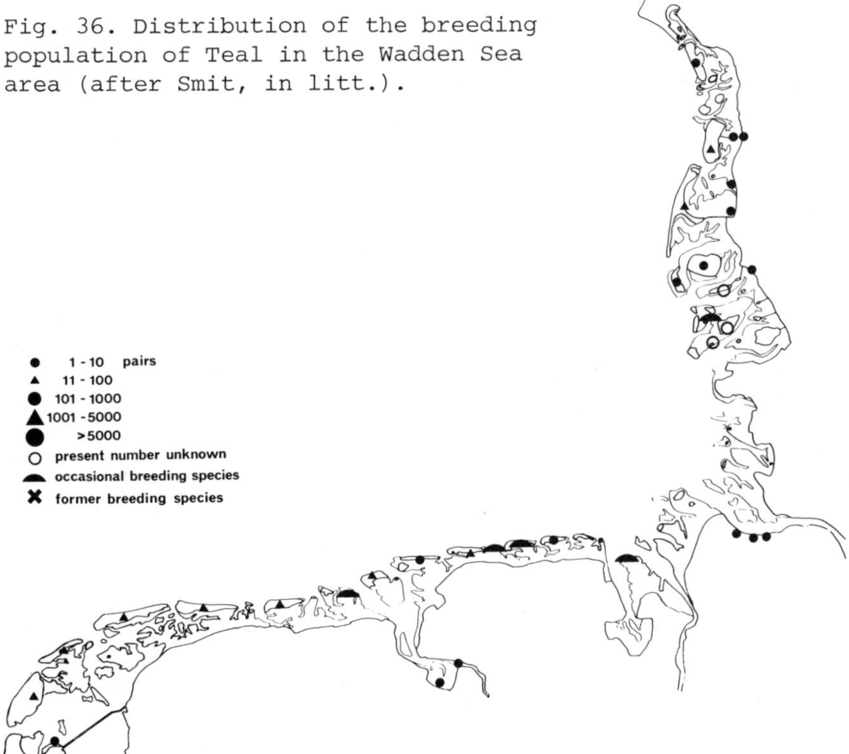

Fig. 36. Distribution of the breeding population of Teal in the Wadden Sea area (after Smit, in litt.).

NW European flyway population, 750,000 to the Black Sea/Mediterranean population and the Middle East Caspian Sea/Persian Gulf and Turkestan/Pakistan populations together to at least 1,500,000 (Atkinson-Willes, 1976). The size of the breeding population in the Wadden Sea area amounts to about 250-350 pairs. Its distribution is shown in fig. 36 (Lauwersmeer and inland polder areas on the mainland excluded).

3.7.3.2 Number per area

In the Danish Wadden Sea numbers are highest from September to November. Peak numbers occur in November. About 11,000 have been counted in November 1968, 9000 in November 1969 and 7000 in October 1971. From December to February few remain in the area. Numbers in spring are less than 5000. Occurrence in the Wadden Sea is more or less confined to sheltered coastal areas (Joensen, 1974 and unpubl. data Game Biology Station, Kalø.

Busche (1980) estimates numbers in Schleswig-Holstein in September at 20,000-25,000 and in mild winters in December-January at 3000. In the cold January of 1970 11 Teal have been counted. Fig. 34 shows fluctuation in numbers throughout the year.

In Niedersachsen peak numbers occur from September-November. In the Eastern part of the area numbers in winter become very small. Generally in the whole area numbers in spring are small as well. In this time of the year larger concentrations of Teal can only be found in the Dollard.

Population fluctuations in the Netherlands are shown in fig. 35. Most important areas for Teal are the Dollard where up to 20,000 have been counted regularly in October, the Frisian coast (though numbers in January/February are small in these areas) and the islands of Terschelling and Texel. Since the endikement of the Lauwerszee large numbers of Teal can be found here. Up to 65,000 have been counted in November (Ebbinge, pers. comm.). Fig. 37 shows regular peak occurrences per area.

3.7.4 Food

3.7.4.1 Food composition

In the Dollard Teal feed on mudflats, often along the edges of the salt marshes. Food mainly consists of seeds of *Aster tripolium*, *Scirpus maritima*, *Atriplex hastata* and *Phragmites australis* (Glas et al., 1974). In Teal killed on Vlieland and Terschelling in winter seeds were found of *Heleocharis palustris*, *Salicornia herbacea*, Gramineae, *Spergularia marina* and *marginata*, *Ranunculus flammula*, *Aster tripolium* and *Suaeda maritima* and berries of *Empetrum nigrum* (De Vries, 1939), but they may also take seeds from *Atriplex hastata Potamogeton* spec., *Carex* spec., small bivalves, insects and various plant material. In the Lauwersmeer Teal mainly take seeds from *Spergularia* and larvae of Diptera *(Chironomus)* (Prop, pers. comm.). In Schleswig-Holstein they have been observed to feed on the uppermost mud-layers containing especially diatoms, meiofauna and perhaps some seeds (König, in litt.). Rather often they prefer *Salicornia-* and

Fig. 37. Regular peak occurrences of Teal
per census area in autumn. Based on Joensen
(1974), Smit (1977) and Busche (1980).

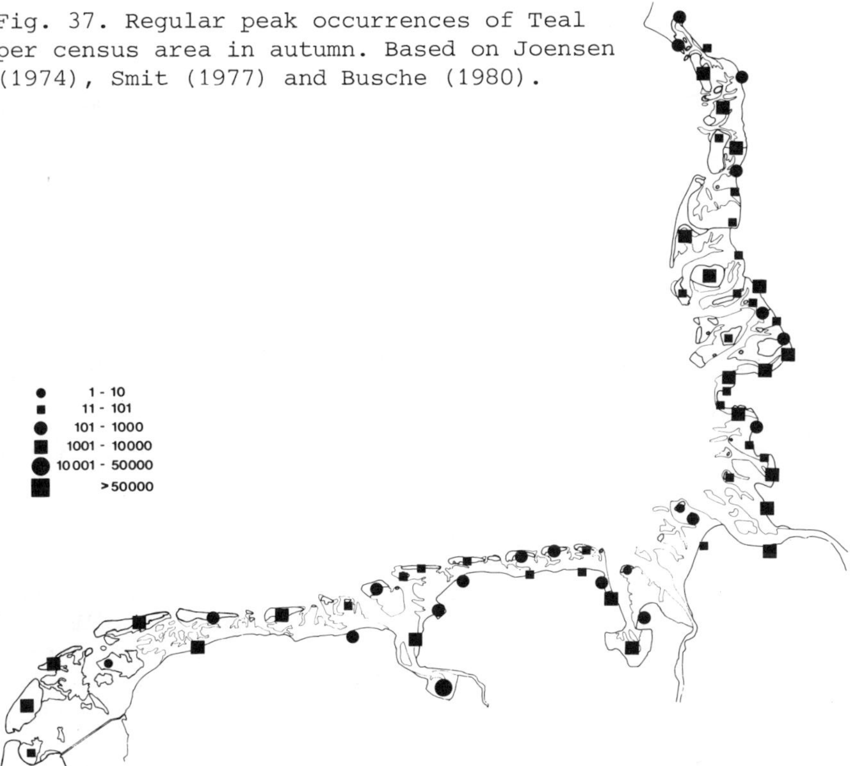

Spergularia fields to feed in (Busche, 1980). Moreover they feed in
marshy inland polder areas. Detailed information on the food choise
in British estuaries can be found in Olney (1963), on the food in
the Camargue in Tamisier (1971a, 1971b).

3.7.4.2 Feeding activities

Generally feeding activities start after sunset. Daytime haunts
are densily vegetated lakes and sheltered coastal marshland, in the
Danish Wadden Sea often *Spartina* areas (Joensen, 1974) or the edges
of the salt marshes (Glas et al., 1974). Moreover the Wadden Sea
acts as a roosting place. Food may be taken by dabbling or by fil-
tering it from the mudflats. Seeds may also be stripped from the
plants. Roosting places may be situated at some distance from feeding places (Fog, 1968).

Drinking water is taken from ditches, pools, duck decoys etc.

3.7.4.3 Total food consumption

No information from the Wadden Sea available.

References

Atkinson-Willes, G.L., 1976. The numerical distribution of ducks, swans and coots as a guide in assessing the importance of wetlands in midwinter. In: M. Smart (ed.), Proc. Int. Conf. Conserv. Wetlands and Waterfowl, Heiligenhafen 1974. IWRB, Slimbridge:p. 199-254.

Bauer, K.M. & U.N. Glutz von Blotzheim, 1968. Handbuch der Vögel Mitteleuropas, Vol. 2. Akademische Verlagsgesellschaft, Frankfurt/Main: 535 pp.

Bergh, L.M.J. van den, 1979. Verslag van de watervogeltellingen in januari en maart 1978. Watervogels 4: p. 48-72.

Bergh, L.M.J. van den & B.E. Schäffner, 1977. Verslag van de watervogeltellingen in de jaren 1972 t/m 1974. Watervogels 2 (extra nummer): p. 119-143.

Bergh, L.M.J. van den, B.E. Schäffner & J.J. Smit, 1978. Verslag van de watervogeltellingen in de jaren 1975-1977. Watervogels 3 (extra nummer): p. 43-71.

Boere, G.C. & P.M. Zegers, 1974. Wadvogeltelling in het Nederlandse Waddengebied in juli 1972. Limosa 47: p. 23-28.

Boere, G.C. & P.M. Zegers, 1975. Wadvogeltellingen in het Nederlandse Waddengebied in april en september 1973. Limosa 48: p. 74-81.

Boere, G.C. & P.M. Zegers, 1977. Watervogeltellingen in het Nederlandse Waddenzeegebied in 1974 en 1975. Watervogels 2: p. 161-173.

Busche, G., 1980. Vogelbestände des Wattenmeeres von Schleswig-Holstein. Kilda, Greven (in press).

Fog, J., 1968. Krikandens (Anas crecca) spredning under foureringstogter fra en rasteplads (Albuebugten vildtreservat, Fanø). Dansk Orn. Foren. Tidsskr. 62: p. 32-36.

Glas, P., J.B. Hulscher & S.T. Tjallingii, 1974. Terreingebruik van de vogels in de Dollard. In: Werkgroep Dollard, Dollard, portret van een landschap, Harlingen: p. 47-55.

Joensen, A.H., 1974. Waterfowl populations in Denmark 1965-1973. A survey of the non-breeding populations of ducks, swans and coot and their shooting utilization. Dan. Rev. Game Biol. 9(1): p. 1-206.

Lebret, T., 1947. The migration of the Teal, Anas crecca crecca L., in western Europe. Ardea 35: p. 79-130.

Olney, P.J.S., 1963. Food and feeding habits of Teal, Anas crecca. Proc. Zool. Soc. London 140: p. 169-210.

Rooth, J., 1966. Vogeltelling in het hele Nederlandse Waddengebied in augustus 1963. Limosa 39: p. 175-181.

Smit, C.J., 1977. On the occurrence of 32 bird species in the Danish, German and Dutch Wadden Sea. Unpubl. report Intern. Wadden Sea Working Group, part 2: 171 pp.

Spaans, A.L., 1967. Wadvogeltelling in het gehele Nederlandse Waddengebied in december 1966. Limosa 40: p. 206-215.

Tamisier, A., 1971a. Les biomasses de nourriture disponible pour les sarcelles d'hiver Anas crecca crecca en Camargue. La Terre et la Vie 1971: p. 344-377.

Tamisier, A., 1971b. Régime alimentaire des sarcelles d'hiver Anas

crecca L. en Camargue. Alauda 39: p. 261-310.
Voous, K.H., 1960. Atlas of European birds. Nelson, London: 284 pp.
Vries, V. de, 1939. Bijdrage tot de voedselbiologie van een viertal
    eendensoorten, naar aanleiding van materiaal, afkomstig van Vlie-
    land en Terschelling. Limosa 12: p. 87-98.
Wolff, W.J., 1966. Migration of Teal ringed in the Netherlands. Ar-
    dea 54: p. 230-270.

3.8 MALLARD *(ANAS PLATYRHYNCHOS L.)*
   M. Fog

Da: Graaand; G: Stockente; Du: Wilde Eend

3.8.1 Distribution

3.8.1.1 Breeding area
   The species has a holarctic breeding distribution in the boreal,
temperate, Mediterranean, steppe, desert as well as in the tundra
climatic zones (Voous, 1960). Generally four subspecies are distinguished (fig. 38). Mallards breed in every type of wetland habitat:
ponds, brooks, moors, on salt marshes and islands and along coasts
and streams. It is a highly adaptable species, common in cultivated
areas.

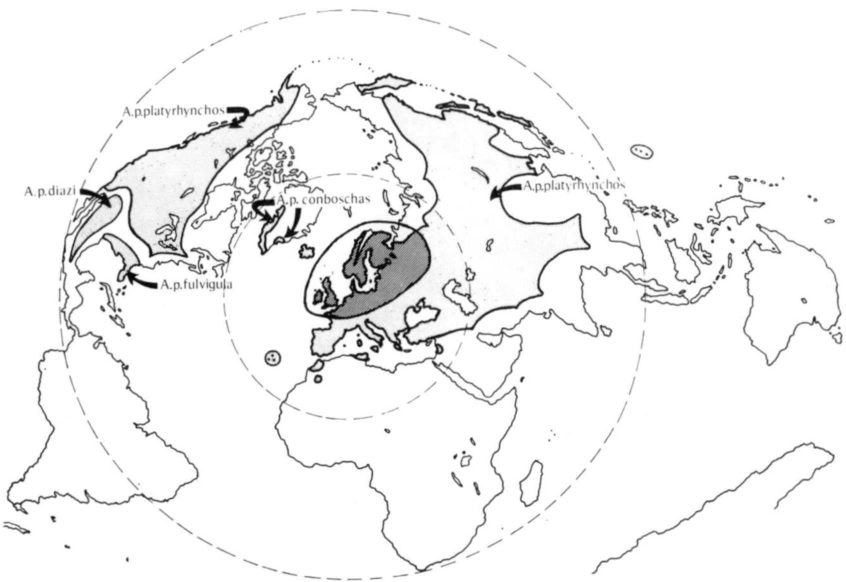

Fig. 38. Breeding area of Mallard (after Voous 1960, modified). The
dark shaded area indicates approximately the region where birds visiting the Wadden Sea originate from.

Fig. 39. Recoveries of Mallards, ringed at the island of Texel from 1965-1973 (after unpublished data Vogeltrekstation, Arnhem).

Fig. 40. Recoveries of Mallards ringed at the island of Fanø from 1965-1973 (after Fog, 1974).

3.8.1.2 Migration routes

Only the northernmost populations are migratory. Populations breeding in W Europe are sedentary or disperse somewhat after the breeding season. According to ringing data the population passing or wintering in the Wadden Sea originates from NW USSR, Fennoscandinavia and Denmark (figs. 39 and 40).

3.8.1.3 Wintering area

According to data acquired by ringing Mallards in the duck decoys of Texel (fig. 39) and Fanø (fig. 40) the Wadden Sea population winters in a wide area all over NW Europe.

3.8.1.4 Moulting areas

Moulting Mallards can be found in most types of wetland habitat, generally in small flocks, often near the breeding places. Precondition is that enough cover must be present. In some lakes or marsh areas flocks of some hundreds gather to moult. In the Wadden Sea area small concentrations (some hundreds) of moulting birds are known from the Rantum Becken (Sylt) (Busche, 1980), Hauke Haienkoog, mainland coast of Friesland (Timmerman, 1974), Schorren (Texel) (Swennen, pers. comm.) and the Dutch part of the Dollard.

3.8.2 Annual cycle

3.8.2.1 Moult

In the Netherlands and Germany males start moulting in June, in Denmark in July. In August males are able to fly again, females finish moult of flight feathers in September. In the Dutch Wadden Sea Mallards moulting flight feathers are predominantly males. Body feathers are renewed from June-July until September-October (males) or October-November (females).

3.8.2.2 Migration

The peak in June (fig. 41) in numbers registered in the Dutch Wadden Sea will be due to migration towards moulting areas and because the area acts as a moulting area itself. By the time males are able to fly again (August) numbers in the Wadden Sea increase (figs. 41 and 42). Females follow in September. In October and November Fennoscandinavian and Russian birds arrive in the Wadden Sea area. At the same time a departure takes place to wintering places elsewhere in W Europe. In February Mallards already leave for the breeding places.

3.8.2.3 Weight changes

Fifteen female Mallards from Denmark in December had a mean weight of 1254 g (range 1095-1520 g), 13 males 1039 g (940-1200 g) (Schiøler, 1925). Unfortunately there is no information from the Wadden Sea area available.

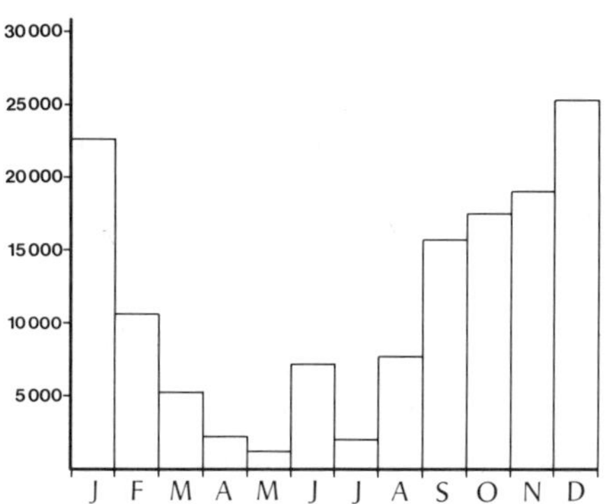

Fig. 41. Mean number of Mallards per month in the Dutch part of the Wadden Sea (excluding Lauwersmeer) (after Smit, 1977).

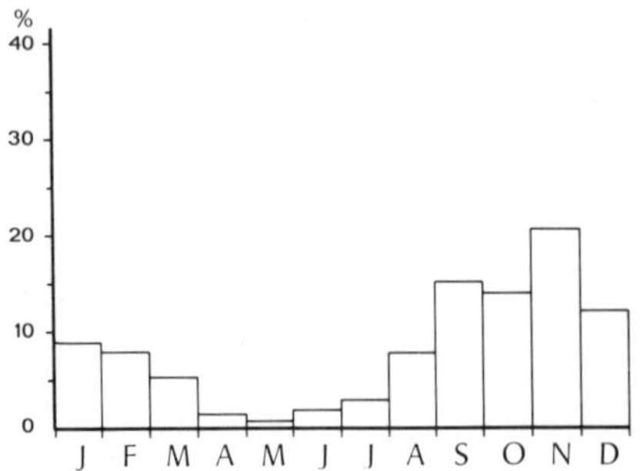

Fig. 42. Occurrence of the Mallard in the Wadden Sea area in Schleswig-Holstein. Numbers for each month are expressed as a percentage of the total number of birds observed in all months (after Busche, 1980).

3.8.3 Numbers

3.8.3.1 Population size

The size of the Western palaearctic population is estimated at 4 to 5 million. Of these at least 1½ million belong to NW European flyway population, 1½ million to the Black Sea/Mediterranean population and 1 million to the Middle East/Caspian population (Atkinson-Willes, 1976). In the NW European population bastardizing occurs with domesticated ducks. The size and distribution of the breeding population in the Wadden Sea area are shown in fig. 43. The population breeding on islands and salt marshes along the mainland coast counts 3900-4600 pairs (excluding Lauwersmeer and inland polders on the mainland).

3.8.3.2 Numbers per area

Numbers in the Wadden Sea area in Schleswig-Holstein and the Netherlands are shown in figs. 41 and 42. Numbers regularly occurring in autumn per census area are shown in fig. 44. Probably these numbers are underestimated since during high tide large numbers may be present on the Wadden Sea out of sight from the shore.

Numbers determined by aerial surveys in the Danish part of the Wadden Sea are shown in Table 10. During low tide, like Wigeon and Shelduck, Mallards are scattered over the tidal flats. During high tide, especially during westerly storms, they concentrate near the Rømø-dam, Mandø, Højer, the West coast near Ribe, Skallingen and in the Ho bugt (Joensen, 1974). The January population in Denmark (including the Wadden Sea) amounts to about 9,000 - 150,000 birds (Joensen, 1974).

In Schleswig-Holstein the West coast population in winter varies from 3500 to 19,000. In cold winters (January, 1970) there may be less than 1000. The average November population amounts to about 50,000 (Busche, 1980).

In Niedersachsen fluctuations in numbers are more or less comparable with those in the Dutch part of the Wadden Sea. The species is most numerous along the mainland coast. Numbers in winter however generally are somewhat lower as compared to the Netherlands.

Mean numbers per month in the Dutch part are shown in fig. 41, actual counting data in Tables 11 and 12.

Fig. 43. Distribution of the breeding population
of Mallard in the Wadden Sea area (after Smit,
in litt.).

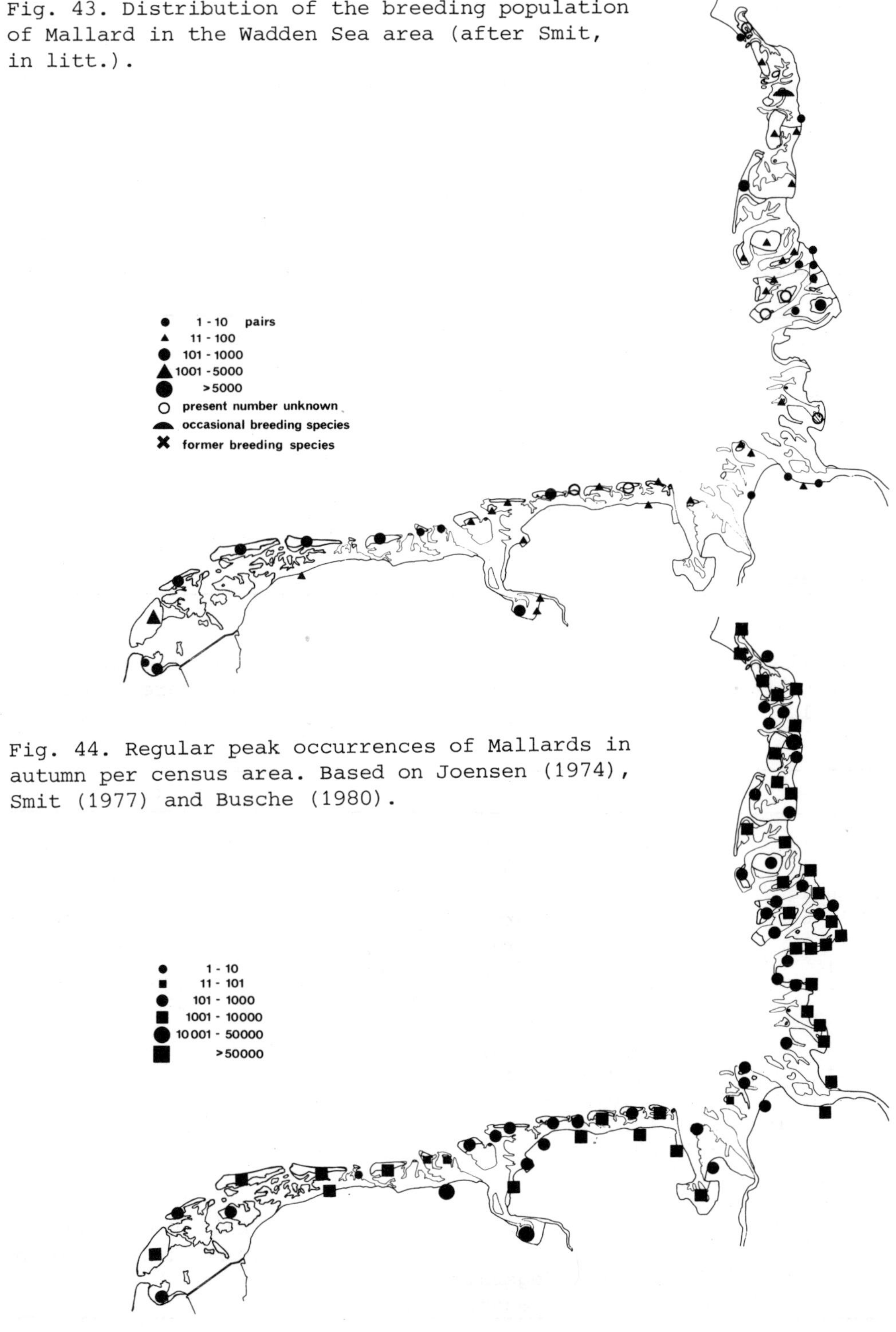

Fig. 44. Regular peak occurrences of Mallards in
autumn per census area. Based on Joensen (1974),
Smit (1977) and Busche (1980).

3.8.4 Food

3.8.4.1 Food composition

Composition of the food is extremely varied and dependent on what can be found. An extensive survey from English estuarine and coastal areas is given by Olney (1964). In Danish waters seeds (especially

Table 10. Approximate number of Mallards determined by aerial surveys in the Danish Wadden Sea (data: Game Biology Station, Kalø).

Month	Year	Number	Remarks
January	1969	30,000	
January	1970	2,000	Area totally covered with ice.
January	1971	11,000	
March	1969	25,000	
September	1970	7,000	
October	1969	7,000	
October	1971	8,000	
November	1968	32,000	
November	1969	19,000	
December	1972	38,000	

Table 11. Number of Mallards determined by counts from the shore in the Dutch Wadden Sea. Results of recent January counts are given in Table 12.

Date	Year	Number	Source	Remarks
March 12	1977	10,022	Van den Bergh et al., 1978	Lauwersmeer incl.
March 13	1976	6,844	"	"
April 6	1973	3,100	Boere & Zegers, 1975	
April 19	1975	1,630	Boere & Zegers, 1977	
May 1	1976	1,675	Zegers, in litt.	
May 11	1974	1,260	Boere & Zegers, 1977	
July 29	1972	1,690	Boere & Zegers, 1974	
August 22	1963	5,000	Rooth, 1966	
August 30	1975	6,990	Boere & Zegers, 1977	
September 1	1973	12,020	Boere & Zegers, 1975	
October 19	1974	10,315	Boere & Zegers, 1977	
November 16	1974	20,793	Van den Bergh & Schäffner, 1977	"
November 17	1973	27,692	"	"
November 25	1972	14,908	"	Lauwersmeer incl. Area partly covered
November	1976	25,400	Zegers, in litt.	
December 29	1966	15,000	Spaans, 1967	

Table 12. Number of Mallards in January counts in the Netherlands. Source 1972-1974: Van den Bergh & Schäffner (1977); 1975-1977: Van den Bergh et al. (1978); 1978: Van den Bergh (1979). Wadden Sea data include Lauwersmeer.

Area	1973	1974	1975	1976	1977	1978	Mean 1973-1978	%
Delta area	40,754	43,845	103,245	33,290	36,143	58,324	53,000	22
Duck decoys (whole country)	35,336	50,962	53,557	49,480	32,103	41,548	44,000	18
IJsselmeer	35,126	56,139	30,199	27,874	37,177	53,898	40,000	16
Wadden Sea	21,449	16,267	26,373	26,497	17,601	43,641	25,000	10
Rivers Rhine, Waal, Meuse, IJssel	15,287	19,012	24,579	31,590	28,764	27,220	24,000	10
Rest of the country	44,788	45,330	54,779	73,286	45,066	84,241	58,000	24
Total	192,740	231,555	292,732	242,017	196,854	308,872	244,000	100

Scirpus, *Potamogeton*, grains) were found in 96% of the stomachs, grass and plant leaves in 22% and animal food (crustaceans, snails, insects) in 27% (Spärck, 1958). Out of 14 stomachs from Vlieland (October-December 1964) in two, considerable amounts of seeds were found. Ten contained animal food (especially *Gammarus locusta*, *Carcinus maenas*, *Hydrobia ulvae*, *Macoma balthica* and *Mytilus edulis*). Besides small amounts of *Balanus* spec., *Littorina littorea*, *Cerastoderma edule* (in six) and *Tellina tenuis* (in one). Highest number of *Hydrobia* in a stomach was 8275, *Gammarus* 42, *Mytilus* 21, *Carcinus* 8 and *Macoma* 6 (Swennen, pers. comm.). In the Dollard seeds of salt marsh plants seem to be favoured though Mallards also forage in the polders here (Glas et al., 1974). Stomachs of Mallards shot on Vlieland and Terschelling in November and December especially contained seeds of *Heleocharis palustris*, *Scirpus maritimus*, *Potentilla anserina*, several *Carex* species, *Empetrum nigrum* and *Galeopsis* spec. (De Vries, 1939). In this time of the year the birds were mainly foraging in agricultural areas bordering the duck decoys.

3.8.4.2 Feeding activities

Mallards prefer places where good foraging and resting possibilities occur together. The largest concentrations can be found in places where shallow parts of the Wadden Sea or mudflats are bordered by salt marshes. Inland foraging generally occurs in the dark. In the Dollard most of the feeding- and resting places lie close to the salt marsh border (Glas et al., 1974). During high tide the Wadden Sea may be used as a resting place as well. The daily rhythm of activities in the Wadden Sea area has not been investigated properly.

After foraging Mallards often come to freshwater ponds and ditches to drink.

3.8.4.3 Total food intake

As food composition and food intake have not been investigated thoroughly, nothing is know about total food consumption and average food composition per area or season.

References

Atkinson-Willes, G.L., 1976. The numerical distribution of ducks, swans and coots as a guide in assessing the importance of wetlands in midwinter. In: M. Smart (ed.), Proc. Int. Conf. Conserv. Wetlands and Waterfowl, Heiligenhafen 1974. IWRB, Slimbridge: p. 199-254.

Bergh, L.M.J. van den, 1979. Verslag van de watervogeltellingen in januari en maart 1978. Watervogels 4: p. 48-72.

Bergh, L.M.J. van den & B.E. Schäffner, 1977. Verslag van de watervogeltellingen in de jaren 1972 t/m 1974. Watervogels 2 (extra nummer): p. 119-143.

Bergh, L.M.J. van den, B.E. Schäffner & J.J. Smit, 1978. Verslag van de watervogeltellingen in de jaren 1975-1977. Watervogels 3 (extra nummer): p. 43-71.

Boere, G.C. & P.M. Zegers, 1974. Wadvogeltelling in het Nederlandse Waddengebied in juli 1972. Limosa 47: p. 23-28.

Boere, G.C. & P.M. Zegers, 1975. Wadvogeltellingen in het Nederlandse Waddengebied in april en september 1973. Limosa 48: p. 74-81.

Boere, G.C. & P.M. Zegers, 1977. Watervogeltellingen in het Nederlandse Waddenzeegebied in 1974 en 1975. Watervogels 2: p. 161-173.

Busche, G., 1980. Vogelbestände des Wattenmeeres von Schleswig-Holstein. Kilda, Greven (in press).

Fog, J., 1974. Graaand. In: Weitemeyer & Hansen (eds.). Nyt dansk Jagtleksikon, Vol. 3. Brenner & Korch, København: p. 784-803.

Glas, P., J.B. Hulscher & S.T. Tjallingii, 1974. Terreingebruik van de vogels in de Dollard. In: Werkgroep Dollard, Dollard, portret van een landschap. Harlingen: p. 47-55.

Joensen, A.H., 1974. Waterfowl populations in Denmark 1965-1973. A survey of the non-breeding populations of ducks, swans and coot and their shooting utilization. Dan. Rev. Game Biol. 9(1): p. 1-206.

Olney, P.J.S., 1964. The food of the Mallard collected from coastal and estuarine areas. Proc. Zool. Soc. London 142: p. 397-418.

Rooth, J., 1966. Vogeltelling in het hele Nederlandse Waddengebied in augustus 1963. Limosa 39: p. 175-181.

Schiøler, E.L., 1925. Danmarks fugle, Vol. 1. Gyldendahl, København.

Smit, C.J., 1977. On the occurrence of 32 bird species in the Danish, German and Dutch Waddensea. Unpubl. report Int. Wadden Sea Working Group, part 2: 171 pp.

Spaans, A.L., 1967. Wadvogeltelling in het gehele Nederlandse Waddengebied in december 1966. Limosa 40: p. 206-215.

Spärck, R., 1958. An investigation of the food of swans and ducks in Denmark. Transaction third Congr. Int. Union Game Biol., Aarhus 1957. Dan. Rev. Game Biol. 3(3): p. 45-47.

Timmerman, Azn., A., 1974. Over de betekenis van de Friese waddenkust
    tussen Harlingen en Lauwersoog voor wad- en watervogels. Report
    Wadvogelwerkgroep Friesland, Oenkerk: 54 pp.
Voous, K.H., 1960. Atlas of European birds. Nelson, London: 284 pp.
Vries, V. de, 1939. Bijdrage tot de voedselbiologie van een viertal
    eendensoorten, naar aanleiding van materiaal, afkomstig van Vlie-
    land en Terschelling. Limosa 12: p. 87-98.

3.9 PINTAIL *(ANAS ACUTA L.)*
   M. Fog

Da: Spidsand; G: Spiessente; Du: Pijlstaart

3.9.1 Distribution

3.9.1.1 Breeding area
   The nominate form of the species has mainly a holarctic breeding
distribution, occurring in the boreal, temperate, Mediterranean,
steppe and to a lesser extent in the tundra climatic zone. Two distinct subspecies breed on the Kerguelen and Crozet Islands in the
Southern Indian Ocean (fig. 45). It breeds along freshwater pools,
in marshes, moors (Voous, 1960) and on salt marshes.

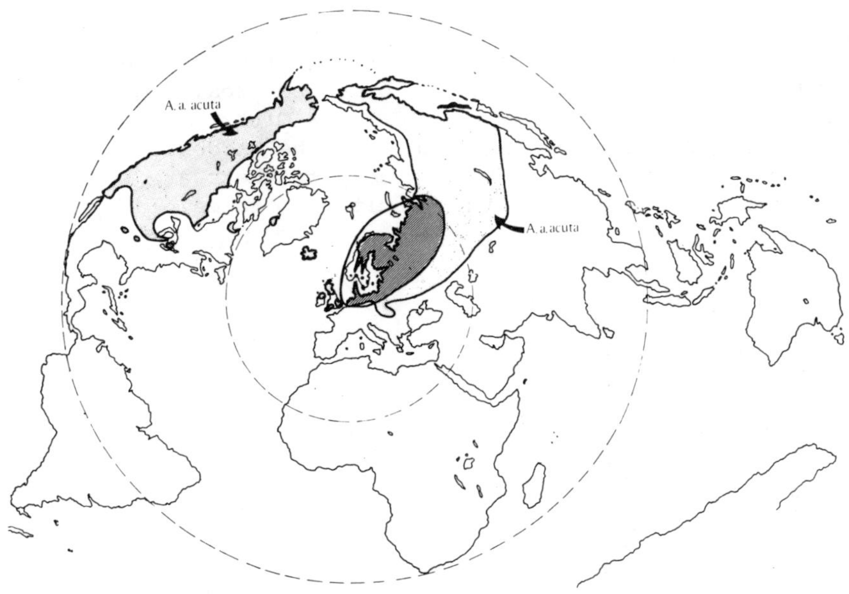

Fig. 45. Breeding area of the nominate subspecies of Pintail (after
Voous, 1960, modified). The dark shaded area indicates approximately
the region where birds visiting the Wadden Sea originate from. Two
subspecies breeding on small islands in the extreme Southern part of
the Indian Ocean are not indicated on the map.

3.9.1.2 Migration routes

According to ringing data the population passing through or wintering in the Wadden Sea originates from Sweden, Finland and the Northern part of the European USSR. After the breeding season (males already during the breeding season) Pintail migrate towards the moulting areas. On their way to the wintering quarters they concentrate more or less along the coasts of North Sea and Atlantic Ocean (Bauer & Glutz, 1968). Population fluctuations in the Wadden Sea area give some support for the suggestion in Bauer & Glutz (1968) that Pintail carry out "Schleifenzug". This suggestion is also supported by recent data shown by Busche (1980) for Schleswig-Holstein where numbers in autumn exceed those in spring by far and by data from the area along the Dutch rivers Rhine, Waal, Meuse and IJssel where the opposite occurs (Stichting VWG Grote rivieren, 1979). Other data from the Wadden Sea however indicate that the situation might be more complicated (details in 3.9.3.2).

3.9.1.3 Wintering areas

Recoveries of Pintail ringed at Texel (fig. 46) show that wintering takes place on the British Isles, in France, along the coasts of the Iberian peninsula and in Morocco. Recoveries of the Fanø duck decoy come from the British Isles, France, the Netherlands, and the Spanish eastcoast. Especially the Dutch part of the Wadden Sea may act as a wintering area. The Delta area however is more important in this respect (Tables 13 and 14).

3.9.1.4 Moulting area

Pintail gather in large concentrations to moult flight feathers on rivers and lakes near or south of the breeding area. Body feathers are moulted in the wintering areas (Bauer & Glutz, 1968). Small numbers of moulting Pintail probably occur in the Wadden Sea area in Schleswig-Holstein (Busche, 1980). As body feathers are renewed in autumn, winter and even spring this moult also takes place in the Wadden Sea.

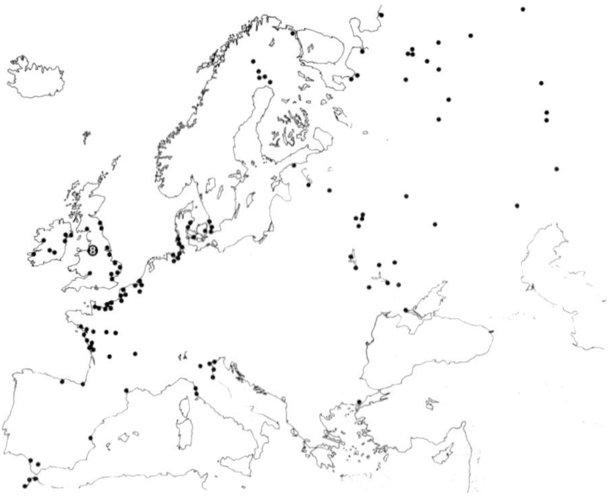

Fig. 46. Recoveries of Pintail ringed on the island of Texel from 1965-1973 (after unpubl. data Vogeltrekstation, Arnhem).

Table 13. Number of Pintail in January counts in the Netherlands.
Source 1972-1974: Van den Bergh & Schäffner (1977); 1975-1977: Van
den Bergh et al. (1978); 1978: Van den Bergh (1979). All data for
the Wadden Sea include the Lauwersmeer.

Area	1973	1974	1975	1976	1977	1978	Mean 1973-1978	%
Delta	12,697	8,425	26,537	7,399	5,818	10,878	11,959	73
Wadden Sea	2,340	1,895	3,365	3,174	1,095	5,680	2,925	18
Rivers Rhine, Meuse, Waal, IJssel	585	779	2,363	492	179	433	805	5
Duck decoys	229	412	262	445	289	838	413	3
Rest of counting	157	210	318	250	320	454	285	2
Total	16,008	11,721	32,845	11,760	7,701	18,283	16,386	100

Table 14. Number of Pintail determined by counts from the shore in
the Dutch Wadden Sea. Results of recent January counts are given in
Table 14.

Date	Year	Number	Source	Remarks
March 12	1977	1,507	Van den Bergh et al., 1978	Lauwersmeer incl.
March 13	1976	1,279	"	
April 6	1973	1,040	Boere & Zegers, 1975	
April 19	1975	210	Boere & Zegers, 1977	
May 1	1976	70	Zegers, in litt.	
May 11	1974	17	Boere & Zegers, 1977	
July 29	1972	0	Boere & Zegers, 1974	
August 22	1963	150	Rooth, 1966	
August 30	1975	410	Boere & Zegers, 1977	
September 1	1973	220	Boere & Zegers, 1975	
October 19	1974	2,390	Boere & Zegers, 1977	
November 16	1973	9,000	Van den Bergh & Schäffner, 1977	Lauwersmeer incl.
November 17	1974	3,263	"	"
November 25	1972	7,298	"	Lauwersmeer incl. Area partly covered
November	1976	5,150	Zegers, in litt.	
December 29	1966	5,000	Spaans, 1967	

3.9.2 Annual cycle

3.9.2.1 Migration

Males and unsuccessfully breeding females leave the breeding area by the end of May or in the first part of June. They gather and migrate towards the moulting areas shortly afterwards. Females which have raised young successfully, follow around mid-July. Moulting areas are left from mid-August until October. The first Pintail arrive in the Wadden Sea as early as August (Joensen, 1974 and fig. 47), larger numbers however arrive in September and October. Peak numbers in Britain and Ireland occur from November to January. On most places in the Wadden Sea, numbers in spring increase only slightly.

3.9.2.2 Moult

Males moult flight feathers from the end of June on, females generally follow later in the season. During the moulting period the birds are unable to fly for 20-28 days. By mid-August most Pintail have completed moult and begin to leave the moulting areas. Moult of body feathers is completed in most of the males in the first part of January, for females this may last until May (Bauer & Glutz, 1968).

3.9.2.3 Weight changes

Weights of Pintail from Denmark in August-October were: adult males (40) 746-1111 g (mean 928 g); adult females (17) 601-845 g (762 g); first c.y. males (32) 710-1096 g (mean 882 g); first c.y. females (15) 665-912 g (mean 710 g) (Schiøler in Bauer & Glutz, 1968). Data from the Wadden Sea are not available.

3.9.3 Numbers

3.9.3.1 Population size

The breeding population of the Western palaearctic amounts to about 320,000 pairs, 316,000 of these breeding in the USSR. The NW European flyway population counts 50,000, the Black Sea/Mediterra-

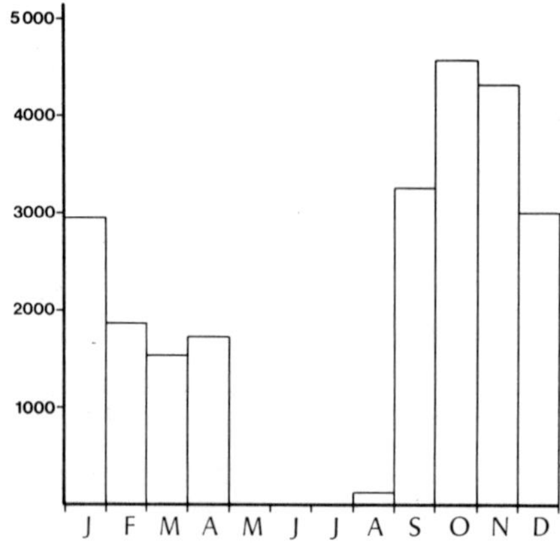

Fig. 47. Mean number of Pintail per month in the Dutch part of the Wadden Sea (excluding Lauwersmeer). After Smit (1977).

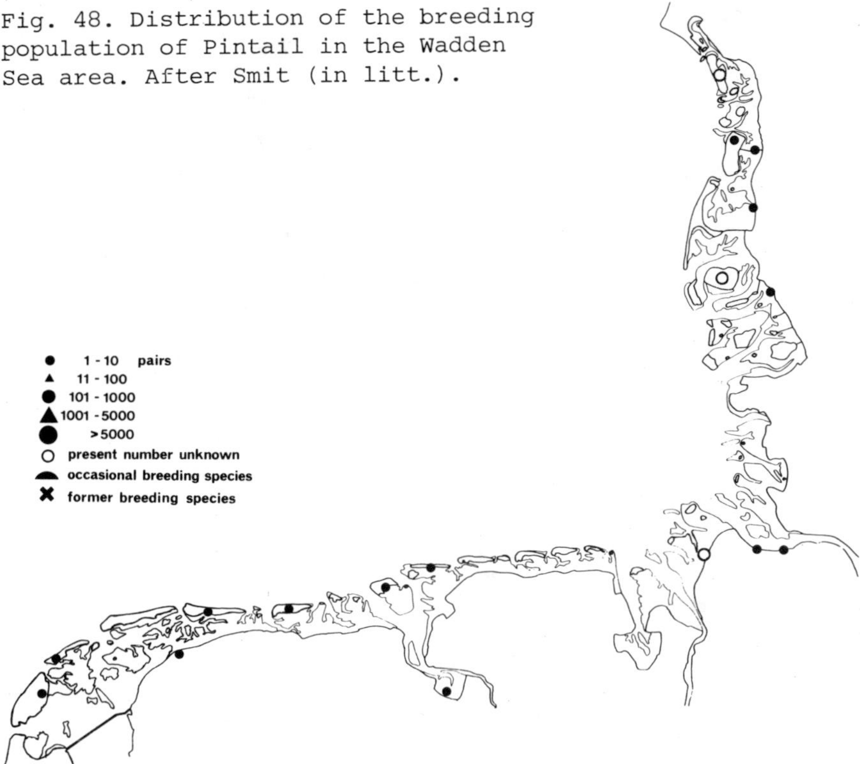

Fig. 48. Distribution of the breeding population of Pintail in the Wadden Sea area. After Smit (in litt.).

nean population 250,000, the Caspian/Iranian population 150,000, the Turkestan population 55,000 and the Senegal/Mali population 200,000 birds (Atkinson-Willes, 1976). The breeding distribution of the Wadden Sea population is shown in fig. 48. It should be noted that part of the breeding population originates from semi-domesticated Pintail from duck decoys. The total size of the Wadden Sea breeding population amounts to only 30-50 pairs (Smit, pers. comm.).

3.9.3.2 Numbers per area

In the Danish Wadden Sea peak numbers occur in late September and during October. Only few are seen from December to February. From early March on numbers increase, after mid-May only few remain. Up to 8000 have been recorded near the Rømø-dam in October 1969 (Joensen, 1974), up to 7000 have been counted on the tidal flats west of Højer (Jepsen, 1975). Occurrence in the Wadden Sea is more or less confined to sheltered coastal areas.

In Schleswig-Holstein about 80% of all Pintail present per year are observed from September to November. Numbers in this time of the year in the area are estimated to amount to about 15,000. In mild winters 200-1500 are present, in cold winters none (Busche, 1980).

Along the NW part of the mainland coast of Ostfriesland numbers occurring in spring exceed those in autumn by far (Rettig, 1977).

Fig. 49. Regular peak occurrences of
Pintail per census area in autumn.
Based on Joensen (1974), Smit (1977)
and Busche (1980).

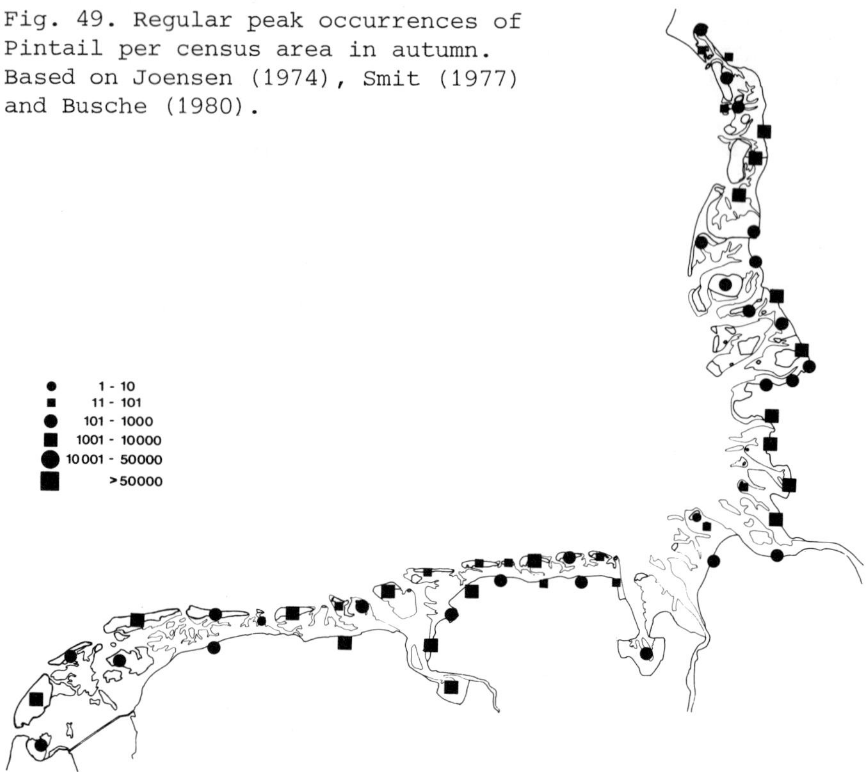

This appears not to be true for the Eastern part of the mainland
coast and for most of the islands (Grosskopf, 1968; Smit, 1977).
Numbers in most coastal areas however are seldomly larger than a
few hundreds.

An estimate of the mean numbers per month in the Dutch Wadden Sea
is shown in fig. 47. The role of the area as compared to the Delta
is shown in Tables 13 and 14. On all Wadden Sea islands numbers drop
suddenly in March. In coastal areas however, including the Dollard,
numbers rise in March and April, sometimes exceeding autumn- and midwinter numbers (Smit, 1977).

Fig. 49 illustrates the distribution of Pintail in the Wadden Sea
area.

3.9.4 Food

3.9.4.1 Food composition

Generally food in autumn and winter consists of vegetable matter
(seeds, leaves, roots). In spring and summer a rather important part
of food may consist of small molluscs, crustaceans and insects (Bauer & Glutz, 1968). From the Wadden Sea only one study of the food
exists. In 14-16 stomachs of Pintail killed between December and Fe-

bruary on Terschelling 2764 seeds of 33 plant species were found. Seeds of Carex species, *Rumex acetosella*, grasses, *Triglochin maritima*, *Heleocharis palustris*, *Salicornia herbacea*, several dicotylodonous plants and berries of *Empetrum nigrum* were most common. Besides large amounts of *Hydrobia ulvae* were found (De Vries, 1939).

3.9.4.2 Feeding activities

Generally feeding takes place after sunset. Food is taken by dabbling or, whenever meadows or marshes are partly flooded, by grazing. Salt marshes as well as the Wadden Sea may act as a roosting area.

Though not studied it may be assumed that Pintail drink fresh water from pools, ditches etc.

3.9.4.3 Total food consumption

No information available.

References

Atkinson-Willes, G.L., 1976. The numerical distribution of ducks, swans and coots as a guide in assessing the importance of wetlands in midwinter. In: M. Smart (ed.), Proc. Int. Conf. Conserv. of Wetlands and Waterfowl, Heiligenhafen 1974. IWRB Slimbridge: p. 199–254.

Bauer, K.M. & U.N. Glutz von Blotzheim, 1968. Handbuch der Vögel Mitteleuropas, Vol. 2. Akademische Verlagsgesellschaft, Frankfurt/Main: 535 pp.

Bergh, L.M.J. van den, 1979. Verslag van de watervogeltellingen in januari en maart 1978. Watervogels 4: p. 48–72.

Bergh, L.M.J. van den & B.E. Schäffner, 1977. Verslag van de watervogeltellingen in de jaren 1972 t/m 1974. Watervogels 2 (extra nummer): p. 119–143.

Bergh, L.M.J. van den, B.E. Schäffner & J.J. Smit, 1978. Verslag van de watervogeltellingen in de jaren 1975–1977. Watervogels 3 (extra nummer): p. 43–71.

Boere, G.C. & P.M. Zegers, 1974. Wadvogeltelling in het Nederlandse Waddengebied in juli 1972. Limosa 47: p. 23–28.

Boere, G.C. & P.M. Zegers, 1975. Wadvogeltellingen in het Nederlandse Waddengebied in april en september 1973. Limosa 48: p. 74–81.

Boere, G.C. & P.M. Zegers, 1977. Watervogeltellingen in het Nederlandse Waddenzeegebied in 1974 en 1975. Watervogels 2: p. 161–173.

Busche, G., 1980. Vogelbestände des Wattenmeeres von Schleswig-Holstein. Kilda, Greven (in press).

Grosskopf, G., 1968. Die Vögel der Insel Wangerooge. Abhandl. Vogelk. 5. Inst. für Vogelforschung, Wilhelmshaven:293 pp.

Jepsen, P.U., 1975. Vadehavet vildtreservat med øen Jordsand. Danske Vildtundersøgelser 24: 80 pp.

Joensen, A.H., 1974. Waterfowl population in Denmark 1965–1975. A survey of the non-breeding populations of ducks, swans and coot and their shooting utilization. Dan. Rev. Game Biol. 9: p. 1–206.

Rettig, K., 1977. Zum Durchzug der Spiessente (Anas acuta) an der
    niedersächsischen Nordseeküste. Vogelk. Ber. Nieders. 9(1): p. 1-3.
Rooth, J., 1966. Vogeltelling in het hele Nederlandse Waddengebied in
    augustus 1963. Limosa 39: p. 175-181.
Spaans, A.L., 1967. Wadvogeltelling in het gehele Nederlandse Wadden-
    gebied in december 1966. Limosa 40: p. 206-215.
Smit, C.J., 1977. On the occurrence of 32 bird species in the Danish,
    German and Dutch Waddensea. Unpubl. report Intern. Wadden Sea
    Working Group, part 2: 171 pp.
Stichting VWG Grote Rivieren, 1979. Vogels van de Grote Rivieren.
    Spectrum, Utrecht: 328 pp.
Voous, K.H., 1960. Atlas of European birds. Nelson, London: 284 pp.
Vries, V. de, 1939. Bijdrage tot de voedselbiologie van een viertal
    eendensoorten, naar aanleiding van materiaal, afkomstig van Vlie-
    land en Terschelling. Limosa 12: p. 87-98.

3.10 EIDER *(SOMATERIA MOLLISSIMA L.)*
C. Swennen

Da: Ederfugl; G: Eiderente; Du: Eidereend

3.10.1 Distribution

3.10.1.1 Breeding area
The species has a discontinuous holarctic distribution. It is a
coastal breeding bird in the arctic, boreal and temperate climatic
zones (Voous, 1960). A number of subspecies, generally with broad
zones of intermediate forms, can be distinguished. The most important are shown in fig. 50. In the Wadden Sea only the nominate form
occurs.
Ringing results indicate the Baltic Sea as the origin of the Eiders moulting and/or wintering in the Wadden Sea, and also of most
of the immature summer visitors (fig. 51). Moreover the Eider breeds
in the Wadden Sea area.

3.10.1.2 Migration routes
The Baltic migrants migrate over sea and often cross land in
Skaane (South Sweden), Sjaelland, South Jutland and Schleswig-Hol-
stein (Swegen, 1972; Joensen, 1973b; Alerstam et al., 1974).

3.10.1.3 Wintering areas
The birds breeding around the Baltic have their main wintering
grounds in the South-Western part of the Baltic Sea, the Kattegat
and the whole Wadden Sea.
The birds breeding on the Wadden Sea islands are resident. Only
a part of the yearlings leave the Wadden Sea in their first autumn
and disperse mainly into a SW direction. They winter around the
coasts of the North Sea, in the Channel and along the Atlantic coast
of France (Swennen, 1976; fig. 52).

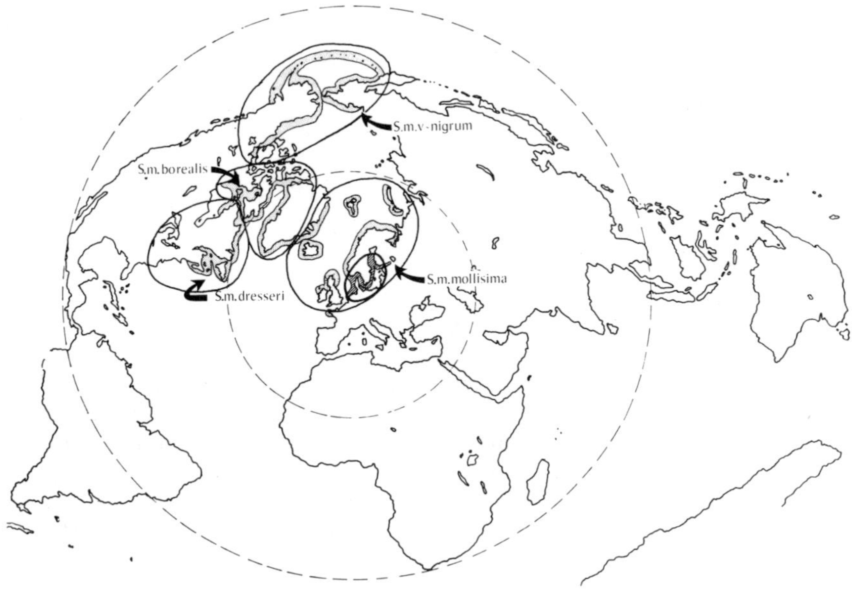

Fig. 50. Breeding area of the Eider (after Voous, 1960; Bauer & Glutz, 1969). The encircled dark shaded area indicates a region where birds visiting the Wadden Sea originate from.

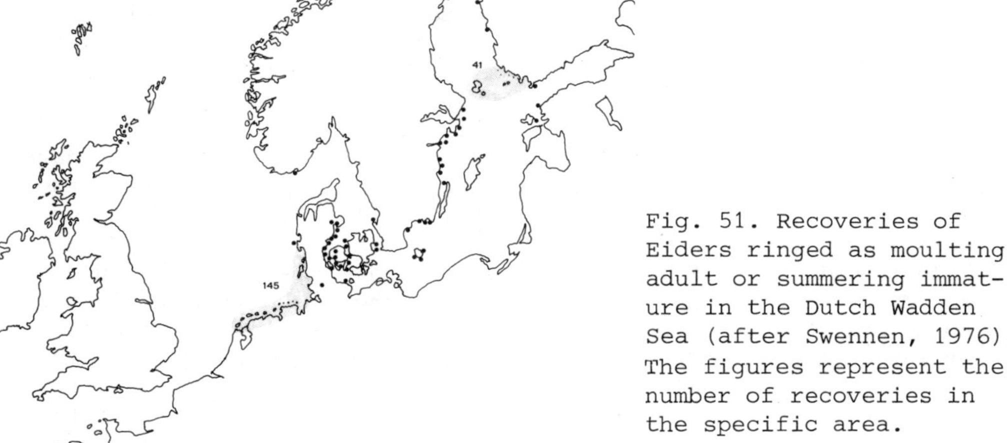

Fig. 51. Recoveries of Eiders ringed as moulting adult or summering immature in the Dutch Wadden Sea (after Swennen, 1976). The figures represent the number of recoveries in the specific area.

3.10.1.4 Moulting areas

The Wadden Sea functions as the moulting area for the local birds and for large numbers from the Baltic Sea (fig. 51) (Joensen, 1973b; Swennen, 1976). For more detailed information see 3.10.2.2 and 3.10.3.2.

Fig. 52. Recoveries of Eiders ringed as ducklings in the Dutch Wadden Sea and recovered as sub-adults (after Swennen, 1976). The figure represents the number of recoveries in the Dutch part of the Wadden Sea.

3.10.2 Annual cycle

3.10.2.1 Migration

The first flocks of adults gather at sea close to the breeding islands during March-April, in some years already in February. In the colonies on the Wadden Sea islands laying begins in the first half of April. In the same time, (mid-February until mid-April), the winter visitors return to the Baltic Sea. A small part of them, mainly immatures, stays as non-breeding summer visitors. The first ducklings leave their nests in the first half of May (earliest date May 4, Vlieland) to grow up in sea. Most of the young hatch in the second half of May, small numbers still follow in June and in some years a few in July.

In the middle of May the adult males leave the resting grounds around the colonies and concentrate near the moulting areas in the Wadden Sea. On the Dutch Wadden Sea islands a part of the females returns to the dunes where they have been breeding after deserting or loosing their ducklings.

In the second half of June the first Baltic birds arrive at the moulting places in the Wadden Sea. The adult males arrive mainly by the end of June and early July, the females by the end of July and the first half of August. The winter visitors from the Baltic arrive in October and November, sometimes even later, and join the birds already present.

3.10.2.2 Moult

In the first half of June the males get their dark eclipse plumage. From the 7th of July on the first ones shed their flight feathers. Of 152 males caught near Vlieland on July 26, 1967 exactly 50%

had lost flight feathers (Swennen, unpubl.). In the Danish part of the
Wadden Sea between July 23 and July 26, 1970 65-75% of the males was
flightless (Joensen, 1973b). Early September practically all males
are able to fly again. On Vlieland the first females shed wing feathers in the last week of July. On August 2, 1967 40% of 80 females
caught off Terschelling had lost the flight feathers. The females
moult their wing feathers later and over a longer period than the
males. In the second part of September most females are able to fly
again (Swennen, unpubl.).

3.10.2.3 Weight changes

Adults are heavier than sub-adults, and during the greater part
of the year males are heavier than females. For example: of Eiders
caught in the Wadden Sea off Terschelling and Vlieland in August
1968, males had a weight of 2243 g $\pm$ s.d. 149 (n=274), females of
2068 g $\pm$ 126 (n=155). Just before the breeding period the weight of
the adult females increases noteworthy. Birds trapped on Vlieland
between April 30 and May 4, 1974 showed the following figures: adult
males 2188 g $\pm$ 103 (n=52), adult females 2566 g $\pm$ 204 (n=17), but
immature females 1845 g $\pm$ 142 (n=35). Females in search of a nesting
place weighed 2713 g $\pm$ 126 (n=3), after laying their eggs at the
start of the incubation 2404 g $\pm$ 98 (n=7). After the 4 weeks fasting
period during incubation the weight of the females is strongly de-

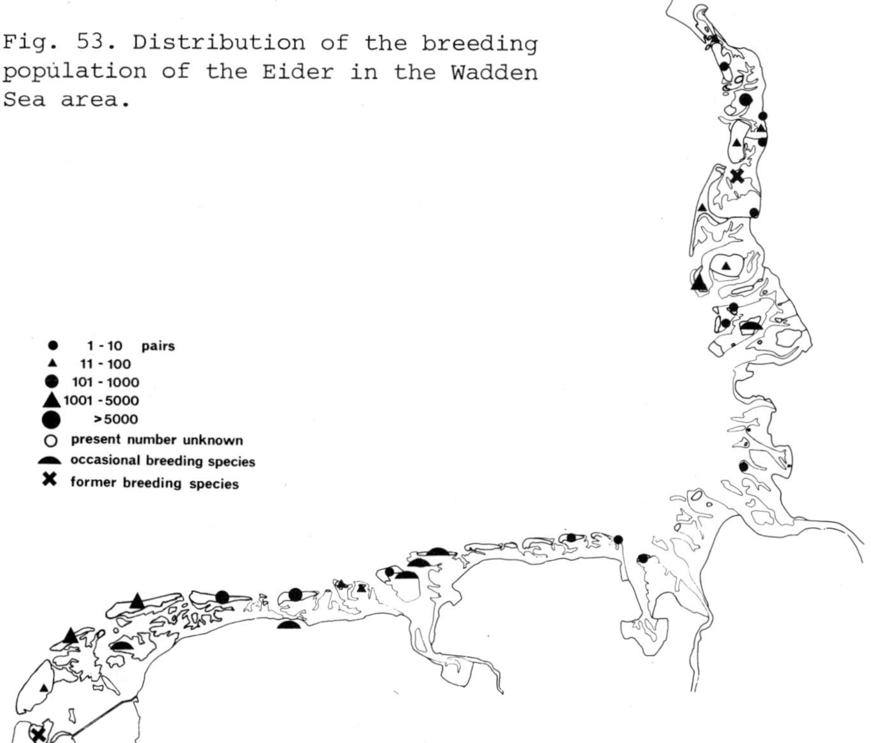

Fig. 53. Distribution of the breeding
population of the Eider in the Wadden
Sea area.

creased: 1543 g ± 102 (n=126). The conditions of foraging shortly
before the beginning of the breeding-season will be decisive whether the birds are able to breed or not.

3.10.3 Numbers

3.10.3.1 Population size

The size of the Baltic/Wadden Sea population amounts to at least
650,000 birds (Almkvist et al., 1975; this paper).

The distribution of the breeding population in the Wadden Sea
area is shown in fig. 53. The breeding population on the Wadden Sea
islands in Denmark and Schleswig-Holstein (Langli, Fanø, Mandø, Rømø, Sylt, Amrum and Föhr, along the Rømø-dam and along the coast
near Ballum and Højer) comprises at least about 1500 pairs (Joensen,
1973a; Quedens, 1974; Bauer & Glutz, 1969; Dansk Orn. Forening, in litt.;
Smit, in litt.). On the Wadden Sea islands and along the mainland coast
of Niedersachsen the Eider is only breeding irregular or in small numbers, the total population amounting only to about 20 pairs (König,
in litt.; Smit, in litt.). On the Dutch Wadden Sea islands (Rottum,
Schiermonnikoog, Ameland, Terschelling, Vlieland and Texel) about
4000 pairs are nesting (Swennen, 1976). The whole breeding population of the Wadden Sea area can be estimated at about 5500-6500 pairs.

3.10.3.2 Numbers per area

In the Wadden Sea a mixture is found of birds actually breeding
there and of birds breeding in the Baltic Sea. Both populations are
morphologically identical. The best way to say something on the occurrence of Eiders in the Wadden Sea is to divide the year into periods.

From mid-April until mid-June apart from the breeding population
non-breeders (mainly immatures) of the Baltic breeding population
are present. The numbers differ substantially from year to year, but
usually there seem to be more non-breeding summer visitors than Wadden Sea breeding birds.

From mid-June until October substantial numbers of Eiders from
the Baltic population arrive at the moulting places in the Wadden
Sea. In the Danish part 30,000 moulting birds have been recorded in
successive years (Joensen, 1973b). In the German part in one year
up to 40,000 moulting birds have been recorded off Schleswig-Holstein, but considerably less were counted in other years. Near Scharhörn and Knechtsand 1000-2000 birds have been recorded, in other
places however only very small numbers (Smit, 1977). In two surveys in the Dutch part the number of moulting birds amounted to more
than 21,000 and 36,000 respectively (Swennen, 1976).

From November until March or early-April the following groups of
Eiders are present:
1) birds belonging to the breeding population of the Wadden Sea;
2) non-breeding Baltic birds that were already present in summer;
3) Baltic moult migrants that arrived in summer and autumn;
4) a group of winter visitors from the Baltic arriving from November
   on.

The numbers in winter occurring in the Danish part range from
42,000 to 74,000 (Joensen, 1974), in the German part from 25,000 to
40,000 off Schleswig-Holstein (Busche, 1980). Small numbers occur
in other areas (Smit, 1977). In the Dutch part 104,000 (January 1978,
Swennen unpubl.) to 168,000 (Swennen, 1976) are present. These numbers show that 26-43% of the Baltic population (Almkvist et al.,
1975) stays in the Wadden Sea in winter.

In the sixties of this century on the Dutch islands a severe decline set in. This was caused by huge mortality among breeding females due to poisoning by pesticides that were discharged into the
river Rhine (Swennen, 1972). In 1965 the discharge of the pesticides
was stopped and after 1968 a gradual recovery set in. Compared to
the situation in 1950 the number of breeding, moulting and wintering
birds has increased.

3.10.4 Food

3.10.4.1 Food composition

The diet mostly consists of molluscs (chiefly *Cerastoderma edule*,
Mytilus edulis) and crustaceans (chiefly *Carcinus maenas*). No clear
seasonal variation in the main food components was found in the Dutch
part of the Wadden Sea. By faecal analysis it was found that *Cerastoderma* and *Mytilus* each contribute 40% of the diet and *Carcinus* 6.5%
(Swennen, 1976).

3.10.4.2 Feeding activities

Feeding areas in the Wadden Sea are situated between half tidal
level and a depth of 5 meter. In the breeding period they forage also on higher flats near the breeding places.

During high tide in the breeding period Eiders rest ashore near
the nesting places. During moult they rest mainly on high sand ridges
during high tide, but during quiet weather they rest at sea. During
high tide in winter Eiders often congregate into flocks of hundreds,
sometimes up to several thousands, at sea. In some places they rest
on the flats emerging during low tide. In winter only sick birds come
ashore.

Outside the breeding season Eiders drink seawater. Also by eating
seawater is ingested (in cockles about half of the weight consists of
it). During and shortly after breeding females prefer to drink fresh
or brackish water and move to pools ashore or freshwater seepage
places along the shoreline. Recently hatched ducklings need fresh-
or slightly brackish water before they can use seawater. Until fledging they prefer to drink water with a low salt content.

3.10.4.3 Total food consumption

On the basis of determinations with captive birds the total food
consumption was estimated at 138 g ash-free dry weight, which is
750 Kcal (=3150 KJoule) per bird per day (Swennen, 1976). No clear
seasonal fluctuation in the consumption was found, but in winter,
due to the low flesh content of molluscs in that time, the birds
eat twice as much specimens as in summer.

References

Alerstam, T., C.A. Bauer & G. Roos, 1974. Spring migration of Eiders
    Somateria mollissima in southern Scandinavia. Ibis 116: p. 194-
    210.
Almkvist, B., Å. Andersson, A. Jogi, M.K. Pirkola, M. Soikkeli & J.
    Virtanen, 1975. The number of adult Eiders in the Baltic Sea. Wild-
    fowl 25: p. 89-94.
Bauer, K.M. & U.N. Glutz von Blotzheim, 1969. Handbuch der Vögel Mit-
    teleuropas, Vol. 3. Akademische Verlagsgesellschaft, Frankfurt/
    Main: 504 pp.
Busche, G., Vogelbestände des Wattenmeeres von Schleswig-Holstein.
    Kilda, Greven (in press).
Joensen, A.H., 1973a. Ederfuglen (Somateria mollissima) som yngle-
    fugl i Danmark. Danske Vildtundersøgelser 20: p. 1-36.
Joensen, A.H., 1973b. Moult migration and wing-feather moult of Sea-
    ducks in Denmark. Danish Rev. Game Biol. 8 (4): p. 1-42.
Joensen, A.H., 1974. Waterfowl populations in Denmark 1965-1973. Dan-
    ish Rev. Game Biol. 9 (1): p. 1-206.
Quedens, G., 1974. Das Tierleben der Insel Amrum. Nordfriesland 29:
    p. 49-56.
Smit, C.J., 1977. On the occurrence of 32 bird species in the Danish,
    German and Dutch Wadden Sea. Unpubl. report Intern. Wadden Sea
    Working Group, part 2: 171 pp.
Swegen, H., 1972. Ejderns (Somateria mollissima) träck over land i
    söndre Sverige. Vår Fågelvärld 31: p. 183-190.
Swennen, C., 1972. Chlorinated hydrocarbons attacked the Eider popu-
    lation in the Netherlands. TNO-nieuws 27: p. 556-560.
Swennen, C., 1976. Populatie-structuur en voedsel van de Eidereend
    Somateria m. mollissima in de Nederlandse Waddenzee. Ardea 64: p.
    311-371.
Temme, M., 1974. Zugbewegungen der Eiderente (Somateria mollissima)
    vor der Insel Norderney unter besonderer Berücksichtigung der Wet-
    terverhältnisse. Vogelwarte 27: p. 252-263.
Voous, K.H., 1960. Atlas of European birds. Nelson, London: 284 pp.

3.11 RED-BREASTED MERGANSER (*MERGUS SERRATOR* L.)
   C.J. Smit & C. Swennen

Da: Toppet Skallesluger; G: Mittelsäger; Du: Middelste Zaagbek

3.11.1 Distribution

3.11.1.1 Breeding area
   The species has a discontinuous holarctic distribution breeding
mainly in the boreal, but also in the tundra and temperate climatic
zones (fig. 54). Nesting places are situated along lakes, rivers, sea
coasts and estuaries. Generally Red-breasted mergansers breed in more
northern regions than the Goosander (Voous, 1960).

There is only limited information where birds visiting the Wadden Sea originate from. The few available ringing recoveries suggest at least a part comes from Fenno-Scandinavia and Iceland (Bauer & Glutz, 1969).

3.11.1.2 Migration routes
After the breeding season both males and females migrate to coastal areas near the breeding places. From here they disperse to the wintering areas. Migration routes are not known.

3.11.1.3 Wintering areas
Red-breasted mergansers of the populations visiting W Europe winter mainly along the coasts of the North Sea and the Western part of the Baltic.

3.11.1.4 Moulting areas
Moult of the primaries takes place shortly after the breeding season in coastal areas not far from the breeding grounds. Birds summering in the Wadden Sea moult primaries there, especially in August.

3.11.2 Annual cycle

3.11.2.1 Migration
In the first part of July males leave the breeding places and migrate to nearby coastal areas to moult. Females leave the breeding

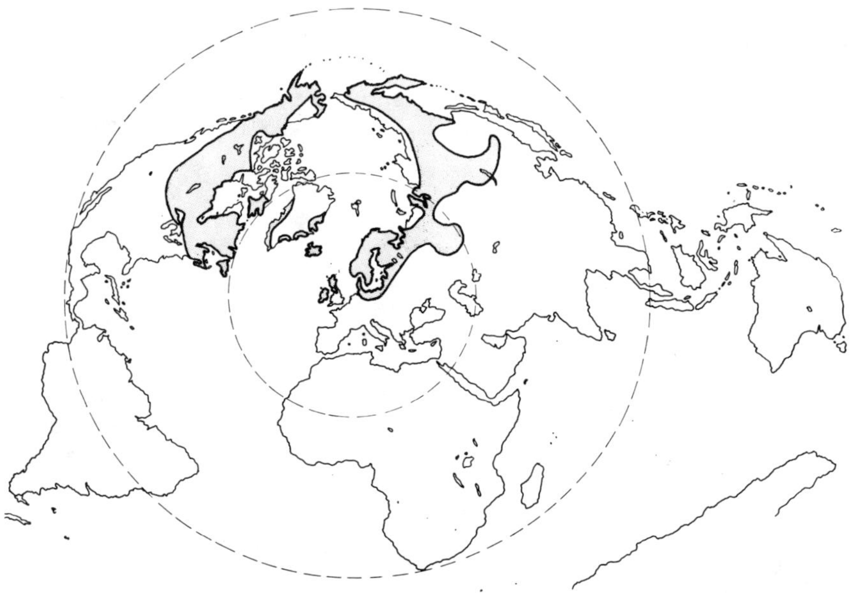

Fig. 54. Breeding area of the Red-breasted merganser (after Voous, 1960; Bauer & Glutz, 1969).

places from the second part of July on. Departure from the areas where primary moult has taken place begins in the course of September (Bauer & Glutz, 1969).

At Blaavandshuk autumn migration starts in September. Spring migration of adults starts in March and peaks in April. Migration of mostly immature birds however continues until early July (Petersen, 1974).

They generally arrive on the breeding places around the Baltic Sea from the second part of April on.

3.11.2.3 Weight changes

No information available.

3.11.3 Numbers

3.11.3.1 Population size

Atkinson-Willes (1972) estimates the size of the population wintering in NW Europe at 75,000 birds.

The species occasionally breeds in the Wadden Sea area. Breeding has been recorded on Texel, Rottumerplaat, Föhr, Sylt, Langli, the Hauke Haienkoog and probably on Schiermonnikoog and Rømø. From 1973 on 1-3 pairs breed on Amrum every year (Bauer & Glutz, 1969; Quedens, unpubl.).

3.11.3.2 Numbers per area

In the Danish Wadden Sea the species is scarce. With aerial surveys a maximum of 1000-3000 has been recorded in March. Generally, however, numbers are much lower and in winter hardly any birds are present. Usually the largest numbers occur in the Ho bugt (Joensen, 1974).

Some aerial surveys and ground counts showed that in Schleswig-Holstein they are most numerous in March, April and October. The es-

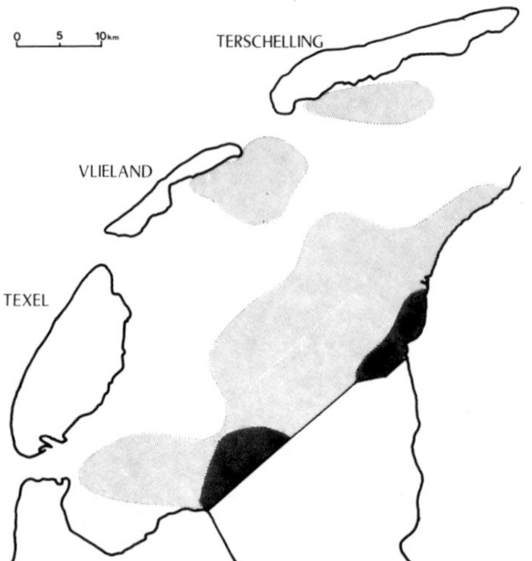

Fig. 55. Distribution of the Red-breasted merganser in the Western part of the Dutch Wadden Sea determined by counts from ships. In the dark shaded area numbers per 7 km stretch exceeded 256 birds; in the light shaded area numbers exceeded 16 birds per 7 km. Numbers are average data from counts from November-March from 1964-1969 (Swennen, unpubl.).

timated numbers in April however amounts to only 600, scattered all
over the area (Busche, 1980).

In Niedersachsen only counts from land have been carried out.
Speculations on numbers can not be made.

In the Dutch part the species is most numerous in the Western
part of the Wadden Sea along the Afsluitdijk and in the somewhat
sheltered gullies south of the islands (fig. 55). Unlike the Goosander
only relatively small numbers occur on the IJsselmeer (Swennen, 1970).

Numbers computed from boat surveys in the Western part of the Wadden Sea show an increase from October (about 800) until January (about
12,000) and a decrease till April (5000). In May only small numbers
are left (Swennen, unpubl.). A small number is staying during summer.

The peak numbers show that the Danish part is used by maximally
4%, the part in Schleswig-Holstein by 1% and the Dutch part by 15%
of the NW European flyway population.

3.11.4 Food

3.11.4.1 Food composition

In general a large variety of fishes is taken. Less dominant are
crustaceans and sometimes polychaetes and molluscs (Bauer & Glutz,
1969). From the Wadden Sea no data on food composition are available.

In the Dutch part of the Wadden Sea the largest concentrations of
Red-breasted mergansers coincide with that of Smelt, *Osmerus eperlanus*.

3.11.4.2 Feeding activities

Red-breasted mergansers catch fish by diving. At high tide the
birds occur all over the area, during low tide in the gullies, ashore
or on the tidal flats along the edges of the gullies.

No high- or low tide roosts can be distinguished. Red-breasted mergansers rest and sleep on the water, to a lesser extent ashore.

As drinking water they probably take salt as well as brackish water. No special drinking flights are recorded.

3.11.4.3 Total food consumption

No data available.

References

Atkinson-Willes, G.L., 1972. In: M. Smart (ed.): The international
    wildfowl censuses as a basis for wetland evalution and hunting
    rationalization. Proc. Intern. Conf. Conservation Wetlands and
    Waterfowl, Iran 1971: p. 87-110.
Bauer, K.M. & U.N. Glutz von Blotzheim, 1969. Handbuch der Vögel
    Mitteleuropas, Vol. 3. Akademische Verlagsgesellschaft, Frank-
    furt/Main: 504 pp.
Busche, G., 1980. Vogelbestände des Wattenmeeres von Schleswig-
    Holstein. Kilda, Greven (in press).

Joensen, A.H., 1974. Waterfowl populations in Denmark 1965-1973: A survey of the non-breeding populations of ducks, swans and coots and their shooting utilization. Danish Rev. Game Biol. 9(1): p. 1-206.

Petersen, F.D., 1974. Traekket af aender Anatidae ved Blaavand 1963-71. Dansk Orn. Foren. Tidsskrift 68: p. 25-37.

Swennen, C., 1970. Vogelwaarnemingen op het IJsselmeer. Limosa 43: p. 1-10.

Voous, K.H., 1960. Atlas of European birds. Nelson, London: 284 pp.

3.12 GOOSANDER (*MERGUS MERGANSER* L.)
C. Swennen

Da: Stor Skallesluger; G: Gänsesäger; Du: Grote Zaagbek

3.12.1 Distribution

3.12.1.1 Breeding area

The species has a discontinuous holarctic distribution breeding mostly in the boreal but also in the temperate climatic zone and in some places in mountain regions. Generally it breeds further south than the Red-breasted merganser. It occurs along lakes, rivers, estuaries, brackish fjords and bays (Voous, 1960). Three subspecies are distinguished (fig. 56). In the Wadden Sea area the subspecies *M. m. merganser* occurs as a winter visitor.

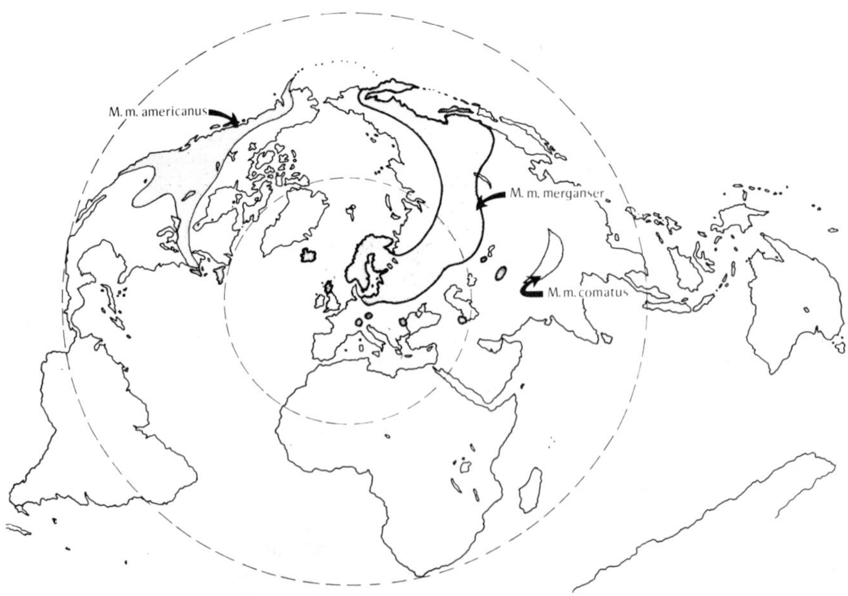

Fig. 56. Breeding area of the Goosander (after Voous, 1960; Bauer & Glutz, 1969).

It is not known exactly from which part of the breeding range birds
visiting the Wadden Sea originate. At least a part of the Goosanders
present in the Wadden Sea originate from Fenno-Scandinavia (Bauer &
Glutz, 1969).

3.12.1.2. Migration routes
Migration routes are not known.

3.12.1.3 Wintering area
Goosanders winter on many fresh- or brackish water lakes and rivers throughout Europe, but also along the coasts of Northern and Western Europe and the Baltic. In severe winters they reach the Northern
coasts of the Mediterranean (Bauer & Glutz, 1969). Large numbers, in
some winters maybe up to 22,000 birds have been counted on the Northern part of the IJsselmeer (Swennen, 1970).

3.12.1.4 Moulting areas
Wing moult takes place in the surrounding of the breeding places.
Only a part of the body feathers still has to be moulted when the
birds arrive in the Wadden Sea.

3.12.2 Annual cycle

3.12.2.1 Migration
After the moult of flight feathers Goosanders gather in many fresh-
and brackish water areas. In South Sweden they depart from mid-October on. The last birds leave the Eastern part of the Baltic by early
November (Bauer & Glutz, 1969). The Wadden Sea acts as wintering
place from the moment the large inland waters in Northern Europe and
the coastal waters of the Baltic Sea become ice-covered. The first
birds arrive in the Wadden Sea from the first of October on (fig. 57).
Numbers increase gradually in the area and peak numbers are recorded
in February and March. In April a rapid decline sets in. Arrival of
the birds in the breeding area is dependent on the thawing of the
ice in these areas. This occurs for example in Southern Finland in
early April (Bauer & Glutz, 1969).

3.12.2.2 Moult
Moult of males starts already in May or June. Flight feathers are
moulted from mid-June to mid-July, by mid-August most birds are able

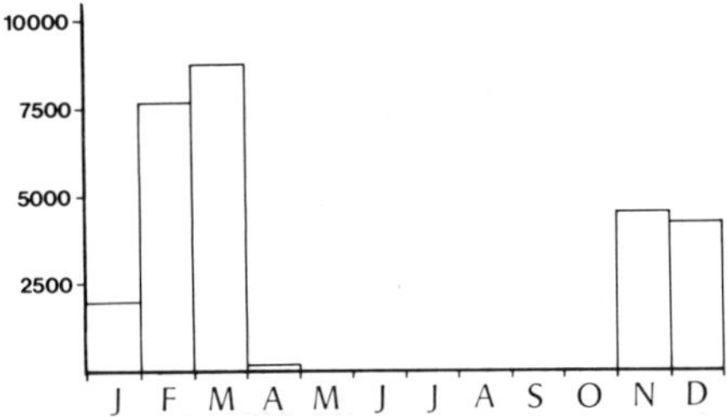

Fig. 57. Mean number of Goosanders per month in the Dutch
part of the Wadden Sea. Numbers
are based on counts from ships
in the years 1964-1970 (Swennen,
unpubl.).

to fly again. A moult of body feathers follow afterwards and is completed in December. Females moult somewhat later, generally flight feathers are moulted around mid-July (Bauer & Glutz, 1969).

3.12.3 Numbers

3.12.3.1 Population size

The total size of the NW European flyway population is estimated at 40,000 birds (Atkinson-Willes, 1972). The species does not breed in the Wadden Sea area. No data on the population structure are known. Field observations near Kornwerderzand suggest a juvenile percentage of less than 25% and males predominate among the adults.

3.12.3.2 Numbers per area

The distribution in the Wadden Sea is practically limited to shallow parts which do not emerge during low tide. Unlike the Red-breasted merganser clear preference is shown for areas with freshwater influence.

In the Netherlands large numbers only have been discovered by aerial and boat surveys. Also in Denmark the maxima have been counted by aerial surveys. So the possibility exists that in places with only observations from the shore, as in the German part, numbers are underestimated.

In the Danish part of the Wadden Sea the Goosander is most numerous from December to February. In January 1970 a maximum number of 3400 has been counted. The largest concentrations have been found in the Ho Bugt and near the Rømø-dam (Joensen, 1974).

In Schleswig-Holstein counts showed a maximum of about 500 Goosanders, in severe winters they apparently leave the area. Generally peak numbers appear to occur in February and March (Busche, 1980).

In Niedersachsen in winter Goosanders are most numerous in the Elbe (with an estimated maximum of about 100), Weser (100), Jade (250), and Ems estuary (200) (Smit, 1977).

Table 15. Numbers of Goosanders in aerial surveys over the Western part of the Dutch Wadden Sea between Harlingen and Den Oever (after Swennen, unpubl.).

December	1968	4,300
January	1968	3,900
February	1965	9,100
February	1967	9.800
February	1967	5,100
February	1968	6,500
March	1965	2,500
March	1965	9,500
March	1968	3,500

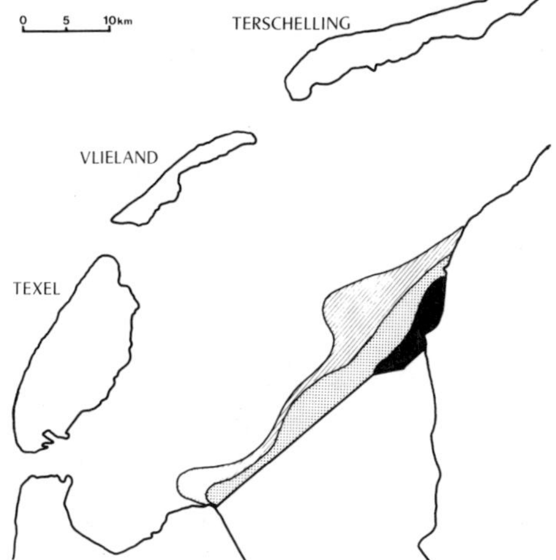

Fig. 58. Distribution of the Goosander in the Western part of the Dutch Wadden Sea determined by counts from ships. In the dark shaded area numbers per 7 km stretch exceeded 256 birds, in the light shaded area 625, in the hatched area 25. Numbers are average data from counts from November-March from 1964-1969 (Swennen, unpubl.).

Numbers in the Dutch part of the Wadden Sea, as recorded by counts from ships are shown in fig. 57. Numbers recorded in aerial counts are listed in Table 15. The distribution in the Western part of the Wadden Sea is shown in fig. 58.

In all areas summer records (June-July) do not occur. There are only irregular observations with low numbers in August and September. The peak numbers show that the Danish part may be used by 12% and the Dutch part by 25% of the estimated flyway population wintering in NW Europe (Atkinson-Willes, 1972).

3.12.4 Food

3.12.4.1 Food composition

Direct observations point to Smelt *(Osmerus eperlanus)* as dominating in the food. However, stomachs have not been analized to prove such.

3.12.4.2 Feeding activities

Goosanders catch their prey by diving. In the Wadden Sea Goosanders forage in water with high contents of suspended matter. On the foraging places in the Western part of the Wadden Sea, the water depth amounts to 1-5 m, rarely more. Flock feeding is usual. Foraging flocks are often accompanied by flocks of Black-headed gulls. Flying Goosanders often descend near flocks of plunge-diving gulls. In comparison with the Red-breasted merganser Goosanders in the Wadden Sea normally stay in less exposed but more turbid water. This contradicts the statement in Bauer & Glutz (1969) that Goosanders have a strong preference for clear water. Goosanders rest on water during the day. Exceptionally a few birds come ashore to preen. Usually they leave the foraging places in the afternoon, it is unknown where they stay at night.

Though the distribution in the Wadden Sea seems to indicate a preference for water with a low salinity, drinking flights to inland waters have never been noticed.

3.12.4.3 Total food consumption
   No data available

References

Atkinson-Willes, G.L., 1972. In M. Smart (ed.): The international wildfowl censuses as a basis for wetland evaluation and hunting rationalization. Proc. Intern. Conf. Conservation of Wetlands and Waterfowl, Iran 1971: p. 87-110.
Bauer, K.M. & U.N. Glutz von Blotzheim, 1969. Handbuch der Vögel Mitteleuropas, Vol. 3. Akademische Verlagsgesellschaft, Frankfurt/ Main: 504 pp.
Busche, G., 1980. Vogelbestände des Wattenmeeres von Schleswig-Holstein. Kilda, Greven (in press).
Joensen, A.H., 1974. Waterfowl Populations in Denmark 1965-1973. Danish Rev. Game Biol. 9(1): p. 1-206.
Smit, C.J., 1977. On the occurrence of 32 bird species in the Danish, German and Dutch Wadden Sea. Unpublished report Intern. Wadden Sea Working Group, part 2: 171 pp.
Swennen, C., 1970. Vogelwaarnemingen op het IJsselmeer. Limosa 43: p. 1-10.
Voous, K.H., 1960. Atlas of European birds. Nelson, London: 284 pp.

3.13 OYSTERCATCHER (*HAEMATOPUS OSTRALEGUS* L.)
   J.B. Hulscher

Da: Strandskade; G: Austernfischer; Du: Scholekster

3.13.1 Distribution

3.13.1.1 Breeding area
   The Oystercatcher has a discontinuous semi-cosmopolitan distribution, breeding in all climatic zones (Voous, 1960). Three subspecies are distinguished in the Palaearctic (fig. 59).
   In NW Europe two populations of the nominate form can be distinguished with more or less separated wintering areas: an Atlantic population breeding on Iceland, the Färöer islands and the British Isles and wintering mainly on the British Isles and a continental population, breeding from the Petsjora river in Siberia to Bretagne in France, wintering mainly along the coast from S Norway to Morocco including the S and SE coast of England (Dare, 1970).
   An analysis of recoveries of ringed birds (figs. 60 and 61) shows that the Wadden Sea derives its Oystercatchers from the continental population, breeding between 53° 30' N and 70° N and 4° E and 38° E.

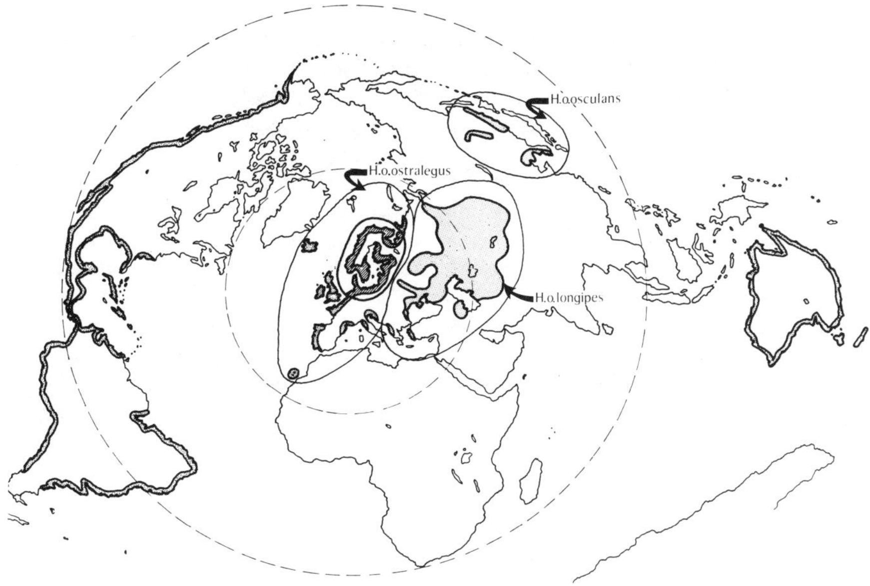

Fig. 59. Breeding area of the Oystercatcher (after Voous, 1960). The dark shaded area represents the region where birds visiting the Wadden Sea originate from.

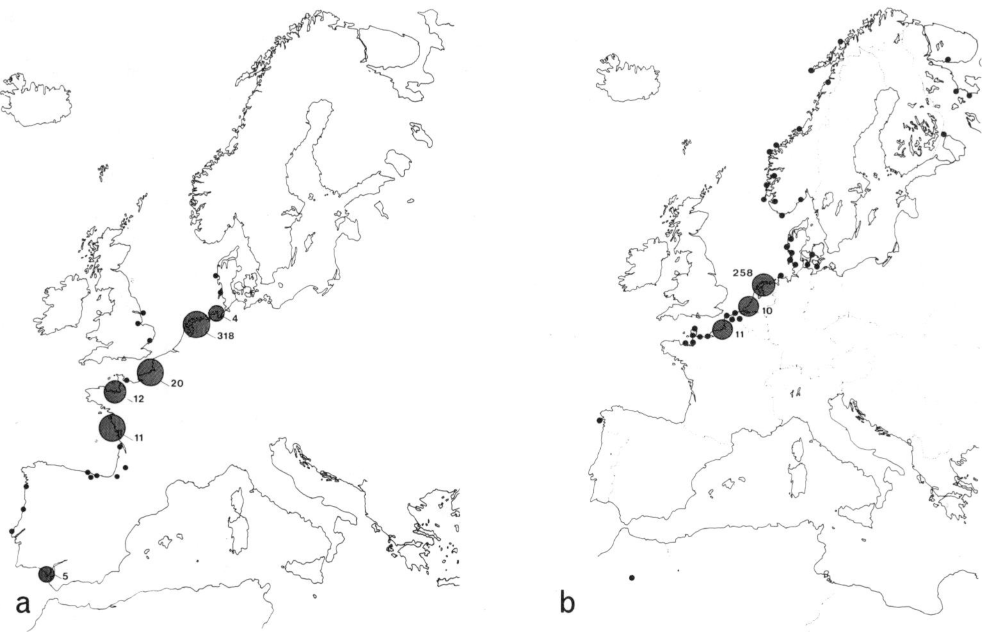

Fig. 60. Recoveries of Oystercatchers ringed as pullus (a) and full-grown (b) in the Dutch Wadden Sea area (data: Vogeltrekstation, Arnhem). The figures represent the number of recoveries in the specific area.

3.13.1.2 Migration routes

North Russian birds reach the Wadden Sea via a direct route over land to the Danish Westcoast, or via a more southerly route, along the South coast of the Baltic Sea (Glutz et al., 1975; Bianki, 1977).

Those Norwegian Oystercatchers wintering in the Wadden Sea first follow the West coast of Norway to the south (Folkestad, 1975) and leave the South coast somewhere between Revtangen and Egersund in a SSE direction. They cross the North Sea and Skagerrak, reach the West coast of Jutland between Hansholm and Blaavandshuk and follow the Danish Westcoast to the Wadden Sea (Thelle, 1970). The "Baltic" migration is about three weeks in advance of the "Norwegian" migration, and the latter is three times as strong (Meltofte et al., 1972; Meltofte & Rabøl, 1977).

Adult birds of the inland population in the Northern part of the Netherlands reach the Wadden Sea from the south. Oystercatchers leaving the Wadden Sea to the southwest follow the North Sea coast. In spring the birds follow reverse routes.

3.13.1.3 Wintering area

Birds visiting the Wadden Sea either winter in the Wadden Sea itself, or along the West coast of the Netherlands, Belgium and France, the South and SE coast of England including the Wash, or the Atlantic coast of the Iberian peninsula south to Cadiz (figs. 60 and 61).

3.13.1.4 Moulting areas

Moulting is common everywhere in the Dutch Wadden Sea area, on the islands as well as along the mainland coast (Boere, 1976; Zegers, pers. comm.), also in the German (Glutz et al., 1975) and probably also in the Danish part. There are some indications that moulting Oystercatchers concentrate mainly on the islands and not along the coast (Glutz et al., 1975).

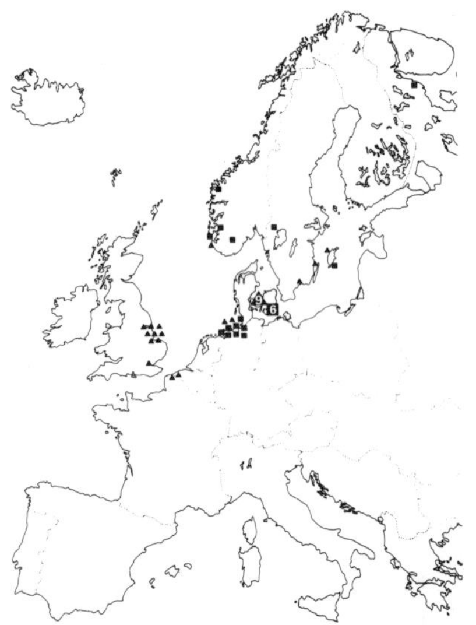

Fig. 61. Ringing places of Oystercatchers recovered in the Dutch Wadden Sea area and ringed as pullus (■) or fullgrown (▲) elsewhere (data: Vogeltrekstation, Arnhem). The figures represent the number of recoveries in the specific area.

3.13.2 Annual cycle

3.13.2.1 Migration

In the Kandalaksha Gulf (White Sea) the breeding grounds are left from the second August decade to the first October decade (Bianki, 1977). Birds pass through Ottenby (Öland) between the beginning of June and the first September decade, with a peak from the second July decade till the first August decade (Edelstam, 1972). On the East coast of Schleswig-Holstein migration starts mid-June, maximum numbers pass through in July (Glutz et al., 1975). The Dutch province of Friesland is left between the end of June and the end of August. Maximum numbers arrive in the first August decade (Hulscher, 1977).

The number of Oystercatchers starts to increase in the whole Wadden Sea area somewhere between the last week of June and the last week of July. The start of the influx of birds may vary from year to year, probably linked with differences in the timing and success of breeding. On the island of Griend for instance in 1967 the first birds arrived in the last June decade, in 1969 however not until July 21 (Veen, pers. comm.).

In the Northern part of the Wadden Sea between Esbjerg and the Weser peak numbers occur from August until October followed by a decline in November. From then on numbers remain more or less constant until January (compare diagrams of numbers for Jordsand (fig. 62) and Schleswig-Holstein (fig. 63)). In the Western part of the Dutch Wadden Sea between Den Helder and Ameland numbers of Oystercatchers increase steadily till September, then decrease somewhat in October and November and increase again till the highest numbers are reached in December/January (figs. 64 and 65). Apparently there is a shift of Oystercatchers from the Eastern to the Western part of the Wadden Sea in September-October. Possibly this annual distribution pattern in the Wadden Sea is linked with the decrease in mean winter temperatures from west to east.

Birds leave the Wadden Sea probably not before September because recoveries of Oystercatchers ringed on the Dutch Wadden Sea islands and recovered in the same season further south do not occur before the first half of September (Table 16).

Mass movements of Oystercatchers in the Wadden Sea occur when severe frost lasts for weeks, especially in combination with snowfall (van Eerden, 1977). In such a case the feeding areas are turned into icefields. However large groups of Oystercatchers remain despite the icy spell: on February 22, 1956 a few hundred were counted in the Western part of the Wadden Sea (Verwey, 1956), at the same time 5000 to 10,000 on Wangerooge (Grosskopf, 1968). During the cold spell in 1962-1963 (December to March) 5000-15,000 were counted in different parts of the Western Wadden Sea (Spaans & Swennen, pers. comm.; Boer & van Orden, 1963). Oystercatchers huddle close together without moving in such periods, but many succumb if the icy weather lasts too long.

Numbers of Oystercatchers in the Wadden Sea decline from the end of January or the beginning of February onwards till June. Along the Dutch North Sea coast visible spring migration also continues till in June (van Dijk, 1976).

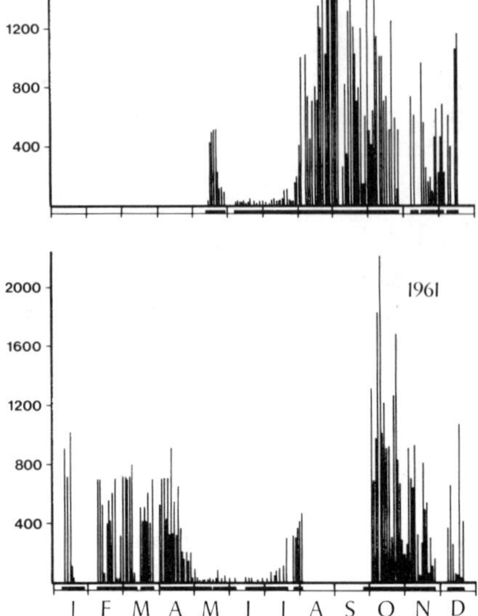

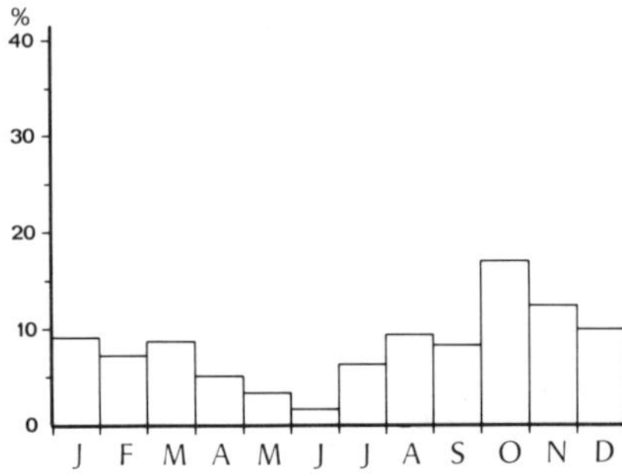

Fig. 62. Numbers of Oystercatchers recorded at the island of Jordsand in 1960 and 1961. Periods of observation are indicated by a heavy horizontal line (after Jepsen, 1975).

Fig. 63. Occurrence of Oystercatchers in the Wadden Sea area in Schleswig-Holstein. Numbers for each month are expressed as a percentage of the total number observed in all months (after Busche, 1980).

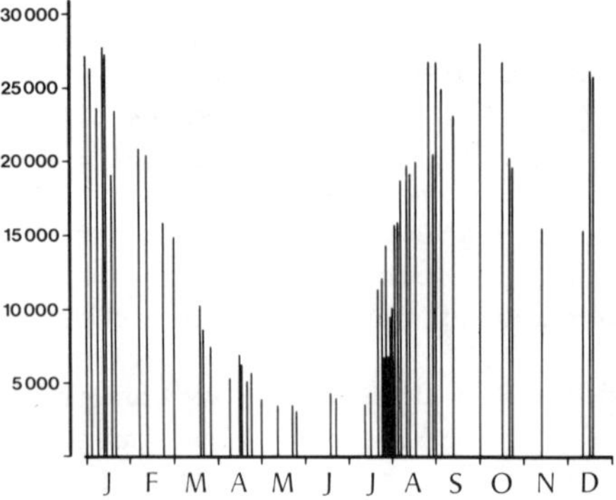

Fig. 64. Numbers of Oystercatchers recorded on the island of Ameland from July 12, 1972 until January 14, 1978) (data: Kersten, pers. comm.).

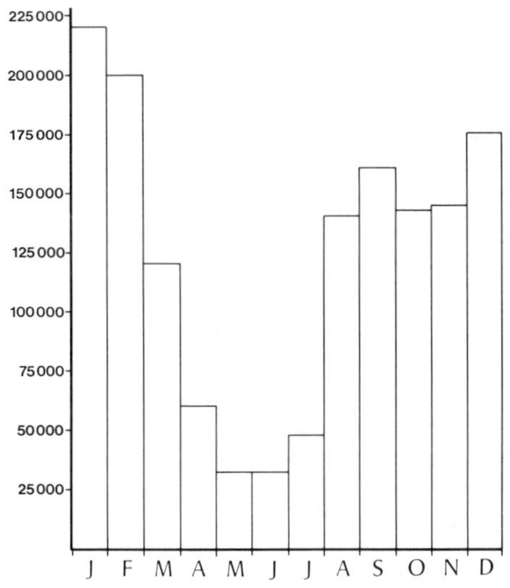

Fig. 65. Mean number of Oyster-catchers per month in the Dutch part of the Wadden Sea (after Hulscher, 1971).

Table 16. Number of recoveries of Oystercatchers ringed on the Dutch Wadden Sea islands (data: Vogeltrekstation, Arnhem, unpubl.). ( )= data unreliable.

locality of recovery	age at ringing	1	2	3	4	5	6	7	8	9	10	11	12	total
N Russia	fullgrown							1	1	1	(1)			4
Norway	fullgrown				2	3	2	4	1				1	13
Denmark	fullgrown					2			3	2	1			8
	pullus							1						1
Germany	fullgrown											1		1
	pullus				2	2	3	2					1	10
England	fullgrown			1										1
	pullus	1					1						1	3
Netherlands, Wadden Sea	fullgrown	26	58	25	23	28	8	21	6	11	13	10	29	258
islands	pullus	17	32	15	19	28	19	66	71	20	15	11	14	327
Zeeland	fullgrown	1	3	5	1									10
	pullus	3		1		1							3	8
Belgium	fullgrown			1							1			2
	pullus	1	1										1	3
France	fullgrown	6	5		1			2			1	3	1	19
	pullus	5	12	1	2	1		3		4	8	4	6	46
Spain	fullgrown												1	1
	pullus	1	3				1				2		2	9
Portugal	pullus	1									1			2
		61	115	49	50	65	34	99	83	38	43	30	59	726

Oystercatchers breeding in Friesland, only some tens of kilometers from the Wadden Sea, arrive at the breeding grounds already from the end of January till the beginning of March, depending upon the weather (Hulscher, 1975). Oystercatchers breeding in the far North along the White Sea arrive from the first May decade on (Bianki, 1977).

In the Wadden Sea flocks of non-breeding Oystercatchers occur in May and June. Most of these birds are juveniles with white collars. The rest are subadults and adults, these mostly have inactive gonads (Van Oordt & Mörzer Bruyns, 1938).

3.13.2.2 Moult

In the Wadden Sea second calendar-year birds start primary moult from the end of April onwards. The primary moulting period for the individual bird lasts for 130-140 days. The last birds complete moult towards the end of September. Adults start primary moult after the breeding season, from the end of June. Their individual primary moulting period lasts for about 100-110 days. The majority of birds has completed it in November, the last birds however not before the end of December. No information is available as to whether summering adults start primary moult in advance of adult breeders (Boere, 1976). Most adults of the inland Frisian population start primary moult on the breeding grounds, reaching the Wadden Sea with actively moulting wings (Hulscher, 1977).

It is not known whether the Scandinavian, Russian and Baltic populations start the initial stages of primary moult on the breeding grounds. Most flight feathers of these populations however, are moulted in the Wadden Sea (Glutz et al., 1975).

3.13.2.3 Weight changes

On average females are heavier than males. Table 17 shows seasonal variations in body weight of birds of different age.

Frost casualties, mainly adults, found from January 31 to February 5, 1976 on the island of Vlieland varied in weights from 270-450 g, mean 341, that is 60% of the normal live weight at the time (Boere et al., in prep.).

Table 17. Mean live weight (in grams) of Oystercatchers caught in the Dutch Wadden Sea. Data a, b & c from Swennen, in Glutz et al., 1975; d & e from Zegers, pers. comm. Numbers in brackets denote the numbers of birds weighed.

	first year	second year	later
a) September-October (males + females)	518.9 (35)	533.2 (37)	545.7 (400)
b) November-February (males + females)	494.7 (97)	529.6 (35)	567.9 (284)
c) March-May (males + females)	487.1 (258)	513.2 (97)	528.5 (359)
d) October-November (males)	532.0 (5)	–	524.0 (10)
e) October-November (females)	552.0 (15)	–	577.0 (23)

3.13.3 Numbers

3.13.3.1 Population size

A simultaneous count in the whole Wadden Sea on September 3, 1972 revealed a total of 474,660 Oystercatchers, of which 1% occurred in the Danish, 48% in the German and 51% in the Dutch part (Prater, 1974).

A total of 500,000 Oystercatchers is probably the maximum number that can be counted on one day and this must also be the total number that frequents the Wadden Sea, because at the beginning of September immigration into the Wadden Sea has nearly come to an end and emigration out of the Wadden Sea has not yet started. This number stresses the importance of the Wadden Sea for the Oystercatcher, because the total winterpopulation in Europe and NW Africa is estimated at 560,000 birds (Prater, 1976).

The distribution of the breeding population of the Wadden Sea area is shown in fig. 66. Information on size and distribution of the breeding populations along the mainland coast of Schleswig-Holstein and Niedersachsen is incomplete. As far as information could be collected the size of the whole breeding population amounts to at least 16,500-19,600 pairs, distributed as 900-1000 pairs in Denmark, about 7000 pairs in Schleswig-Holstein (Busche, 1980), at least 2800-3900 pairs in Niedersachsen and 8400-8800 pairs in the Netherlands (data: Institut für Vogelforschung, Wilhelmshaven; Meltofte, in litt. and Smit, in litt.).

Fig. 66. Distribution of the breeding population of the Oystercatcher in the Wadden Sea area (after Smit, in litt.).

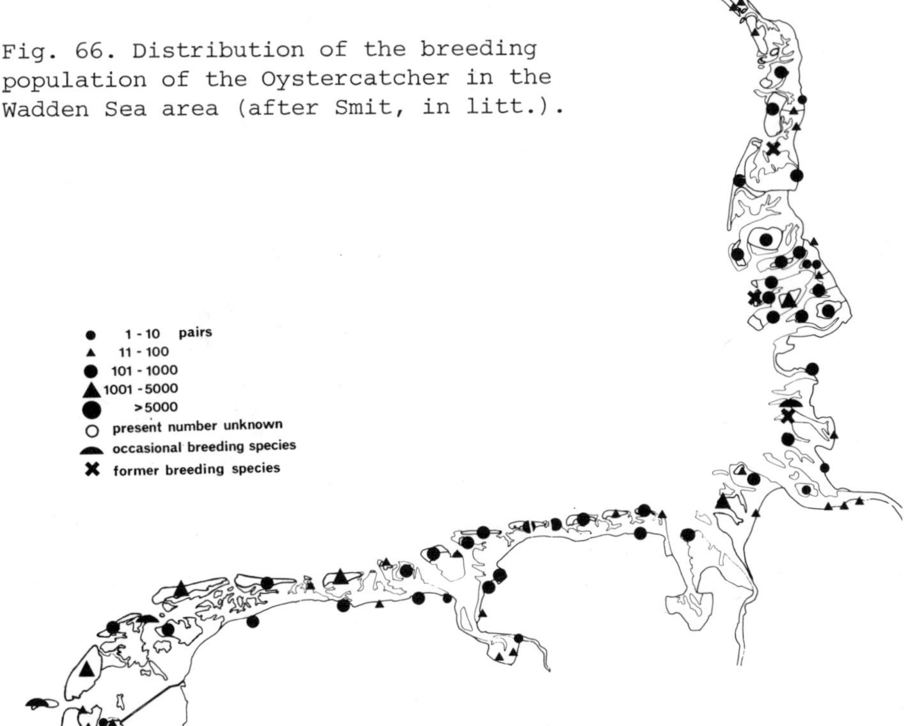

The breeding population in the Wadden Sea area increased during
this century. The breeding population of Mellum increased with 5% per
year over the period 1913-1968 (Schnakenwinkel, 1970). For the increase on Griend see Brouwer (1950), for the increase on Wangerooge
Grosskopf (1968). Recently the breeding area of Oystercatchers in
NW Europe expanded as well to inland areas (Voous, 1960). An increase
of the number of Oystercatchers frequenting the Wadden Sea during
migration and in the winter has not been documented very well because counts in past decennies are lacking. Counts on Schiermonnikoog in the last week of August and first half of September over 1963-
1968 varied from 13,600 to 28,500 (mean 21,100), over 1971-1977 from
21,000 to 36,000 (mean 28,800), an increase of 36%, which statisti-
cally is not significant (Zegers, pers.comm.).

3.13.3.2 Numbers per area
    In the Danish part of the Wadden Sea Oystercatchers are most nume-
rous in the more sandy parts of the area (fig. 67): between Skallingen, Langli and Fanø, and near Kilsand, Mandø, Koresand and Rømø.
Peak numbers occur in September and October when up to 65,000 may
be present, and in March and April with a maximum of 30,000. In mild
winters 20,000-40,000 can be counted, in severe winters numbers drop
to a few thousands (Meltofte, 1980; reports Vadefuglegruppen Dansk
Ornithologisk Forening).

Fig. 67. Regular peak occurrences of
Oystercatchers in autumn per census
area (after Smit, 1977; Busche, 1980;
Meltofte, 1980 and data from Vadefuglegruppen Dansk Ornithologisk Forening
and the Institut für Vogelforschung,
Wilhemshaven).

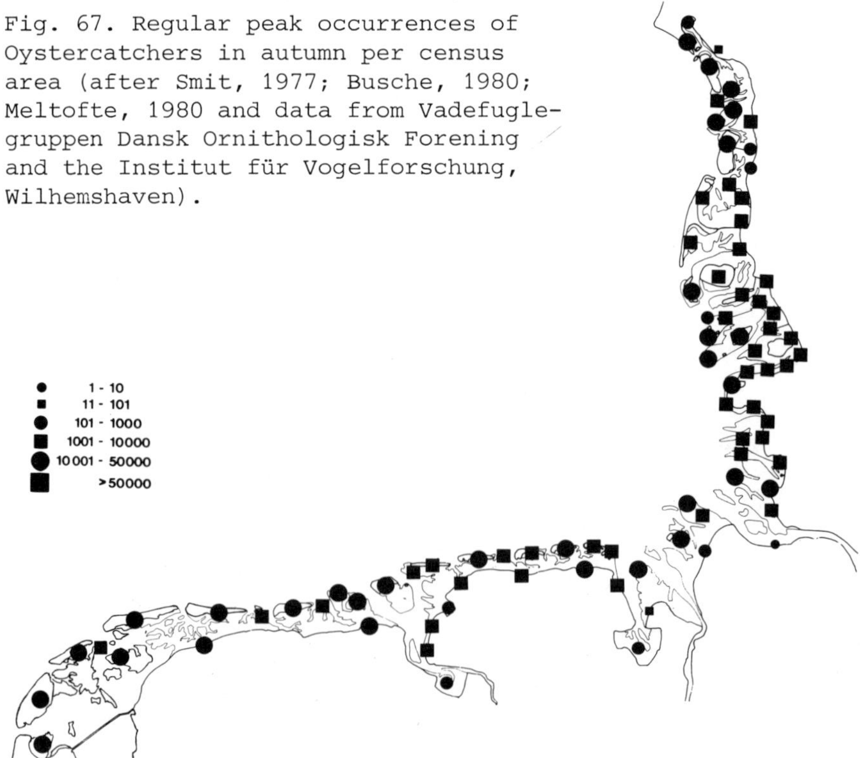

In Schleswig-Holstein the distribution resembles that in Denmark, the largest concentrations occurring in the sandy Western part of the area which is exposed to the North Sea. Peak numbers occur from August to October (fig. 63) when up to 150,000 birds may be present. In normal winters numbers drop to 30,000, in cold winters to 10,000-20,000. More than 10,000 Oystercatchers may occur as summer visitor in May and June (Busche, 1980).

In the Wadden Sea area in Niedersachsen Oystercatchers are most numerous near the islands. Peak numbers in most areas occur from August to October. In 1964, 1965, 1968, 1969 and 1974 January and February counts have been carried out covering the whole area. During these years 13,600, 33,900, 13,700, 21,000 and 18,000-23,000 respectively have been counted (data: Institut für Vogelforschung, Wilhemshaven).

Fig. 65 shows the mean monthly number of Oystercatchers occurring in the Dutch Wadden Sea. The highest number counted in the Dutch Wadden Sea was 243,000 (September 1973), the lowest 22,000 (May 1974) (Boere & Zegers, 1977). Table 18 shows results of counts in the area, fig. 67 the occurrence of Oystercatchers in the whole Wadden Sea area.

3.13.4 Food

3.13.4.1 Food composition

In the Wadden Sea Oystercatchers take mainly molluscs, to a lesser extent worms and crustaceans (Hulscher, 1964a, 1964b, 1964c). Inland mainly insects and earthworms *(Lumbricus)* are taken. Of the molluscs *Cerastoderma edule*, *Mytilus edulis* and *Macoma balthica* are the main prey species, *Macoma* especially when *Cerastoderma* is lacking (Hulscher, 1964b). *Mya arenaria* and *Scrobicularia plana* may be

Table 18. Numbers of Oystercatchers in the Dutch Wadden Sea (after various publications in Watervogels and Limosa and Zegers, in litt.).

Date	Year	Number
January 8	1977	171,270
January 12	1974	181,880
January 17	1976	143,500
January 18	1975	132,640
April 6	1973	46,800
April 19	1975	29,440
May 1	1976	28,000
May 11	1974	21,140
July 29	1972	101,410
August 22	1963	163,000
August 30	1975	208,650
September 1	1973	242,700
October 19	1974	164,620
November 13	1976	163,670
December 29	1966	213,000

important locally, sometimes also *Littorina littorea* (Dircksen, 1932; Höfmann & Hoerschelmann, 1969). Of the annelids *Arenicola marina*, *Nereis diversicolor* and *Harmothoe* spec. (Lange, 1968) are eaten on mudflats, *Scolelepis squamatus (= Nerine cirratulus)* on beaches (Dircksen, 1932; Hulscher, 1964c), earthworms *(Lumbricus)* in polders. Of the crustaceans *Carcinus maenas* (Dircksen, 1932; Hulscher, 1964c) and in small numbers *Crangon crangon*, *Corophium volutator* and *Talitrus saltator* (Hulscher, unpubl.) are taken. *Orchestia* spec. plays a role as food for chicks (Glutz et al., 1975). Of the insects mainly leatherjackets (larvae of Tipulidae), caterpillars (Noctuidae) and beetles (Hulscher, unpubl.) occur. Adult *Calliphora* sometimes serve as food for dune breeders (Glutz et al., 1975).

A small species of sea anemone was seen eaten in large numbers by a first calendar-year bird (Hulscher, unpubl.).

3.13.4.2 Feeding activities

Oystercatchers feed on all suitable exposed flats. Maximum distance to the coast amounts to 13 km. The majority concentrates on musselbeds and cocklefields which are exposed for 0-6 hours per tide, or on places with concentrations of *Mya* or *Scrobicularia*. There are always small numbers of birds feeding on the tidal flats characterized by the lugworm *(Arenicola marina)* where *Arenicola*, *Nereis* and *Macoma* are eaten. The upper part of this zone is particularly exploited during the first 1 to 2 hours after exposure when the birds follow the tide line. When the main feeding areas become immersed the birds immediately fly to the roosts, they usually do not continue feeding on the higher tidal flats that are still exposed.

In winter at high tide, or during bad weather when exposure time of the feeding area on the tidal flats is too short for collecting the required daily ratio of food, also meadows along the mainland coast and on the islands are used as feeding grounds.

During the breeding season birds from the Wadden Sea breeding population may feed in their territories (polders, salt marshes, dunes) as well as on the tidal flats.

High tide roosts can be found anywhere along the coast. The largest flocks are generally found near the largest feeding areas, usually on bare sandflats, along the edges of the salt marsh, especially at the mouths of creeks, in the polders or on small meadows in the dunes. These "inland roosts" are not occupied at night. Along the mainland, roosts are situated on the salt marsh. If no salt marsh is available or if the normal roosts are flooded, birds take refuge behind the dike on bare arable fields or in meadows. Roosting at sea has been reported for Vlieland (Swennen, 1968).

3.13.4.3 Total food consumption

Food intake of birds in captivity amounted in summer to 23.5-36.3 g ash-free dry weight *(Cerastoderma* and/or *Mytilus*, mean 28.8 g)* (Hulscher, 1974). Koene (1978) found with *Mytilus* in March an intake of 29.6-40.7 g ash-free dry weight (mean 37.0 g per bird/24 hrs).

Food intake of birds in the field has not yet been determined

on a 24 hour basis, because intake in darkness is not precisely
known. The intake of an adult bird, feeding under experimental conditions on the mudflat was identical in darkness and in daylight,
3.40 g ash-free dry weight of flesh/hr (Hulscher, 1974, 1976). The
food intake of wild birds per low water period in daylight in summer
varied between 17.1 and 42.2, mean 26.6 ± 8.9 g of *Cerastoderma* or
Macoma flesh per bird (Hulscher in Glutz et al., 1975).

References

Bianki, V.V., 1977. Gulls, shorebirds and alcids of Kandalaksha Bay.
    Keter, Jerusalem: 250 pp.
Boer, P. & C. van Orden, 1963. Enige gegevens over winterslachtof-
    fers, voornamelijk van het strand van Noord-Holland. De Pieper 2
    (4/5): p. 13-26.
Boere, G.C., 1976. The significance of the Dutch Waddenzee in the
    annual cycle of arctic, subarctic and boreal waders. Part. 1. The
    functions as a moulting area. Ardea 64: p. 210-291.
Boere, G.C. & P.M. Zegers, 1977. Wadvogeltellingen in het Nederland-
    se Waddengebied in 1974 en 1975. Watervogels 2: p. 161-173.
Brouwer, G.A., 1950. De vogels, in: Griend, het vogeleiland in de
    Waddenzee. Nijhoff, 's-Gravenhage: p. 243-288.
Busche, G., 1980. Vogelbestände des Wattenmeeres von Schleswig-Hol-
    stein. Kilda, Greven (in press).
Dare, P.J., 1970. The movements of Oystercatchers visiting or breed-
    ing in the British Isles. Fishery Invest. (ser. II), 25 (9); Lon-
    don: 137 pp.
Dircksen, R., 1932. Die Biologie des Austernfischers, der Brandsee-
    schwalbe und der Küstenseeschwalbe nach Beobachtungen und Unter-
    suchungen auf Norderoog. J. Orn. 80: p. 427-521.
Dijk, J. van, 1976. Club van zeetrekwaarnemers, Report 8: 25 pp.
Edelstam, C., 1972. The visible migration of birds at Ottenby, Swe-
    den. Vår Fågelvärld, Suppl. 7.
Eerden, M.R. van, 1977. Vorstvlucht van watervogels door het ooste-
    lijk deel van de Nederlandse Waddenzee op 30 december 1976. Water-
    vogels 2: p. 11-14.
Folkestad, A.O., 1975. Wetland bird migration in Central Norway. Or-
    nis Fennica 52: p. 49-56.
Glutz von Blotzheim, U.N., K.M. Bauer & E. Bezzel, 1975. Handbuch der
    Vögel Mitteleuropas, Vol. 6. Akademische Verlagsgesellschaft, Wies-
    baden: 840 pp.
Grosskopf, G., 1968. Die Vögel der Insel Wangerooge. Abhandl. Vogelk.
    5, Institut für Vogelforschung, Wilhelmshaven: 293 pp.
Höfmann, H. & H. Hoerschelmann, 1969. Nahrungsuntersuchungen bei Li-
    mikolen durch Mageninhaltanalysen. Corax 3 (19): p. 7-22.
Hulscher, J.B., 1964a. Hoe een scholekster strandkrabben ving. De
    Levende Natuur 67: p. 49-52.
Hulscher, J.B., 1964b. Scholeksters en lamellibranchiaten in de Wad-
    denzee. De Levende Natuur 67: p. 80-85.
Hulscher, J.B., 1964c. Scholeksters en wormen. De Levende Natuur 67:
    p. 97-102.

Hulscher, J.B., 1971. De scholekster en de Waddenzee. Waddenbulletin 6 (2): p. 9-13.

Hulscher, J.B., 1974. An experimental study of the food intake of the Oystercatcher in captivity during the summer. Ardea 62: p. 155-171.

Hulscher, J.B., 1975. Het voorkomen in de winter en het begin van de voorjaarstrek van de scholekster in het binnenland van Friesland. Vanellus 28: p. 141-147.

Hulscher, J.B., 1976. Localisation of cockles (Cardium edule L.) by the Oystercatcher (Haematopus ostralegus L.) in darkness and daylight. Ardea 64: p. 292-310.

Hulscher, J.B., 1977. The progress of wing-moult of Oystercatchers Haematopus ostralegus at Drachten, Netherlands. Ibis 119: p. 507-512.

Jepsen, P.U., 1975. Vadehavet Vildtreservat med. øen Jordsand. Danske Vildtundersøgelser 24: 80 pp.

Koene, P., 1978. De Scholekster: aantalseffecten op de voedselopname. Report Zool. Lab. University Groningen: 123 pp.

Lange, G., 1968. Ueber Nahrung, Nahrungsaufnahme, und Verdauungstrakt mitteleuropäischer Limikolen. Beitr. Vogelkd. 13: p. 225-334.

Meltofte, H., 1980. Fugle i Vadehavet. Vadefugletaellingen i Vadehavet 1974-1978. Miljøministreriet, Fredningsstyrelsen, København: 50 pp.

Meltofte, H., S. Pihl & B.M. Sørensen, 1972. Efteraarstraekket af vadefugle (Charadrii) ved Blaavandshuk 1963-1971. Dansk Orn. Foren. Tidsskrift 66: p. 63-69.

Meltofte, H. & J. Rabøl, 1977. Vejrets indflydelse paa efteraarstraekket af vadefugle ved Blaavandshuk, med et forsøg paa en analyse af traekkets geografiske oprindelse. Dansk Orn. Foren. Tidsskrift 71: p. 43-63.

Oordt, G.J. van & M.F. Mörzer Bruyns, 1938. Studien über die Gonaden übersommernder Vögel. IV. Die Gonaden übersommernder Austernfischer (Haematopus ostralegus L.). Z. Morphol. Ökol. Tiere 34: p. 161-172.

Prater, A.J., 1974. Coastal wader counts: Waddensea and Delta Region. IWRB Bull. 37: p. 102-104.

Prater, A.J., 1976. The distribution of coastal waders in Europe and North Africa. In: M. Smart (ed.), Proc. Intern. Conf. Conservation of Wetlands and Waterfowl, Heiligenhafen 1974. IWRB, Slimbridge: p. 255-271.

Schnakenwinkel, G., 1970. Studien an der Population des Austernfischers (Haematopus ostralegus) auf Mellum. Vogelwarte 25: p. 336-355.

Smit, C.J., 1977. On the occurrence of 32 bird species in the Danish, German and Dutch Wadden Sea. Unpubl. report Internat. Wadden Sea Working Group, part 3: 174 pp.

Swennen, C., 1968. Zwemmende scholeksters. Limosa 41: p. 150-151.

Thelle, T., 1970. The migration of Oystercatcher from West Norway to the Wadden Sea. Dansk Orn. Foren. Tidsskr. 64: p. 229-247.

Verwey, J., 1956. De Waddenzee als voedsel-areaal voor vogels bij strenge kou. Ardea 44: p. 218-224.

Voous, K.H., 1960. Atlas of European birds. Nelson, London: 284 pp.

3.14 AVOCET *(RECURVIROSTRA AVOSETTA L.)*
G.C. Boere & C.J. Smit

Da: Klyde; G. Säbelschnäbler; Du: Kluut

3.14.1 Distribution

3.14.1.1 Breeding area
The Avocet has a discontinuous palaearctic-ethiopian distribution, breeding in the temperate, Mediterranean, steppe and desert climate zones (fig. 68; Voous, 1960; Glutz et al., 1977). The NW European population which is visiting the Wadden Sea, is quite isolated from the main breeding area (Tjallingii, 1970).

3.14.1.2 Migration routes
Migration of the Avocet is mainly coastal though the species is quite regularly observed at suitable places inland.

3.14.1.3 Wintering areas
The wintering area of the NW European breeding population ranges from the Wadden Sea as far as NW Africa (e.g. Bub, 1967; Edelstam, 1971; Cadbury & Olney, 1978). Especially areas in the Western part of the Mediterranean are important. For instance 11,000 birds have been counted in Portugal and 7000 in Tunesia. Other wintering areas

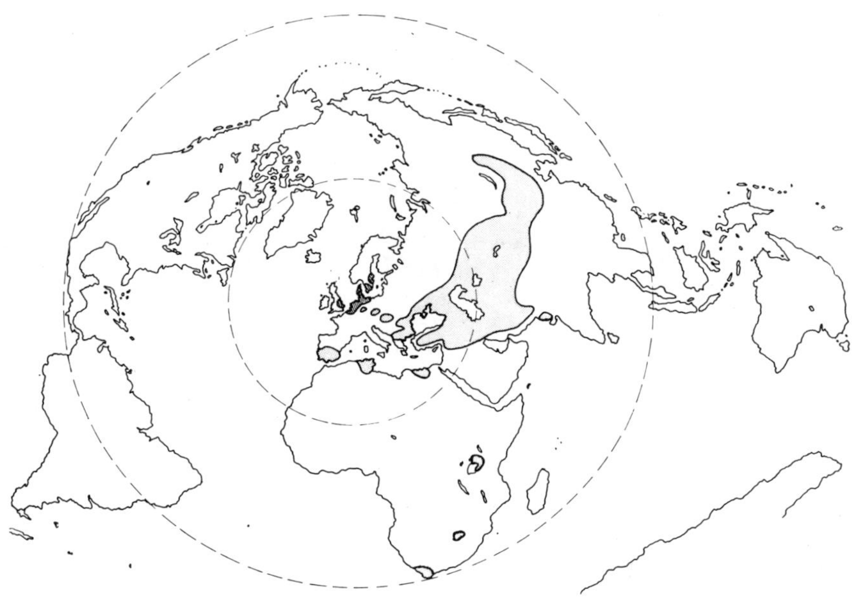

Fig. 68. Breeding area of the Avocet (after Voous, 1960). The dark shaded area represents the region where birds visiting the Wadden Sea originate from.

are coastal lagoons in Morocco and the Banc d'Arguin and coastal as
well as inland areas in W Africa, for instance in Nigeria and Senegal.

3.14.1.4 Moulting areas

Nothing is known from direct observations, but moult probably occurs all over the wintering area. In some places in the Wadden Sea
area thousands stay for several months just after the breeding season. Examples are the Dollard, Leybucht and Jadebusen (Bub, 1967)
and the area along the Rømø-dam. It is quite probable that these
large concentrations of Avocets are in moult. During the last few
years inland wetlands also became very important as moulting areas
(Van Poelgeest & Osieck, 1974).

3.14.2 Annual cycle

3.14.2.1 Migration

Breeding birds from Sweden, the most notherly breeding area of
the population visiting the Wadden Sea area, leave early July to arrive in the Wadden Sea about that time. In the same period breeding
birds of coastal NW Europe (probably excluding those breeding in England and the Dutch Delta area), migrate towards the Wadden Sea. Small
numbers already migrate to Southern Europe. From early August till
mid- or late-September nearly the whole NW European breeding population is present in the Wadden Sea area. From mid-September onwards
numbers decrease due to futher migration to Southern Europe. Depending
on the weather conditions, thousands of birds can be present in the

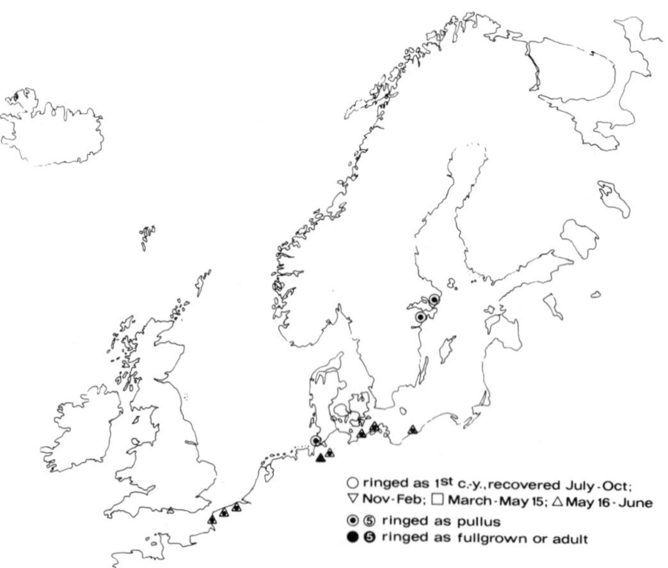

Fig. 69a. Ringing places of Avocets recovered in the Dutch part of the
Wadden Sea (based on published records from 1911-1979).

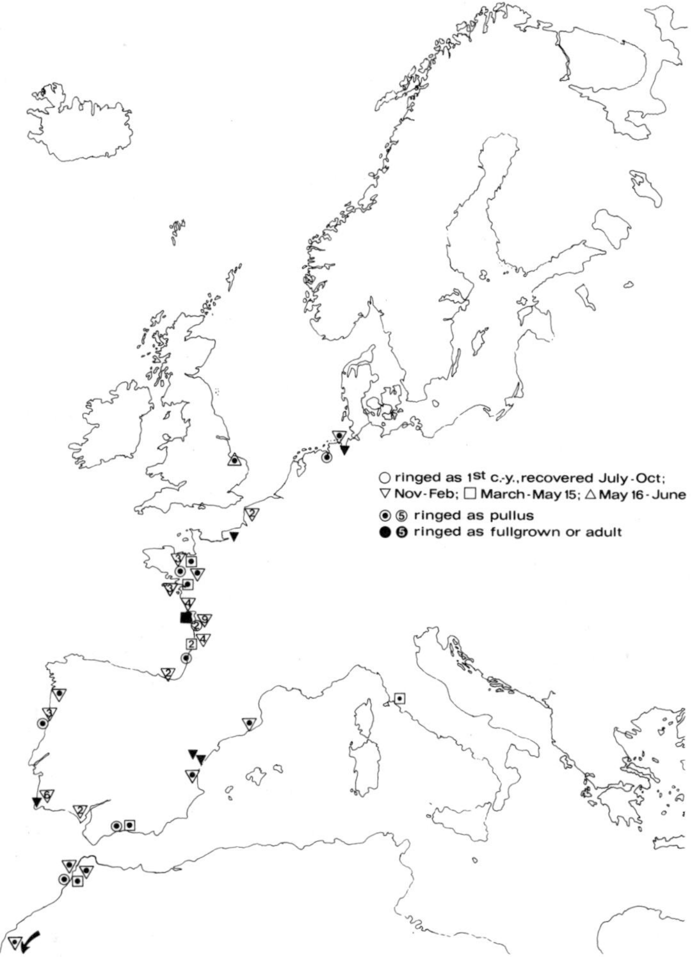

Fig. 69b. Recoveries of Avocets ringed in the Dutch part of the Wadden Sea (based on published records from 1911-1979). The figures represent the number of recoveries in the specific area.

Wadden Sea until early November. Groups of sometimes a hundred or more birds have been observed in December and January. Numbers increase in the Wadden Sea from the end of February onwards, but are not as large as during autumn. Avocets arrive on the breeding places along the Baltic and South Sweden by mid-March (Glutz et al. 1977). Ringing results are shown in figs. 69a and 69b.

3.14.2.2 Moult

Nearly nothing is known on moult progress. According to Glutz et al. (1977) however it is more or less comparable with *Himantopus*. Post-nuptial moult therefore probably takes place from early July till the end of December. Two adult birds caught on the island of

Vlieland in October and November had recently finished moult (Boere, 1976).

3.14.2.3 Weight changes
No information available from the area.

3.14.3 Numbers

3.14.3.1 Population size
Prater (1976) estimates the size of the population wintering in W Europe and NW Africa at 22,300. He estimates the size of the NW European breeding population at somewhat more than 9200 pairs. In the past years however the number of breeding pairs increased a little.
Table 19 shows that in the seventies the Wadden Sea area contributed to about 50% of the NW European breeding population. Fig. 70 shows the distribution of this population.

3.14.3.2 Numbers per area
From the beginning of July on breeding birds from Sweden arrive in Denmark. In the Danish Wadden Sea numbers reach maxima during August. Up to 7700 have been observed in the area. In September and October still 3000-4000 Avocets are present, in some years this number may even stay until December. In January and February only a few have been recorded up to now. Numbers in spring reach maxima of 1400, in some years numbers in April and May are less than 1000. The largest concentrations are observed along the Rømø-dam, smaller near Højer and in the Ho Bugt (Meltofte, 1980; Dansk Ornithologisk Forening).
Fig. 71 shows that in the Wadden Sea area in Schleswig-Holstein

Table 19. Number of breeding pairs of the Avocet in NW Europe (after Ekelöf, 1970; Prater, 1976; Timmerman, 1979; Glutz et al., 1977 and unpublished data from Dansk Ornithologisk Forening, Institut für Vogelforschung, Wilhelmshaven and Smit).

		Wadden Sea area
Sweden	600	
Denmark ($\pm$ 1975)	3000	650
Schleswig-Holstein (1969)	1600	1500
Niedersachsen ($\pm$ 1977)	1600-1900	1600-1900
Netherlands (1975-1977):		
Wadden Sea area	2200	2200
IJsselmeer area	1100	
Delta area	1300	
Rest of the country	100	
Belgium (1969)	180	
England ($\pm$ 1974)	130	
	11810-12110	5950-6250

Fig. 70. Distribution of the breeding
population of the Avocet in the Wadden
Sea area. After Dahl & Heckenroth (1978),
Glutz et al. (1977) and various publis-
hed and unpublished data from Dansk
Ornithologisk Forening, Institut für
Vogelforschung, Wilhelmshaven, König
(in litt.) and Smit (in litt.).

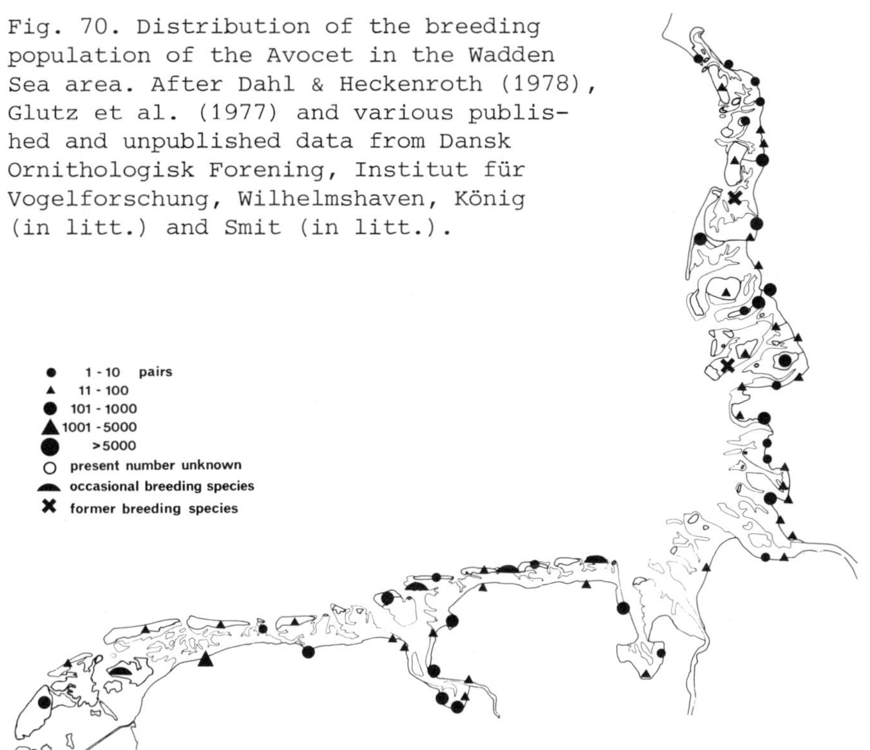

numbers already reach maxima in July. Probably these are birds
from the population breeding along the West coast in Schleswig-
Holstein as well as birds from Swedish and Baltic populations. Relatively high numbers in June probably are mainly birds that have been breeding unsuccesfully. Numbers in July may be up to 5000 birds. Relative
ly few Avocets are present from November until March. In mid-winter
only a few occur in December and January (Busche, 1980).

Large numbers occur in the Wadden Sea area in Niedersachsen, especially in the Leybucht, Jadebusen and Dollart. In all of these areas
large mudflats are present. Numbers in the Leybucht in spring and autumn vary from some hundreds to just over 1000, though numbers of up
2500 sometimes have been recorded (Dahl & Heckenroth, 1978). In the
Jadebusen in the second half of September and the first half of
October up to 4000 are present, though once about 10,000 have been
recorded in October (Bub, 1967). Fig. 72 and 73 show that the species
is even more numerous in the Dollart, both the German and the Dutch
part.

The most important areas for Avocets in the Dutch part of the Wadden Sea are the Dollard and the mainland coast of Friesland. The former Lauwerszee (where up to 5800 have been counted in September) lost
most of its importance. Like in the other parts of the Wadden Sea the
species shows a strong preference for muddy areas. Figure 74 shows its
occurrence in the Dutch part. The mean number of Avocets in the Dol-

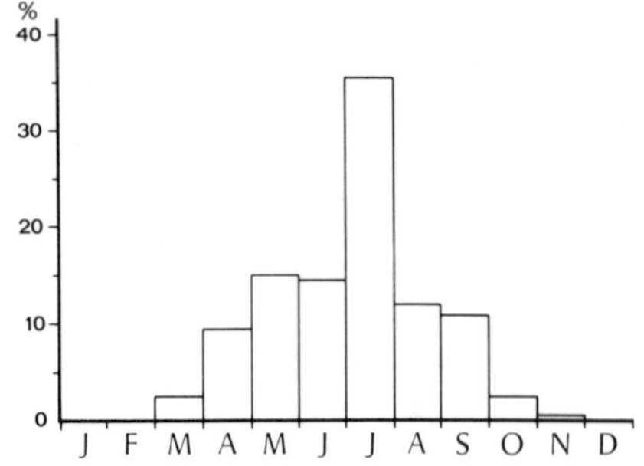

Fig. 71. Occurrence of Avocets in the Wadden Sea area in Schleswig-Holstein. Numbers for each month are expressed as a percentage of the total number observed in all months (after Busche, 1980).

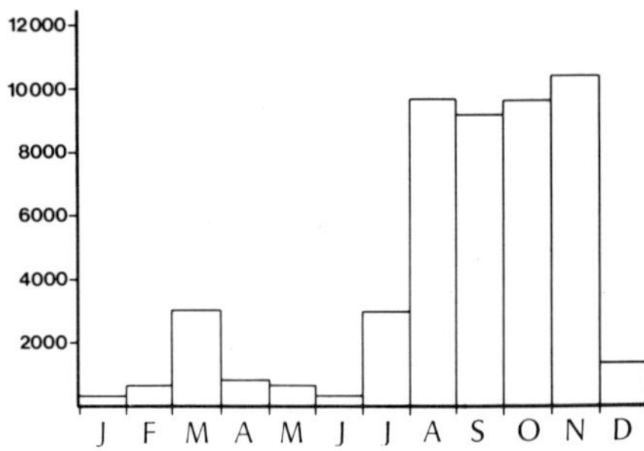

Fig. 72. Mean number of Avocets per month in a 50 km long stretch along the mainland coast in Niedersachsen (after data from Smit, 1977 and the Institut für Vogelforschung, Wilhelmshaven).

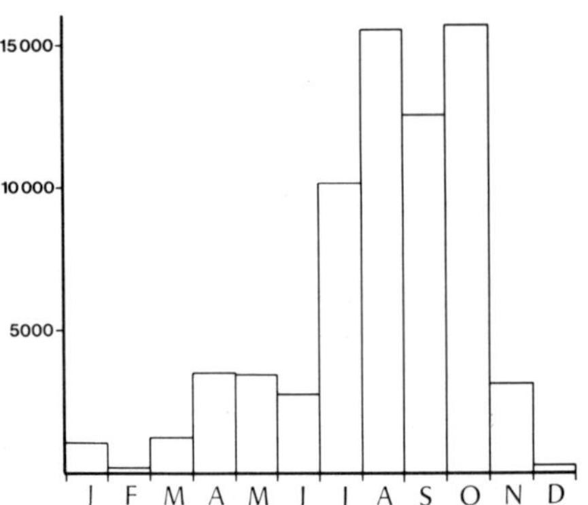

Fig. 74. Mean number of Avocets per month in the Dutch Wadden Sea (after data from Tjallingii, 1969 and Smit, 1977).

lard in August and October is about 10,000. In the Dutch part of the Dollard peak numbers in these months amount to about 12,000 to 15,000 (Tjallingii, 1971). Somewhat smaller numbers occur along other parts of the mainland coast. Numbers from July until September vary from 4000 till 10,000 (Timmerman, 1974). Numbers counted in the Dutch part of the Wadden Sea are listed in Table 20.

On September 1, 1973 also the Danish and German part of the Wadden Sea have been counted. In the Danish part the number was 6576, in the German 23,100. This brings the total number of Avocets to 45,326, representing probably the great majority of the NW European breeding population of the species, including first calendar-year birds.

Fig. 73. Regular peak occurrences of
Avocets per census area in autumn.
After Smit (1977), Busche (1980),
Meltofte (1980) and data from Vade-
fuglegruppen Dansk Ornithologisk
Forening and Institut für Vogel-
forschung, Wilhelmshaven.

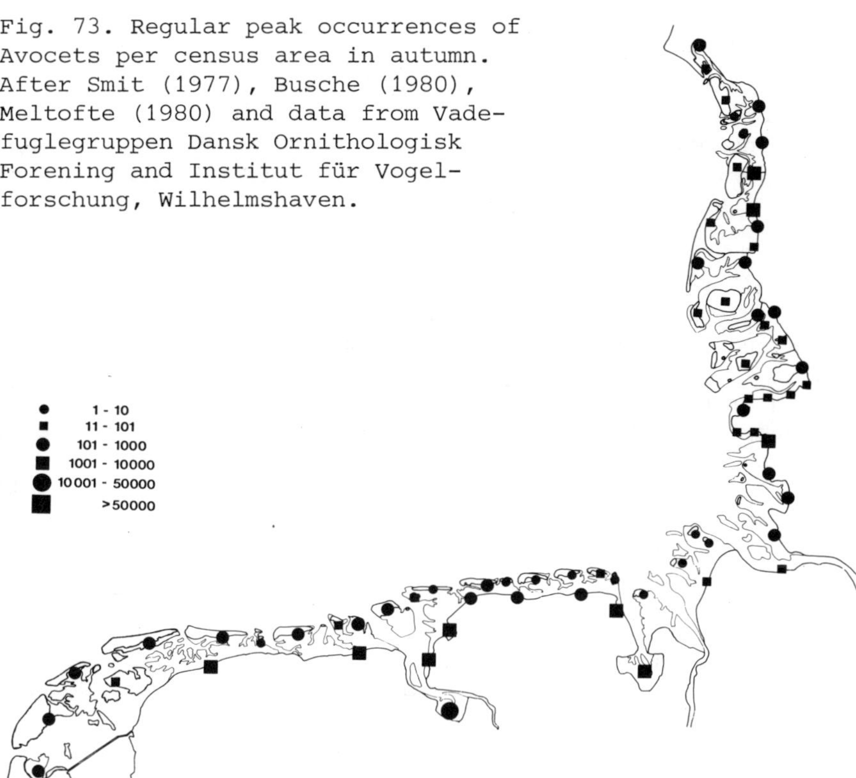

Table 20. Numbers of Avocets in the Dutch Wadden Sea (after various
publications in Watervogels and Limosa and Zegers, in litt.).

Date	Year	Number
January 8	1977	61
January 12	1974	294
January 17	1976	480
January 18	1975	900
April 6	1973	5,040
April 19	1975	3,220
July 29	1972	12,790
August 22	1963	8,900
August 30	1975	6,250
September 1	1973	15,650
October 19	1974	8,390
November 13	1976	4,360
December 29	1966	92

3.14.4 Food

3.14.4.1 Food composition

In marine habitats a great variety of invertebrates are taken, but above all large numbers of *Nereis diversicolor* and *Corophium volutator* (Tjallingii, 1969). Outside the marine habitat the main prey group are chironomid larvae but otherwise also a great variety of other invertebrate species (Glutz et al., 1977).

3.14.4.2 Feeding activities

Avocets have a characteristic feeding technique by sweeping the up-curved bill through the upper part of the soft bottom (Tjallingii, 1969, 1971).

Main feeding areas are soft bottom areas close to the coast. The same type of substrate is frequented when feeding along the shores of ponds, ditches and creeks in the salt marshes.

High tide roosts are situated close to the high tide line and often in the water. Otherwise Avocets roost on salt marshes, inland meadows or rural areas close to the shallows. Flocks of several hundreds, sometimes thousands of birds may group together.

3.14.4.3 Total food intake

Unfortunately only scarce information is available. Tjallingii (1969) found a daily intake of about 3300 specimens of *Nereis diversicolor* per day, corresponding with about 142 gram wet weight.

References

Boere, G.C., 1976. The significance of the Dutch Wadden Sea in the annual life cycle of arctic, subarctic and boreal waders. Part 1. The function as a moulting area. Ardea 64: p. 210-291.

Bub, H., 1967. Ueber den Säbelschnäbler und den Grossen Brachvogel im Jadebusen bei Hochwasser. Vogelwarte 24: p. 135-142.

Busche, G., 1980. Vogelbestände des Wattenmeeres von Schleswig-Holstein. Kilda, Greven (in press).

Cadbury, C.J. & P.J.S. Olney, 1978. Avocet population dynamics in England. Brit. Birds 71: p. 102-121.

Dahl, H.J. & H. Heckenroth, 1978. Landespflegerisches Gutachten zu geplanten Deichbaumassnahmen in der Leybucht. Naturschutz und Landschaftspflege Niedersachsen 7. Landesverwaltungsamt, Hannover: 176 pp.

Edelstam, C., 1971. Flytning och dödlighet hos svenska Skärfläckor Recurvirostra avosetta. Vår Fågelvärld 30: p. 168-179.

Ekelöf, O., 1970. Der Brutbestand des Säbelschnäblers an der Westküste Schleswig-Holsteins im Jahre 1969. Corax 3: p. 97-100.

Glutz von Blotzheim, U.N., K.M. Bauer & E. Bezzel, 1977. Handbuch der Vögel Mitteleuropas, Vol. 7. Akademische Verlagsgesellschaft, Wiesbaden: 894 pp.

Meltofte, H., 1980. Fugle i Vadehavet. Vadefugletaellinger i Vadehavet 1974-1978. Miljøministeriet, Fredningsstyrelsen, København: 50 pp.

Poelgeest, R. van & E.R. Osieck, 1974. Enige gegevens over het voorko-
   men van de Kluut in zuidelijk Flevoland. Limosa 47: p. 94-99.
Prater, A.J., 1976. The distribution of coastal waders in Europe and
   North Africa. In: M. Smart (ed.): Proceedings Intern. Conference on
   Conservation of Wetlands and Waterfowl, Heiligenhafen, 1974. IWRB,
   Slimbridge: p. 255-271.
Smit, C.J., 1977. On the occurrence of 32 bird species in the Danish,
   German and Dutch Wadden Sea. Unpubl. report Intern. Wadden Sea Work-
   ing Group, part 3: 174 pp.
Timmerman, A., 1974. De betekenis van de Friese Waddenkust voor wad-
   en watervogels. Oenkerk, 54 pp.
Timmerman, A. Azn., 1979. Kluut. In R.M. Teixeira (ed.), Atlas van de
   Nederlandse broedvogels. Natuurmonumenten, 's Graveland: p. 136-
   137.
Tjallingii, S.T., 1969. Habitatkeuze en -gebruik van de Kluut. Report
   Zool. Lab. University Groningen: 127 pp.
Tjallingii, S.T., 1970. Inventarisatie van de in 1969 in Nederland broe-
   dende Kluten. De Levende Natuur 73: p. 222-229 & 251-255.
Tjallingii, S.T., 1971. De Kluten van de Dollard. Waddenbulletin 6 (1):
   p. 5-9.
Voous, K.H., 1960. Atlas of European birds. Nelson, London: 284 pp.

3.15 RINGED PLOVER *(CHARADRIUS HIATICULA L.)*

P.H. Becker

Da: Stor Praestekrave; G: Sandregenpfeifer; Du: Bontbekplevier

3.15.1 Distribution

3.15.1.1 Breeding area

   The Ringed plover has a holarctic distribution (fig. 75), breeding
in the tundra, boreal and temperate climatic zones (Voous, 1962). Two
subspecies occur: *C.h. hiaticula*, a breeding bird in Greenland, Iceland, Spitsbergen and in Europe along the coasts of the North Sea and
Baltic and *C.h. tundrae*, breeding from Lapland to NE Europe and Siberia. The Northern part of Scandinavia seems to be an area in which the
two subspecies mix (Väisänen, 1969; Glutz et al., 1975). Along the
North Sea numbers of breeding *C.h. hiaticula* decline to the south. In
the Wadden Sea migrants of both subspecies occur.

3.15.1.2 Migration routes

   C.h. tundrae migrates through Northern, Central and Western Europe.
Breeding populations of *C.h. hiaticula* from Greenland, Iceland and
Spitsbergen may reach their wintering areas by migrating across the
British Isles or via SW Norway. Probably part of the birds from the
Greenland population pass via SW Norway as well (Salomonsen, 1979;
Meltofte & Rabøl, 1977).

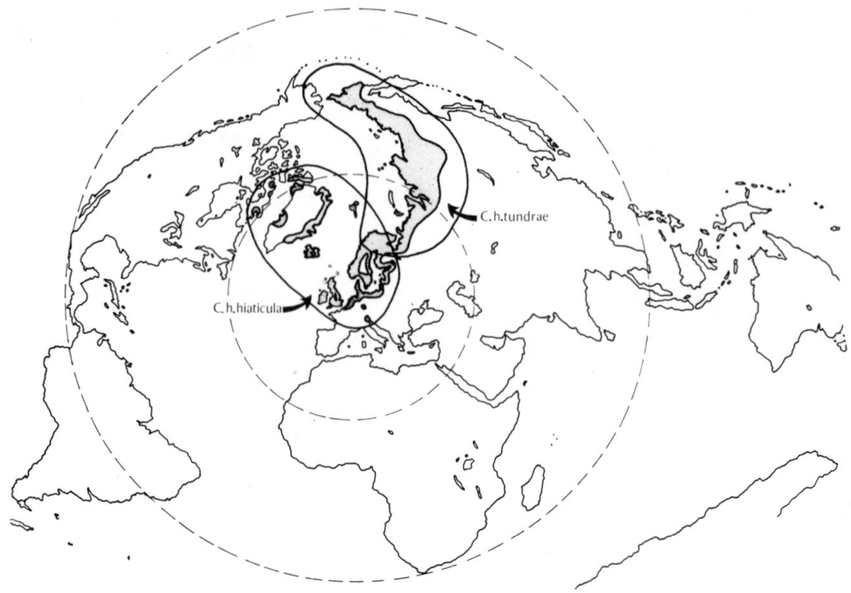

Fig. 75. Breeding area of the Ringed plover (after Voous, 1962, modified).

3.15.1.3 Wintering areas

The species winters along sandy, muddy or gravelly coasts. *C.h. hiaticula* winters in the British Isles, France, the Iberian peninsula and NW and W Africa. Small numbers winter in the Dutch Wadden Sea. The Northern populations winter further south (Prater, 1976; Taylor, 1980; Glutz et al., 1975). *C.h. tundrae* winters on the Canary Islands, in NW and W and E Africa to South-Africa, the Mediterranean region and along the South Caspian Sea (Taylor, 1980). Because only very few recoveries exist of Ringed plovers banded in the Wadden Sea and because of the great differences in wintering areas between Northern and Southern populations nothing can be said with certainty about wintering- and breeding areas of birds visiting the Wadden Sea. Birds belonging to the Wadden Sea breeding population have been recovered on the British Isles and along the French West coast.

3.15.1.4 Moulting areas

In *C.h. hiaticula* postnuptial moult starts in the breeding areas and is finished in the wintering areas. *C.h. tundrae* starts moulting body feathers in Central Europe and completes moult in Africa. First calendar-year birds of both subspecies moult in the wintering areas. Thus the Wadden Sea serves as a moulting station for both subspecies.

3.15.2 Annual cycle

3.15.2.1 Migration

Of *Charadrius h. hiaticula* adult birds generally begin to migrate

before juveniles (Laven, 1940). In the Dutch Wadden Sea small flocks
are formed in June, first migration starts in July. Peak numbers are
counted in August-September, they decrease until mid-October. The number of migrating first calendar-year birds continuously increases from
mid-August to October (Glutz et al., 1975). In France migration com-
mences in mid-July with peak numbers occurring in September. The first
migrating and wintering birds arrive in Portugal, Southern Spain and
Morocco during the mid-August.

Spring migration starts at the beginning of February. The first
birds arrive in the breeding areas in the first decade of March. In
the Wadden Sea the maximum number in spring can be counted in May
(figs. 76 and 77).

In *Charadrius h. tundrae* the departure of Russian birds begins in
the third decade of July. Maximum numbers arrive on Öland over a period from mid-August to the first September decade (for details see Edelstam, 1972). As the subspecies is not easy to distinguish from *C.h.
hiaticula* it is difficult to determine the number present in the
Wadden Sea. The first wintering birds reach Morocco in August, greater
numbers arrive in South Africa from mid-September onwards.

In spring the last birds leave South Africa during the first half
of May. In the second part of May most migrants cross Central Europe
and they reach their breeding areas from the end of May till the beginning of June (Glutz et al., 1975).

3.15.2.2 Moult

First calendar-year birds of both subspecies moult in their wintering areas: *C.h. hiaticula* changes the greater part of the body feathers,
C.h. tundrae changes the body plumage in winter, the tail feathers and
remiges are moulted from January to April. Postnuptial moult of adult
European *C.h. hiaticula* may begin in mid-June with the first primary,
during or after egglaying, but usually after the breeding period. By
the end of September moult of wing feathers is finished. The change
of body plumage starts in June-July.

In August-September adult birds of *C.h. tundrae* moult a large part
of their body feathers while they are still in Central Europe (post-
nuptial moult). In the wintering areas moult is completed when remiges
and rectrices are changed during November-February. Prenuptial body
moult starts in the second half of March and ends before arrival at
the breeding places.

According to Boere (1976) and Pienkowski et al. (1976), the majority of Ringed plovers present in the Wadden Sea in August moults in
Southern Europe and Africa.

3.15.2.3 Weight changes

During moult of remiges weights of British short-distance migrants
goes down from 68 g (n=5, July) to 65.7 g (n=106, September) (Glutz et
al., 1975). Once moult is over weights increase only slightly, maxima
are reached in midwinter. In contrast to that, Arctic Ringed plovers
show a drastic weight increase before leaving their resting places
in Britain: August 63.0 g (n=99), September 69.8 g (n=34), October
71.5 g (n=8) (Minton, 1972). Data from the Wadden Sea are listed in
Table 21.

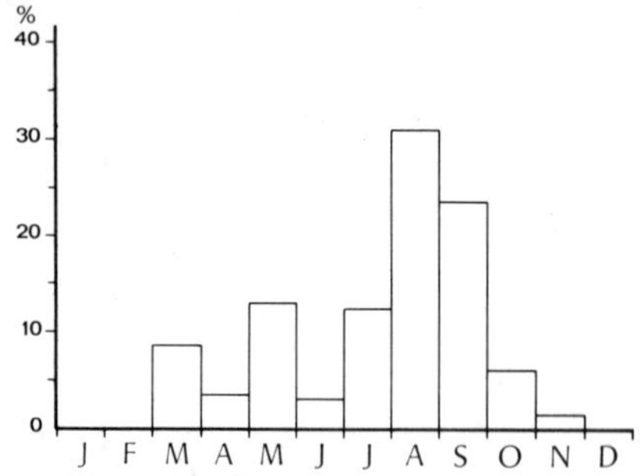

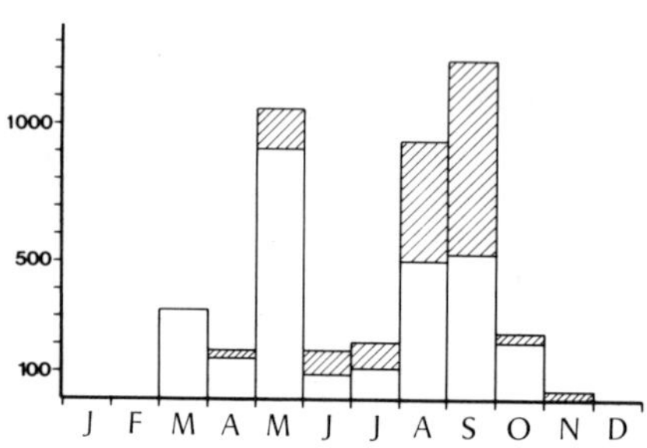

Fig. 76. Occurrence of Ringed plovers in the Wadden Sea area in Schleswig-Holstein. Numbers for each month are expressed as a percentage of the total number observed in all months (after Busche, 1980).

Fig. 77. Mean number of Ringed plovers per month in some areas in Niedersachsen. Coastal areas (white columns) and islands (hatched) have been separated (data: Smit, 1977). In March: peak of C.h. hiaticula, in May: peak of C.h. tundrae. In July-August peak of adults, in September peak of juveniles of both subspecies.

Fig. 78. Distribution of the breeding population of the Ringed plover in the Wadden Sea area. After Smit (in litt.), Dansk Ornithologisk Forening and Institut für Vogelforschung, Wilhelmshaven.

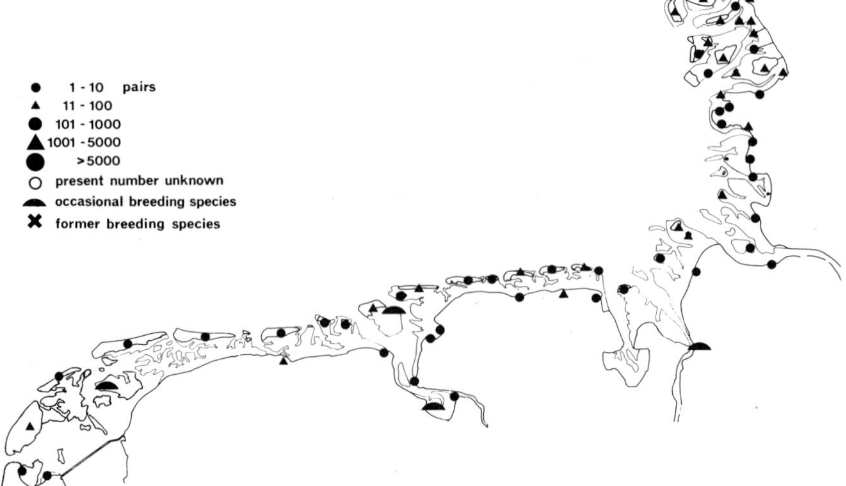

Table 21. Weights (in g) of Ringed plovers caught on the island of Vlieland (Posthuiswad/Kroonspolders: V) and along the mainland coast of Friesland (Westhoek/Bildtpollen: M) (Source: Boere, unpubl.).

month	$\bar{x}$	range	s.d.	n	age	place of catching
August	57.0	42–73	9.1	14	1st, 2nd c.y.	V
August	49.5	49–50	0.7	2	1st, 2nd c.y.	M
August	59.3	41–82	9.5	41	adult	V
August	59.0	53–67	5.4	11	adult	M
September	54.0	43–?	10.2	7	1st, 2nd c.y.	V
September	52.3	48–61	6.1	4	1st, 2nd c.y.	M
September	69.0	66–72	4.2	2	adult	V
September	61.0	58–64	4.2	2	adult	M
October	62.0			1	adult	V

3.15.3 Numbers

3.15.3.1 Population size

The NW European flyway population is estimated by counts in mid-winter to amount to about 42,500 birds. This number probably is an underestimation as breeding birds from N Scandinavia may also winter in W Africa south of Mauritania (Prater, 1976), where no counting has been carried out.

The species is a common breeding bird on the islands in the Wadden Sea and along the mainland coast of Schleswig-Holstein. Fewer Ringed plovers breed along the mainland shores of Niedersachsen and the Netherlands. Distribution of the breeding population is shown in fig. 78. The size of the breeding population in the Wadden Sea area is:
- Denmark: approximately 120–155 pairs (data: Dansk Ornithologisk Forening).
- Schleswig-Holstein: approximately 975–1120 pairs (data: Institut für Vogelforschung, Wilhelmshaven and Smit in litt.). The breeding population of Sylt may amount to more than 600 pairs in some years (Glutz et al., 1975).
- Niedersachsen: approximately 150–300 pairs (data: Glutz et al., 1975 and Institut für Vogelforschung, Wilhelmshaven).
- the Netherlands: approximately 100 pairs (data: Glutz et al., 1975 and Smit, in litt.).
The size of the whole Wadden Sea breeding population therefore amounts to about 1350–1700 pairs.

Table 22 shows examples of population fluctuations resulting from biotope changes. Ringed plovers settle soon on recently formed sandy or gravelly banks, and numbers may decline because of development of vegetation, tourism or biotope destruction.

Table 22. The number of breeding pairs of the Ringed plover, in five areas in nine succeeding years (data: Institut für Vogelforschung, Wilhelmshaven).

year	1970	1971	1972	1973	1974	1975	1976	1977	1978
area									
Leybucht	0	0	0	45	50	50	55	30	18-21
Oldeoog	1	1	1	1	2	2	7	8	≥ 15
Trischen	7	11	9	9	31	22	22-35	2	7-15
Grüne Insel	?	?	?	19	34	47	59	24	32
Juist	11	7	15	?	2	14	18	27	3

3.15.3.2 Numbers per area

A total of 17,472 Ringed plovers was counted in the Wadden Sea on September 1, 1973: 192 in Denmark, 11,200 in Schleswig-Holstein and Niedersachsen and 6080 in the Netherlands (Prater, 1974). In the whole Wadden Sea area numbers may vary strongly from day to day. Large numbers appear to pass through the area without staying long. Therefore peak numbers in a certain area may be much higher than mean numbers. Fig. 79 shows numbers observed regularly in the Wadden Sea area.

In the Danish part of the Wadden Sea Ringed plovers are most numerous in August-September when a few hundred birds are present. Maxima are somewhat more than 700 birds. In winter they are absent, in spring generally 200-300 are present. They are most numerous at the N and S point of Fanø, the S point of Rømø and along the mainland coast near Højer (Meltofte, 1980; Vadefuglegruppen Dansk Ornithologisk Forening).

A few thousand Ringed plovers can be counted in the Wadden Sea area in Schleswig-Holstein. Mean numbers are shown in fig. 76. *C.h. tundrae* passes Schleswig-Holstein in May and returns in August-September. Autumn-migration spreads over a longer period than spring-migration. Sometimes single Ringed plovers stay over winter (Schlenker, 1968).

The peak in numbers during autumn-migration in Niedersachsen (fig. 77), is later than in Schleswig-Holstein. The largest concentrations have been counted on Norderney (1500 in September) and between Knock and Rysum (1500 in May). Differences are found between coastal and island areas (fig. 77) in the timing of the highest monthly means. The largest mean numbers along the coast occur in spring, on the islands in autumn. Probably this difference is due to different food conditions.

Just like in Niedersachsen, peak numbers in the Netherlands in counts along the mainland coast and the islands differ (fig. 80). The largest numbers have been counted along the coast of Groningen and Friesland and on Schiermonnikoog and Vlieland. Numbers per area however generally do not exceed 500. In September more than 6000 birds have been count-

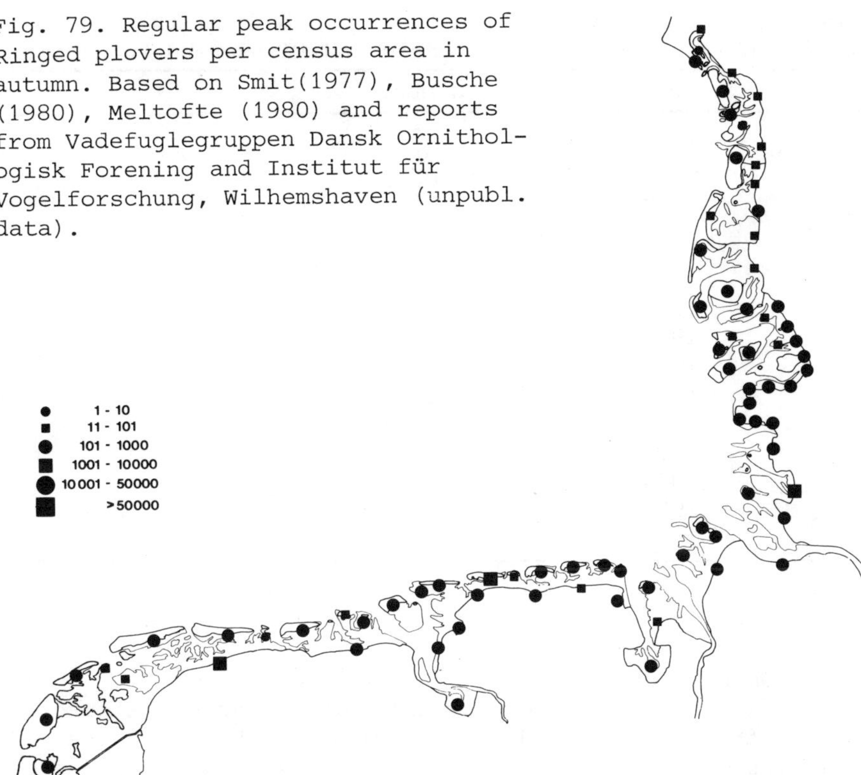

Fig. 79. Regular peak occurrences of Ringed plovers per census area in autumn. Based on Smit(1977), Busche (1980), Meltofte (1980) and reports from Vadefuglegruppen Dansk Ornithologisk Forening and Institut für Vogelforschung, Wilhemshaven (unpubl. data).

ed in the whole area (Boerè & Zegers, 1975). As compared to Schleswig-Holstein and Niedersachsen more birds are present in winter. On several places (Texel, Vlieland) groups of birds, sometimes up to 100 or more, probably belonging to the *tundrae* subspecies, have been observed in May and early June. Table 23 lists the numbers counted in the Dutch part of the Wadden Sea.

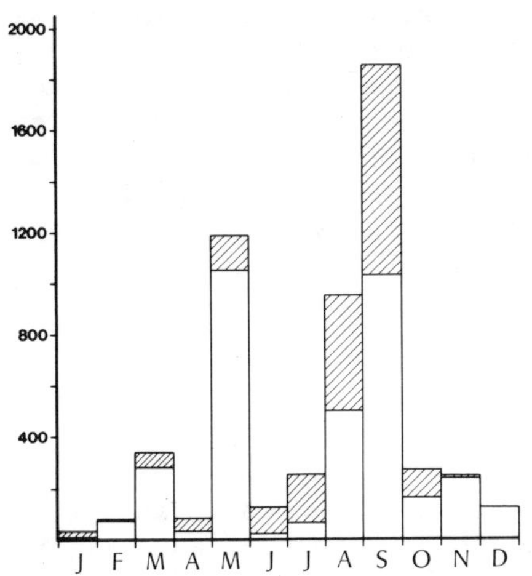

Fig. 80. Mean number of Ringed plovers per month in the Dutch part of the Wadden Sea. Coastal areas (white columns) and islands (hatched) have been separated (after data from Smit, 1977).

Table 23. Number of Ringed plovers in the Dutch part of the Wadden Sea
(after various publications in Watervogels and Limosa and Zegers, in litt.)

Date	Year	Number
January 8	1977	0
January 12	1974	16
January 17	1976	1
January 18	1975	4
April 6	1973	410
April 19	1975	71
May 1	1976	520
May 11	1974	710
July 29	1972	651
August 22	1963	2,600
August 30	1975	2,610
September 1	1973	6,080
October 19	1974	346
November 13	1976	78
December 29	1966	12

3.15.4 Food

3.15.4.1 Food composition

Lange (1968) and Höfmann & Hoerschelman (1969) found in stomachs
of 6 birds from the Wadden Sea: insects (Diptera and larvae), Nereidae
(560 mandibles in one stomach) and molluscs *(Hydrobia)*. Droppings collected on the green beach of Schiermonnikoog contained remains of insects, *Nereis diversicolor* and small specimens of *Carcinus maenas* (Dantuma, 1970). Kersten & Piersma (in litt.) saw Ringed plovers on a muddy substrate near the island of Ameland only take large *Nereis*.

3.15.4.2 Feeding activities

Ringed plovers find their food by sight or acoustically. Another
feeding technique is treading the ground. Birds feed on relatively
hard grounds in the higher eulitoral or supralitoral zone. In the Wadden Sea the species prefers the *Enteromorpha* banks between the *Zostera-*
and *Salicornia* regions, but it also uses salt marshes, which are not
flooded during high tide. Therefore the species is displaced over only short distances during high tide and is able to feed (Glutz et al.,
1975). During high tide Ringed plovers may gather in flocks of tens
of birds as well, but never as concentrated as many sandpiper species.

3.15.4.3 Total food consumption

No information available.

References

Boere, G.C., 1976. The significance of the Dutch Waddenzee in the annual life cycle of arctic, subarctic and boreal waders. Part 1. The

function as a moulting area. Ardea 64: p. 210-291.

Boere, G.C. & P.M. Zegers, 1975. Wadvogeltellingen in het Nederlandse Waddenzeegebied in april en september 1973. Limosa 48: p. 74-81.

Busche, G., 1980. Vogelbestände des Wattenmeeres von Schleswig-Holstein. Kilda, Greven (in press).

Dantuma, R., 1970. Bontbekplevier; verslag over de waarnemingen gedaan tijdens het zomerkamp Schier-3 in 1967. Schierboek 4: p. 41-52.

Edelstam, C., 1972. The visible migration of birds at Ottenby, Sweden. Vår Fågelvärld, Suppl. 7.

Glutz von Blotzheim, U.N., K.M. Bauer & E. Bezzel, 1975. Handbuch der Vögel Mitteleuropas, Vol. 6. Akademische Verlagsgesellschaft, Wiesbaden: 840 pp.

Höfmann, H. & H. Hoerschelmann, 1969. Nahrungsuntersuchungen bei Limikolen durch Mageninhaltsanalysen. Corax 3: p. 7-22.

Lange, G., 1968. Über Nahrung, Nahrungsaufnahme und Verdauungstrakt mitteleuropäischer Limikolen. Beitr. Vogelkunde 13: p. 225-334.

Laven, H., 1940. Beiträge zur Biologie des Sandregenpfeifers. J. Orn. 88: p. 183-287.

Meltofte, H., 1980. Fugle i Vadehavet. Vadefugletaellingen i Vadehavet 1974-1978. Miljøministeriet, Fredningsstyrelsen, København: 50 pp.

Meltofte, H. & J. Rabøl, 1977. Vejrets indflydelse paa efteraarstraekket af vadefugle ved Blaavandshuk, med et forsøg paa en analyse af traekkets geografiske oprindelse. Dansk Orn. Foren. Tidsskr. 71: p. 43-63.

Minton, C.D.T., 1972. Wash Wader Ringing Group. Report 1971/72.

Pienkowski, M.W., P.J. Knight, D.J. Stanyard & F.B. Argyle, 1976. The primary moult of waders on the Atlantic coast of Morocco. Ibis 118: p. 347-365.

Prater, A.J., 1974. Coastal wader counts: Waddensea and Delta Region. IWRB Bull. 37: p. 102-104.

Prater, A.J., 1976. The distribution of coastal waders in Europe and North Africa. In: M. Smart (ed.), Proc. Intern. Conf. Conserv. of Wetlands and Waterfowl, Heiligenhafen 1974. IWRB, Slimbridge: p. 255-271.

Salomonsen, F., 1979. Fra Zoologisk Museum XXV. Dansk Orn. Foren. Tidsskr. 73: p. 191-206.

Schlenker, R., 1968. Uber das Wintervorkommen von Limikolen an der schleswig-holsteinischen Westküste. Corax: p. 92-108.

Smit, C.J., 1977. On the occurrence of 32 bird species in the Danish, German and Dutch Wadden Sea. Unpubl. report Intern. Wadden Sea Working Group, part 2: 171 pp.

Taylor, R.C., 1980. Migration of the Ringed plover Charadrius hiaticula. Ornis Scand. 11: p. 30-42.

Väisänen, R.A., 1969. Evolution of the Ringed plover (Charadrius hiaticula L.) during the last hundred years in Europe. Ann. Ac. Scient. Fennicae Seríes A. IV Biologica. Helsink: p. 1-149.

Voous, K.H., 1962. Die Vogelwelt Europas und ihre Verbreitung. Parey, Hamburg: 284 pp.

3.16 KENTISH PLOVER *(CHARADRIUS ALEXANDRINUS* L.)
P.H. Becker

Da: Hvidbrystet Praestekrave; G: Seeregenpfeifer; Du: Strandplevier

3.16.1 Distribution

3.16.1.1 Breeding area

The Kentish plover is a cosmopolitan bird breeding in all climatic zones except the boreal and tundra zone (Voous, 1962) (fig. 81). In the West Palaearctic Kentish plovers breed on sandy shores, sand dunes, steppes with sparse vegetation, pebble and mud shores, generally along sea coasts but to a lesser extent also inland, for example along rivers. They also breed in sand deserts, natural as well as those created artificially by raising areas with fluid sand.

The most Northern populations breed along the coasts of the North Sea, Wadden Sea and Southwestern Baltic Sea.

3.16.1.2 Migration routes

Because of the absence of inland recoveries and sightings it is to be supposed that the populations of the North- and Baltic Sea probably migrate in autumn and spring mainly along the East coast of the Atlantic. Those birds reaching the Western Mediterranean are supposed to migrate across the mainland back to their breeding areas (Glutz et al., 1975).

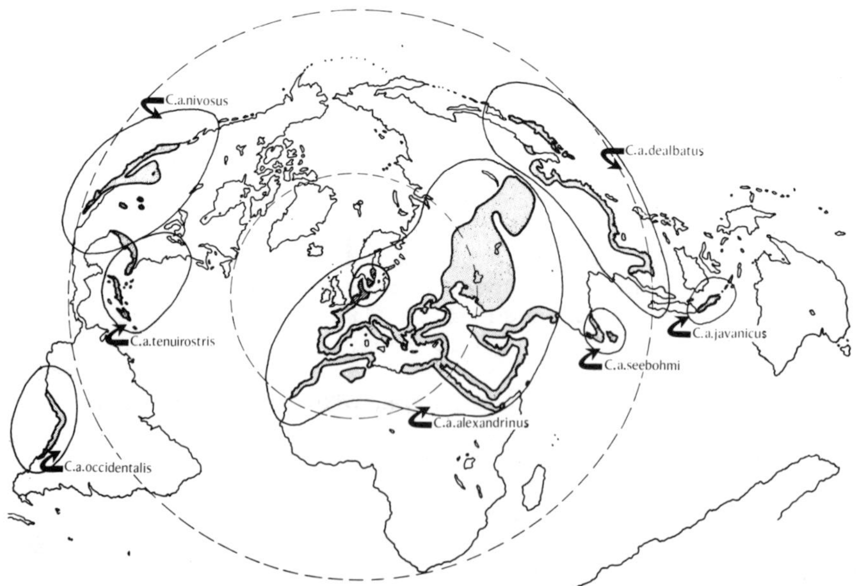

Fig. 81. Breeding area of the Kentish plover, (after Voous, 1962, modified after Glutz et al., 1975). The heavy line encircles the region where birds visiting the Wadden Sea originate from.

3.16.1.3 Wintering areas

The Northern populations winter at the shores of the Mediterranean Sea and the North African coasts.

3.16.1.4 Moulting areas

Birds of the Wadden Sea population moult near their breeding places as well as on their way to and in the wintering areas. Groups of moulting Kentish plovers have been recorded on Fanø, near the mouth of the Eider, near St. Peter, on the Rysumer Nacken and on some of the Dutch Wadden Sea islands. Moulting concentrations are also known from the Dutch Delta area and the European Atlantic coast (Glutz et al., 1975).

3.16.2 Annual cycle

3.16.2.1 Migration

Adult Kentish plovers begin to migrate after fledging of the young but sometimes adult birds and their offspring remain at their breeding places until September (Rittinghaus, 1961). Already by the beginning of June small flocks are formed. Flock formation and departure from the breeding area are maximal in August. In the beginning of September a flock on Texel mainly consisted of juvenile birds (Wolff in Ten Kate, 1967). The last Kentish plovers leave the Wadden Sea in October.

In spring (early March) the first migrants are observed in Bretagne, while they turn up in the North Sea by the end of March. By mid-April most territories are occupied.

3.16.2.2 Moult

First-year birds moult all body-feathers, wing coverts and tertials in August and September. During the first prenuptial moult which starts in November and ends by early April, the birds change parts of the feathers that have been renewed already during the postjuvenal moult.

During the breeding period adult birds change the head- and body feathers, scapulars and tertials. These feathers are changed again during a complete postnuptial moult in the second half of July. This moult is finished by early September. During prenuptial moult the same feathers are changed as in the first-year birds (see above), the procedure however seems to require less time.

3.16.2.3 Weight changes

Data on Kentish plovers from the Camargue (Glutz et al., 1975) show that body weight decreases during postnuptial moult:

Before breeding-moult:	47.7 g (n=15)
During breeding-moult:	45.4 g (n=35)
Change of primary 5-6:	40.0 g (n=105)
Change of primary 10-11:	38.8 g (n=72)
After moult of remiges and tail:	47.8 g (n=43)

Unfortunately no data from the Wadden Sea are available.

3.16.3 Numbers

3.16.3.1 Population size

The distribution of the breeding population of the Wadden Sea is shown in fig. 82. On the West coast of Schleswig-Holstein the species is more numerous along the mainland with its clayey soils as compared to the sandy shores of the islands (König, pers. comm.).

The size of Wadden Sea population amounts to:

Denmark: approximately 30 pairs in 1977-1978, in some years up to 55 pairs (Dansk Ornithologisk Forening);

Schleswig-Holstein: approximately 600-700 pairs (Glutz et al., 1975, and Smit, in litt.);

Niedersachsen: approximately 250-300 pairs (Institut für Vogelforschung, Wilhelmshaven);

The Netherlands: approximately 150-200 pairs (Glutz et al., 1975 and Smit, in litt.).

The total breeding population of the Wadden Sea area therefore amounts to about 1000-1300 pairs. The species prefers areas without vegetation and may settle immediately when such a habitat has been developed. The number of breeding pairs declines if the habitat becomes unfavourable because of growth of vegetation, tourism, or construction works. Some examples of fluctuations in breeding pair numbers are given in Table 24.

Fig. 82. Distribution of the breeding population of the Kentish plover in the Wadden Sea area (after data from Glutz et al., 1975; Institut für Vogelforschung, Wilhelmshaven; Dansk Ornithologisk Forening and Smit, in litt.).

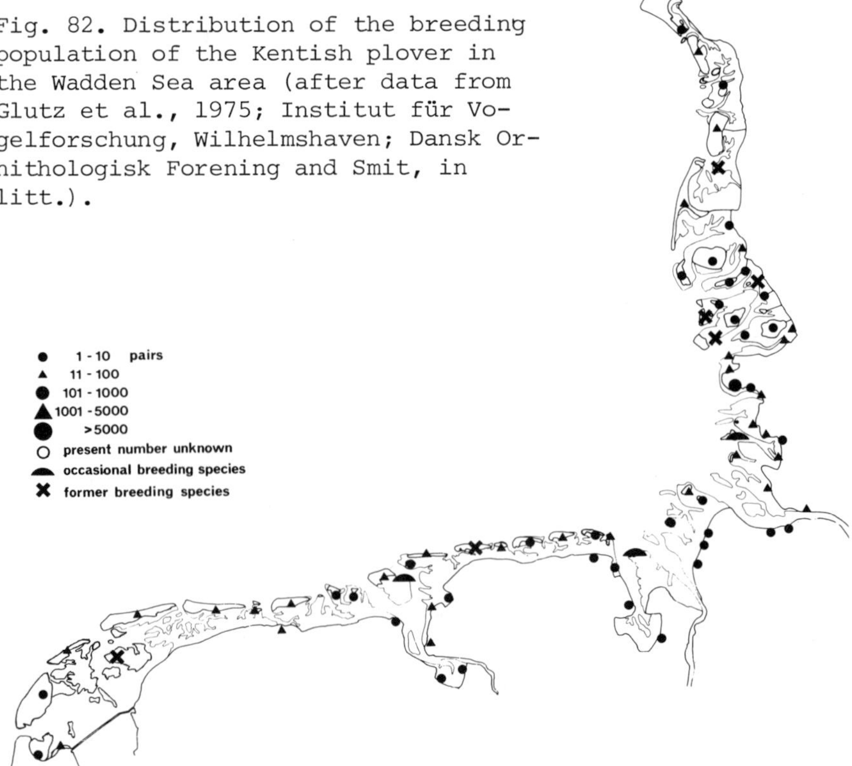

Table 24. The number of breeding pairs of the Kentish plover in five areas in nine succeeding years (data: Institut für Vogelforschung, Wilhelmshaven).

Year	1970	1971	1972	1973	1974	1975	1976	1977	1978
Area									
Leybucht	0	6	120	130	100	80	82	39	23-25
Oldeoog	14	15	9	5	9	5	16	≥ 20	≥ 30
Baltrum	?	?	20-30	3	5	8-10	20	35	30-35
Trischen	?	11	3	8	10	8	8-15	1	3-5
Grüne Insel	?	?	?	6	10	54	61	91	83

3.16.3.2 Numbers per area

Because high tide roosts may have a rather diffuse structure and moreover sometimes are situated on desolate sandbeaches, numbers may have been underestimated in many areas. In the whole area the highest numbers occur in autumn. On September 1, 1973 2880 Kentish plovers were counted of which none in the Danish part, 1900 in Schleswig-Holstein and Niedersachsen and 980 in the Dutch part of the Wadden Sea (Prater, 1974).

In the Danish part of the Wadden Sea during spring and summer only the breeding population is present. In late July and early August post-breeding concentrations occur. Generally numbers are less than 50 birds. On the North point of Fanø, however, up to 137 birds have been counted (Meltofte, 1980; reports Vadefuglegruppen Dansk Ornithologisk Forening).

In Schleswig-Holstein peak numbers occur from July to September and in May (fig. 83). Maximum numbers after the breeding season in the whole area may be up to 2000-2500 birds. Because the breeding population north of Schleswig-Holstein is small spring migration is rather inconspicuous (Busche, 1980).

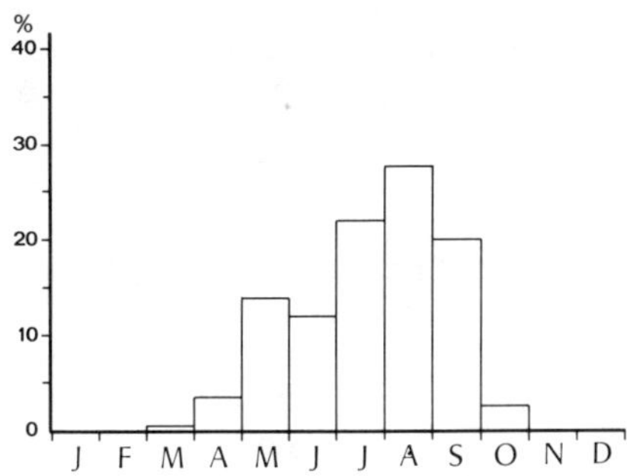

Fig. 83. Occurrence of Kentish plovers in the Wadden Sea area Schleswig-Holstein. Numbers for each month are expressed as a percentage of the total number observed in all months (after Busche, 1980).

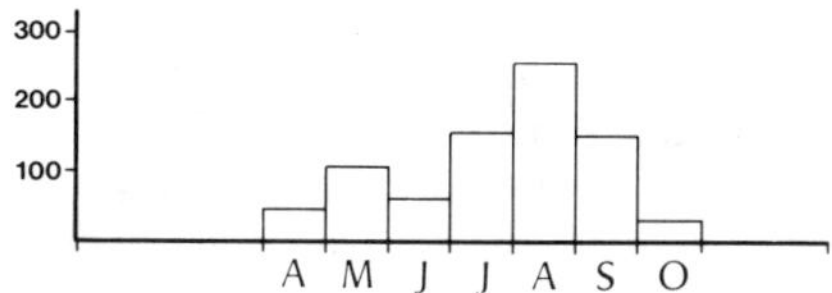

Fig. 84. Mean number of Kentish plovers per month in 6 areas in the Wadden Sea area in Niedersachsen (after data from Smit, 1977).

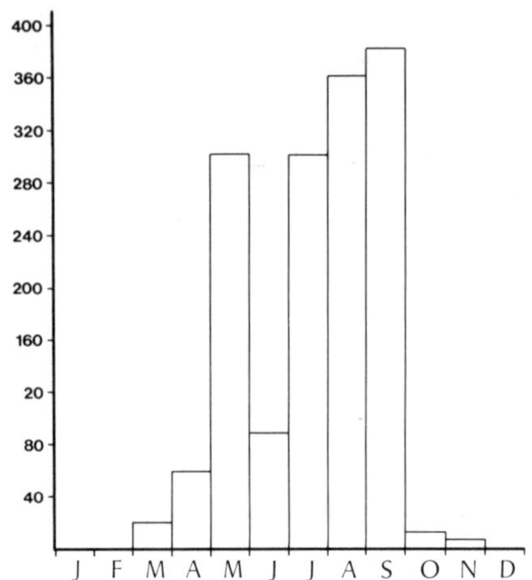

Fig. 85. Mean number of Kentish plovers per month in the Dutch part of the Wadden Sea (after data from Smit, 1977).

In Niedersachsen large numbers have been counted on Scharhörn (300) and in the area between Knock and Rysum (170). Peak numbers coincide with those in Schleswig-Holstein (fig. 84).

In the Dutch part of the Wadden Sea the largest concentrations have been recorded on Ameland, Schiermonnikoog (more than 500), Terschelling (140) and Texel (100). Peak numbers occur in May and September and not in August as in Niedersachsen and Schleswig-Holstein (fig. 85). Numbers counted in the Dutch Wadden Sea in July and August vary from 300 to 350 (Rooth, 1966; Boere & Zegers, 1974; Boere & Zegers, 1977). On September 1, 1973 980 birds have been counted (Prater, 1974). Fig. 86 shows the distribution of Kentish plovers in the Wadden Sea area.

3.16.4 Food

3.16.4.1 Food composition

Stomachs of six birds from the West coast of Schleswig-Holstein contained insects and their larvae, crustaceans *(Carcinus maenas)*, annelids *(Nereis)*, gastropods *(Hydrobia, Littorina)* and a bivalve (Lange, 1968; Höfmann & Hoerschelmann, 1969).

3.16.4.2 Feeding activities

Kentish plovers use shallows in the higher eulittoral or supralittoral zone as feeding areas. According to Glutz et al. (1975) they prefer a moister substrate than Ringed plovers and are found more often on the tidal flats. They find their food by sight and acoustically (Lange, 1968) and peck most prey from the surface. Another feeding technique is probing or treading the ground (Rittinghaus, 1957; Lange, 1968; Glutz et al., 1975).

During high tide Kentish plovers rest in salt marshes and sandy areas along the shore. Flock formation often does not occur.

Data on the choice of drinking water are poor. Occasionally Kentish plovers have been observed to drink fresh water from pools after rainfall (Rittinghaus, pers. comm.).

Fig. 86. Regular peak occurrences of
Kentish plovers in autumn per census
area (based on Smit, 1977; Busche,
1980; Meltofte, 1980; reports Vadefuglegruppen Dansk Ornithologisk Forening and unpubl. data Institut für
Vogelforschung, Wilhelmshaven).

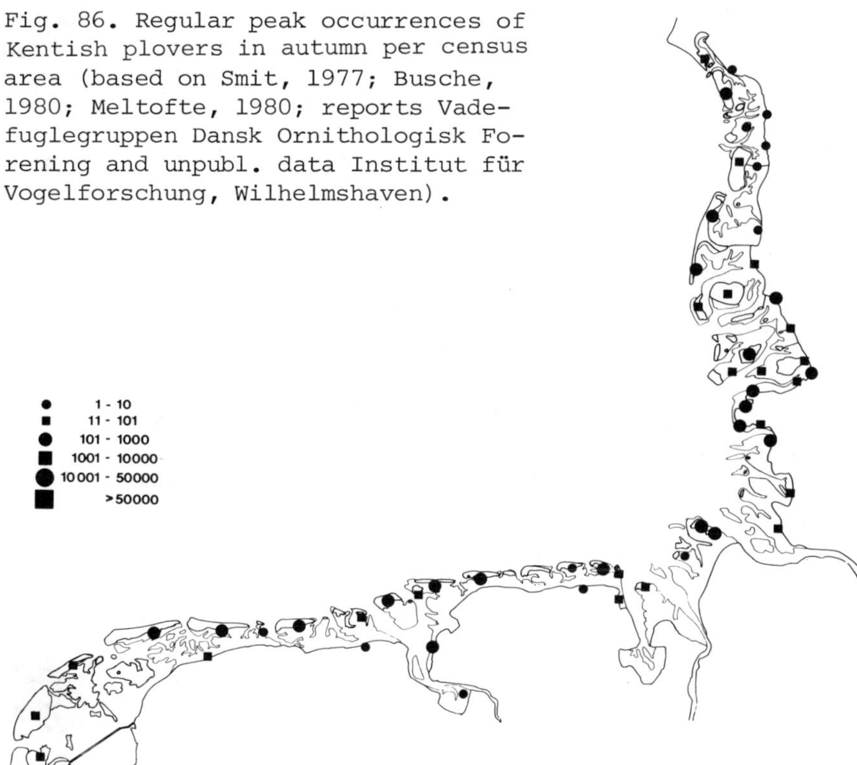

3.16.4.3 Total food consumption

No data available.

References

Boere, G.C. & P.M. Zegers, 1974. Wadvogeltelling in het Nederlandse
    Waddengebied in juli 1972. Limosa 47: p. 23-28.
Boere, G.C. & P.M. Zegers, 1977. Watervogeltellingen in het Nederland-
    se Waddenzeegebied in 1974 and 1975. Watervogels 2: p. 161-173.
Busche, G., 1980. Vogelbestände des Wattenmeeres von Schleswig-Holstein.
    Kilda, Greven (in press).
Dybbro, T., 1970. The Kentish Plover as a breeding bird in Denmark.
    Dansk Orn. Foren. Tidsskr. 64: p. 205-222.
Glutz von Blotzheim, U.N., K.M. Bauer & E. Bezzel, 1975. Handbuch der
    Vögel Mitteleuropas, Vol. 6. Akademische Verlagsgesellschaft, Wies-
    baden: 840 pp.
Höfmann, H. & H. Hoerschelmann. 1969. Nahrungsuntersuchungen bei Limi-
    kolen durch Mageninhaltsanalysen. Corax 3: p. 7-22.
Kate, C.G.B. ten, 1967. Ornithologie van Nederland, 1965. Limosa 40:
    p. 14-58.
Lange, G., 1968. Über Nahrung, Nahrungsaufnahme und Verdauungstrakt
    mitteleuropäischer Limikolen. Beitr. Vogelkunde 13: p. 225-334.

Meltofte, H., 1980. Fugle i Vadehavet. Vadefugletaellingen i Vadehavet 1974-1978. Miljøministeriet, Fredningsstyrelsen, København: 50 pp.

Pienkowski, M.W., P.J. Knight, D.J. Stanyard & F.B. Argyle, 1976. The primary moult of waders on the Atlantic coast of Morocco. Ibis 118: p. 347-365.

Prater, A.J., 1974. Coastal wader counts: Waddensea and Delta region. IWRB Bull. 35: p. 102-104.

Rittinghaus, H., 1957. Charadrius alexandrinus (L.): Nahrungssuche. Encyclopaedia Cinematographica E 135, Göttingen: 3 pp.

Rittinghaus, H., 1961. Der Seeregenpfeifer. Neue Brehm Bücherei 282. Ziemsen, Wittenberg: 126 pp.

Rooth, J., 1966. Vogeltelling in het hele Nederlandse Waddengebied in augustus 1963. Limosa 39: p. 175-181.

Smit, C.J., 1977. On the occurrence of 32 bird species in the Danish German and Dutch Waddensea. Unpubl. report Internat. Waddensea Working Group, Part 3: 174 pp.

Voous, K.H., 1962. Die Vogelwelt Europas und ihre Verbreitung. Pàrey, Hamburg: 284 pp.

3.17 GREY PLOVER *(PLUVIALIS SQUATAROLA L.)*
G.C. Boere & C.J. Smit

Da: Strandhjejle; G: Kiebitzregenpfeifer; Du: Zilverplevier

3.17.1 Distribution

3.17.1.1 Breeding area
The Grey plover has a holarctic distribution, breeding only in the tundra climate zone. The breeding area is almost circumpolar with a disjunction in the North Atlantic region. Birds passing through or wintering in the Wadden Sea come from the breeding areas in Eurasia probably as far as the Taymyr peninsula. There is no proof that birds from the NW Territories (Canada) migrate towards Western Europe (fig. 87; Voous, 1960). No subspecies are distinguished.

3.17.1.2 Migration routes
Probably the first part of migration occurs over land and in a broad front. After arriving in the Western Palaearctic however the species is almost strictly coastal. Individual birds can be observed occasionally in suitable areas inland. These observations concern mainly first calendar-year birds (Glutz et al., 1975). Migration routes from the breeding area to coastal Europe and further south are not well-known, as only very few birds have been ringed.

3.17.1.3 Wintering areas
The wintering area ranges from the North Sea coastal areas and estuaries till the coastal lagoons of South Africa, where several thousands of birds may be present (Summers et al., 1977). Birds wintering in South-Africa are mainly first calendar-year birds.

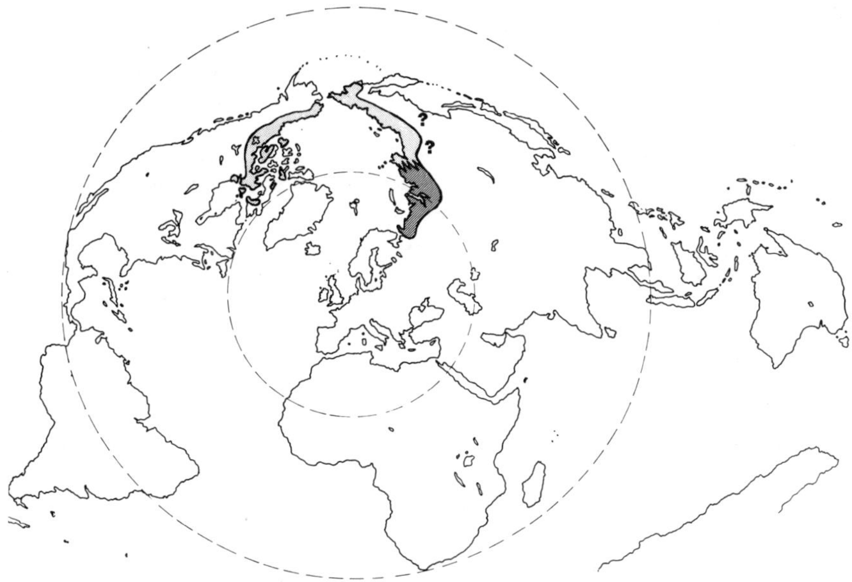

Fig. 87. Breeding area of the Grey plover (after Voous, 1960). The dark shaded area represents the region where birds visiting the Wadden Sea originate from. Question marks indicate a region from which is uncertain if birds breeding there also visit the Wadden Sea.

Important concentrations are found on the British Isles, in some areas in France and Morocco and on the Banc d'Arguin (Mauritania). Several thousands winter in the Wadden Sea area.

Together with the breeding birds of North America, the Grey plover winters almost cosmopolitical.

3.17.1.4 Moulting areas

No particular moulting areas are present in the sense that large numbers stay at one area for a long period to go through a complete moult. There is much evidence (Boere, 1976) that adult post-nuptial moult starts already in the breeding area or soon after departure. It is finished at some place along the migration route, at least as far south as the Banc d'Arguin (Mauritania). Migration in arrested moult has been observed (Boere, 1976).

3.17.2 Annual cycle

3.17.2.1 Migration

Adult birds start to leave the breeding areas by early July. Successfully breeding adults leave by the end of August or early September, juveniles between mid-August and mid-September (Glutz et al., 1975).

First arrivals of Grey plovers in the Wadden Sea take place in the second July decade. Birds which are present earlier are almost all second calendar-year birds summering in the area.

    Peak numbers passing at Blaavandshuk are observed from the end of
July until mid-August (Meltofte et al., 1972). After mid-July numbers
in the Wadden Sea increase rapidly till a mid-August peak. According
to ringing results and field observations there must be a rapid inwards
migration of large numbers in NW Africa in late July and August. First
calendar-year birds arrive in the Wadden Sea from the beginning of September on and leave the area from the last September decade until the
end of November. Table 25 shows, to some extent, the separate migration
of adult and first-year birds based on birds trapped by mist- and cannon netting. Spring migration starts in the first April decade with
peak numbers in the first half of May. Numbers decline very rapidly
towards the end of May and only very small numbers still occur in June.

3.17.2.2 Moult

    The adult post-nuptial moult starts mid-August, sometimes already
in the breeding area. It is finished at the beginning of November, but
many birds do not moult their outer two or three primaries and sometimes some tailfeathers. These remain throughout the winter and are
moulted in March and April. A minor number of birds moult these arrested primaries after a second breeding season which gives a figure of
two moult cycles being in progress in one wing. This is uncommon in
waders, except in plovers (Pienkowski et al., 1976).
    Moult of body feathers of first calendar-year birds takes place
from mid-September until the end of November. As compared to adults
second calendar-year birds moult one month earlier, from the end of
June onwards.
    Pre-nuptial body moult starts by the end of March and is in full
progress during April and early May. Many birds leave the area showing a full breeding plumage but with many body feathers still growing.

3.17.2.3 Weight changes

    There is only information on weight changes available from the is-
land of Vlieland in the Dutch part of the Wadden Sea (Table 25). It
shows relatively low weights during the post-nuptial moult period and
increasing weights towards winter. Weights are highest in May because
of accumulated migratory fat disposals. Figs. 88 and 89 give information on wing- and bill lengths of Grey plovers trapped on the island
of Vlieland. First calendar-year birds are smaller on the average.
Similar data have been obtained in England (Branson & Minton, 1976).

3.17.3 Numbers

3.17.3.1 Population size

    Prater (1976) estimates the number of wintering Grey plovers along
the Atlantic coasts of Europe and NW Africa to about 42,700 birds,
29,200 of these wintering in W Europe. Piersma et al. (1980) counted
23,350 at the Banc d'Arguin while Prater only mentions 3500 for this
area. Praters' figures therefore might be somewhat too low.

Table 25. Mean monthly weights, sexes combined, of adult and first calendar-year Grey plovers, caught on the island of Vlieland, December 1971-December 1975 (Boere, unpubl.).

month	adult birds			first calendar-year birds		
	mean	s.d.	n	mean	s.d.	n
January	241.9	26.4	10			
February	246.9	13.9	11			
March	258.4	25.4	9			
April	251.5	20.7	13			
May	261.9	41.5	32			
July	218		1			
August	229.1	28.6	13			
September	231.1	25.6	46	197.5	13.4	19
October	242.4	25.1	32	219.6	27.7	65
November	251.9	17.3	18	231.1	27.3	10

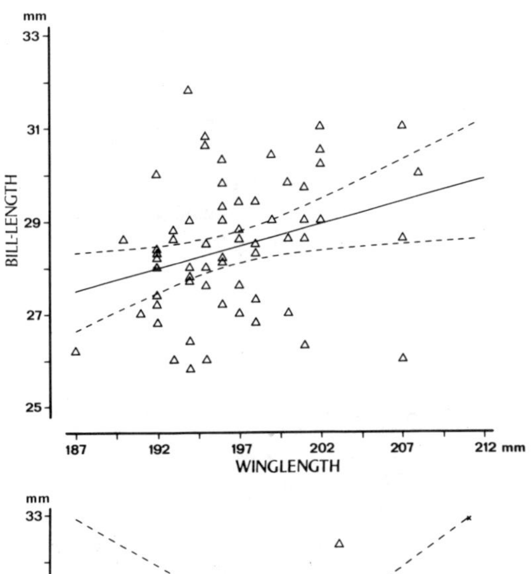

Fig. 88. Exposed bill- (mm) and wing-lenghts (mm) with regression line and 5% interval of first calendar-year Grey plovers caught on the island of Vlieland in October 1972-1975 (n=64) (Boere, unpublished).

Fig. 89. Exposed bill- (mm) and wing lengths (mm) with regression line and 5% interval of adult Grey plovers caught on the island of Vlieland in October-November 1972-1975. Δ denote birds caught in October (n=25), x birds caught in November (n=13).

3.17.3.2 Numbers per area

In the Danish Wadden Sea Grey plovers never occur in large numbers. Maximum numbers are present in August, November and May when 2000-3000 are present. Maximum numbers in June and July are 474 and 900 respectively. From January until April numbers are below 100 (Meltofte, 1980; reports Vadefuglegruppen Dansk Ornithologisk Forening).

Numbers in the Wadden Sea area in Schleswig-Holstein increase in July. Peak numbers occur in August when about 13,000 may be present in the area. Numbers decrease gradually afterwards (fig. 90).

In mild winters about 400 may still be present, in severe winters however only very few still stay. Spring migration begins in March. In June some birds still are present (Schlenker, 1968; Busche, 1980). Some hundreds stay in the area until summer (Heldt, 1968).

In the Wadden Sea area in Niedersachsen numbers in autumn generally peak in August or September (Norderney, 3000; Spiekeroog, 1000; Mellum, 2200; Neuwerk, 650; Knechtsand, 1500). In spring numbers are higher because the birds concentrate in the area before their return to the breeding places. Mean numbers of a coastal area (parts of Dollart and Jadebusen and the area between Westeraccumersiel and Hooksiel, all together forming about 50 kilometers of coastline) are shown in fig. 91.

Phenology of the Grey plover in the Dutch part of the Wadden Sea is shown in fig. 92 and Table 26. Slight differences exist between the occurrence of Grey plovers in the Eastern and Western part of the area. In the Eastern part numbers peak in September, in the Western part in October. In winter Grey plovers are somewhat more numerous in the Western part. Fig. 93 illustrates the distribution in the Wadden Sea.

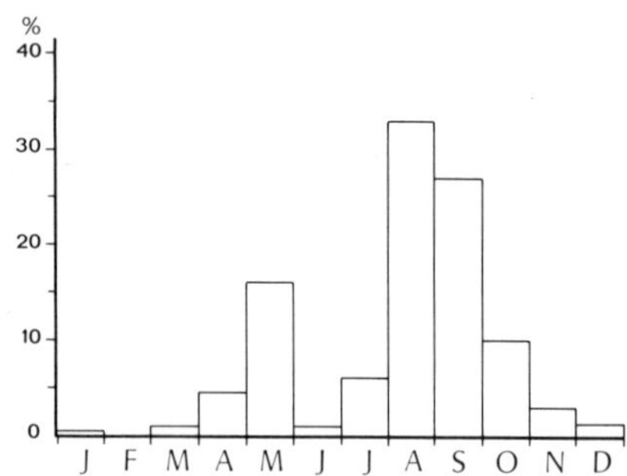

Fig. 90. Occurrence of Grey plovers in the Wadden Sea area in Schleswig-Holstein. Numbers for each month are expressed as a percentage of the total number of birds observed in all months (after Busche, 1980).

Fig. 91. Mean number of Grey plovers per month in a 50 km long stretch along the mainland coast in Niedersachsen (after data from Smit, 1977 and the Institut für Vogelforschung, Wilhelmshaven).

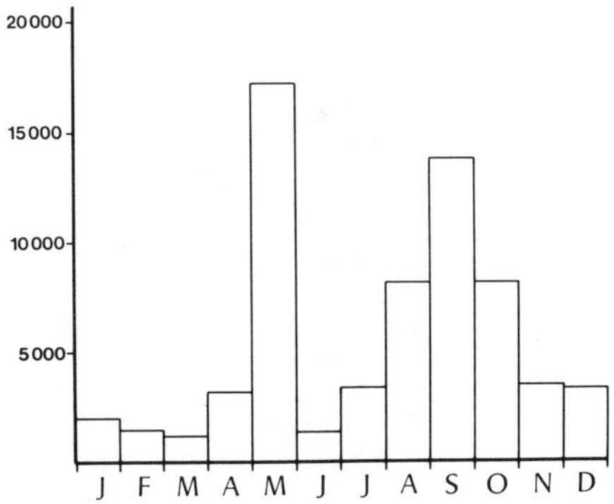

Fig. 92. Mean number of Grey plovers per month in the Dutch part of the Wadden Sea (after data from Smit, 1977).

Table 26. Number of Grey plovers in the Dutch part of the Wadden Sea (after various publications in Watervogels and Limosa and Zegers, in litt.).

Date	Year	Number
January 8	1977	2,300
January 12	1974	2,400
January 17	1976	2,620
January 18	1975	2,110
April 6	1973	3,600
April 19	1975	3,680
May 1	1976	12,760
May 11	1974	13,620
July 29	1972	1,145
August 22	1963	4,500
August 30	1975	14,660
September 1	1973	18,580
October 19	1974	7,145
November 13	1976	5,200
December 29	1966	1,300

Table 27. Peck- and success rate of Grey plovers in two seasons near the island of Ameland (after Kersten & Piersma, in litt.).

	September 1978	May 1979	May 1979
	Mudflat	Mudflat	Waters' edge
Peckrate. min^{-1}	9.9	5.7	6.5
Successful pecks. min^{-1}	3.7	1.8	4.3
Success (%)	39.1	32.6	66.0
Number of birds examined	14	9	1
Minutes of observation	96	60	10

Fig. 93. Regular peak occurrences of
Grey plovers in autumn and spring per
census area (after Smit, 1977; Busche,
1980; Meltofte, 1980 and data from Vadefuglegruppen Dansk Ornithologisk Forening and Institut für Vogelforschung,
Wilhelmshaven).

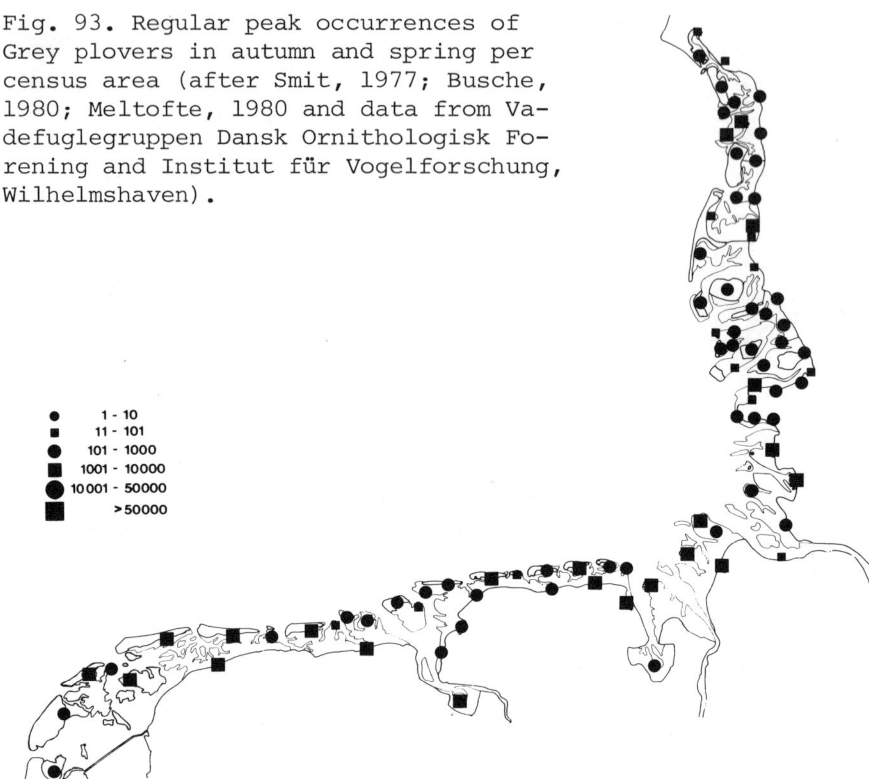

3.17.4 Food

3.17.4.1 Food composition

   Main feeding areas are sandy as well as muddy tidal flats and mus-
sel- and cockle beds. Prey items vary but worms, crustaceans and small
specimens of molluscs, such as the cockle, are most important. On a
muddy flat near the island of Ameland Grey plovers in autumn were feeding on *Nereis* (74%) and small unidentified prey items (probably *Scolop-
los armiger*) (26%), in spring on *Nereis* exclusively (Kersten & Piersma,
in litt.).

3.17.4.2 Feeding activities

   Information on feeding activities is difficult to obtain because of
the shyness of Grey plovers. Table 27 gives the results of some short
observations on feeding activities. High tide roosts of only Grey plovers are rare. They mostly join other species. Mixed flocks of Dunlins,
Knots and Grey plovers, with the latter on the edge of the flock, are
common. The places where they roost during high tide are very much dependent on water levels during high tide (Zwarts, 1969).

3.17.4.3 Total food consumption

   From field observations near the island of Ameland Kersten & Piersma (in litt.) concluded a food intake per emersion period of 12.0 (September 1978) - 16.8 (May 1979) g ash-free dry weight per bird.

References

Boere, G.C., 1976. The significance of the Dutch Wadden Sea in the annual life cycle of arctic, subarctic and boreal waders. Part 1. The function as a moulting area. Ardea 64: p. 210-291.

Branson, H.J.B.A. & C.D.T. Minton, 1976. Moult, measurements and migrations of the Grey plover. Bird Study 23: p. 257-266.

Busche, G., 1980. Vogelbestände des Wattenmeeres von Schleswig-Holstein. Kilda, Greven (in press).

Glutz von Blotzheim, U.N.K.M. Bauer & E. Bezzel, 1975. Handbuch der Vögel Mitteleuropas, Vol. 6. Akademische Verlagsgesellschaft, Wiesbaden: 840 pp.

Heldt, R., 1968. Übersommernde Limikolen an der Westküste von Schleswig-Holstein. Corax 2: p. 108-130.

Meltofte, H., 1980. Fugle i Vadehavet. Vadefugletaellinger i Vadehavet 1974-1978. Miljøministeriet, Fredningsstyrelsen, København: 50 pp.

Meltofte, H., S. Pihl & B.B. Sørensen, 1972. Efteraarstraekket af vadefugle (Charadrii) ved Blaavandshuk 1963-1971. Dansk Orn. Foren. Tidsskr. 66: p. 63-69.

Pienkowski, M.W., P.J. Knight, D.J. Stanyard & F.B. Argyle, 1976. The primary moult of waders on the Atlantic coast of Morocco. Ibis 118: p. 347-365.

Piersma, T., M. Engelmoer, W. Altenburg & R. Mes, 1980. A wader-expedition to Mauritania. Wader Study Group Bull. 29: p. 14.

Prater, A.J., 1976. The distribution of coastal waders in Europe and North Africa. In M. Smart (ed.): Proceedings Intern. Conference on Conservation of Wetlands and Waterfowl, Heiligenhafen 1974. IWRB, Slimbridge: p. 255-271.

Schlenker, R., 1968. Über das Wintervorkommen von Limikolen an der Schleswig-Holsteinischen Westküste. Corax 2: p. 92-108.

Smit, C.J., 1977. On the occurrence of 32 bird species in the Danish, German and Dutch Wadden Sea. Unpubl. report International Wadden Sea Working Group, part 3: 174 pp.

Summers, R.W., J. Cooper & J.S. Pringle, 1977. Distribution and numbers of coastal waders (Charadrii) in the South-Western Cape, South Africa, summer 1975-1976. Ostrich 48: p. 85-97.

Voous, K.H., 1960. Atlas of European birds. Nelson, London: 284 pp.

Zwarts, L., 1969. Zilverplevier '66. Waarnemingen verricht aan het getijritme en de aktiviteit van de Zilverplevier op Schier. Schierboek 3: p. 235-248.

3.18 KNOT *(CALIDRIS CANUTUS L.)*
G.C. Boere & C.J. Smit

Da: Islandsk Ryle; G: Knutt; Du: Kanoetstrandloper

3.18.1 Distribution

3.18.1.1 Breeding area

The Knot has a discontinuous holarctic circumpolar distribution, breeding only in the tundra climate zone.

Within this area three subspecies are described (fig. 94):
C.c. canutus: breeding in Eurasia, Greenland, NE Canada
C.c. rufa   : breeding in North-America
C.c. rogersi: breeding on Wrangel island.

The status of the subspecies *C.c. rogersi* is not generally accepted, and is mostly considered to belong to *C.c. canutus* (Voous, 1960; Glutz et al., 1975).

According to ringing results and biometrical studies the great flocks of Knot migrating through and wintering in the Wadden Sea belong to the breeding populations of NE Canada and Greenland and probably also of Eurasia.

For the occurrence of the latter population however there is no or only small (see fig. 95 and 96) direct evidence from the Wadden Sea. According to the view expressed by Dick et al. (1976) however it is

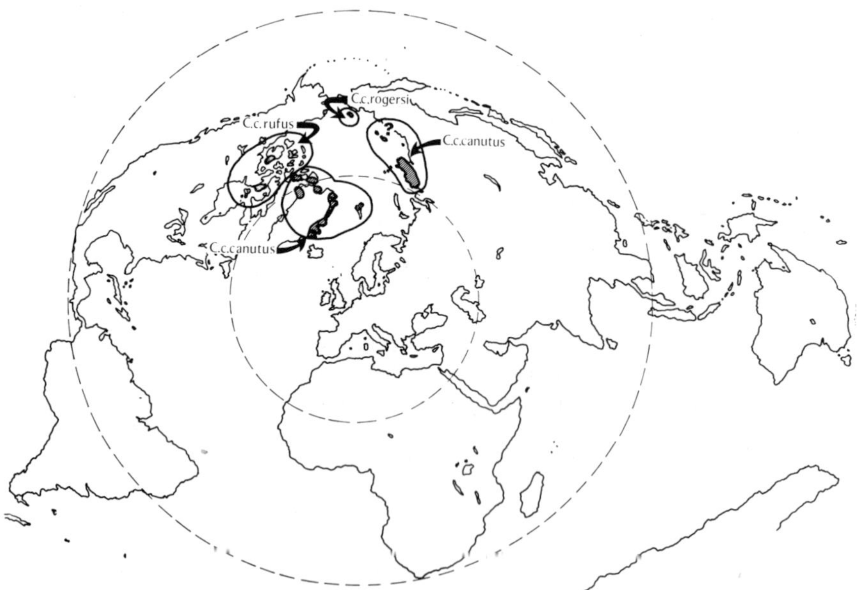

Fig. 94. Breeding area of the Knot (after Voous, 1960). The dark shaded area represents the region where birds visiting the area originate from. Question marks indicate a region from which is uncertain whether birds breeding there also visit the Wadden Sea.

very likely that this population migrates through the area (see also
Fournier & Spitz, 1970).

3.18.1.2 Migration routes

After the breeding season Knots, especially *C.c. canutus* from Eurasia, are almost strictly coastal. After a long distance over land migration the Greenland and NE Canadian populations cross the Atlantic
Ocean by flying directly to Norway and England in autumn.

In spring when departing from NW Europe and England large flocks
have a stop on Iceland. Knots are very rare in inland Europe. Observations inland occur generally during autumn migration and involve mostly first calendar-year birds. Small numbers of the Eurasian population
migrate through Central Europe, the Black- and Caspian Sea area, the
Arabian peninsula and the Rift Valley to South-Africa (Nettlestrøm,
1970; Prater, 1974; Morrison, 1975; Dick et al., 1975; Andreassen &
Raad, 1977).

3.18.1.3 Wintering areas

Wintering Knots build up large flocks of several thousands of birds
and are concentrated at only a small number of suitable coastal shallows and estuaries in NW Europe and North Africa. Most important wintering areas for the Canadian- and Greenland breeding populations are
several estuaries in England (Dee, Ribble, Wash, Morecambe Bay) and
France. The Siberian population winters in Africa and is to a large
extent concentrated on the Banc d'Arguin in Mauritania. Smaller flocks
of this population are spread along the African coast especially in
South-Africa (Dick et al., 1975; Summers et al., 1976). Wintering Knots
are present in the Wadden Sea, but are almost restricted to the Dutch
part. Numbers seem to fluctuate considerably depending on the local
food conditions and, to some extent, the place of the frost line. *C.c.
rogersi* winters locally in Australia and New-Zealand, *C.c. rufa* along
the coast of South-America (Harrington & Morrison, 1980; Morrison et
al., 1980) (see 3.18.2.1).

3.18.1.4 Moulting areas

According to present knowledge two, almost separated areas for the
post-nuptial moult of the two palaearctic subspecies have been described:
1. The North Sea area for the Greenland- and NE Canadian population
   of *C.c. canutus* with the Wadden Sea as the most important part of
   that area (Boere, 1976).
2. North Africa, especially the shallows of the Banc d'Arguin for the
   Siberian population of *C.c. canutus* (Dick et al., 1976).
The North Sea area also acts as the most important area for adult
pre-nuptial moult into the breeding plumage.

3.18.2 Annual cycle

3.18.2.1 Migration

Departure from the high arctic breeding areas starts with the non-breeding birds by late June, followed a few weeks later by the breeding birds. First calendar-year birds generally start to leave the breeding area from early August onwards (Nettleship, 1974).

Numbers of migrating Knots at Blaavandshuk increase by mid-July and peak by the end of this month. A second peak is observed by mid-August (Meltofte et al., 1972).

In the Wadden Sea numbers increase from mid-July and peak numbers occur from mid-August till the end of October.

Approximately half of the Greenland and NE Canadian population performs a loop-migration around the North Sea in autumn, and a smaller part again in spring the opposite way around leaving the German Wadden Sea towards NW over the North Sea by early or mid-May.

Timing and proportion of these migrations are partly influenced by local weather and food conditions.

The onset of spring migration in the Wadden Sea is visible by increasing numbers from the end of March onwards. Numbers reach their peak in the first half of May including both Eurasian and Nearctic birds. The complex migration of these two populations are still not known in detail and subject of an intensive international research project (Dick, 1979; Dick et al., 1980; Haaland & Kaalaas, 1980).

The adult post-nuptial moult is a complete one and starts soon after arrival in the Wadden Sea. The first birds in moult are observed in the first July decade. By the beginning of September all adults are in heavy moult. The adult post-nuptial moult is finished by the end of October. Total moult duration amounts to approximately 100 days. There is some evidence that second calendar-year birds have their moult schedule about two weeks before the adults. First calendar-year birds moult a large part of their body feathers from late August till early November.

Pre-nuptial moult into the breeding plumage takes place from late March till the end of May, but is only partial for summering second calendar-year birds. The birds are concentrated in few areas around the North- and Irish Sea in this period, just before migrating towards the breeding area (Verwey, 1927, 1930; Boere, 1976).

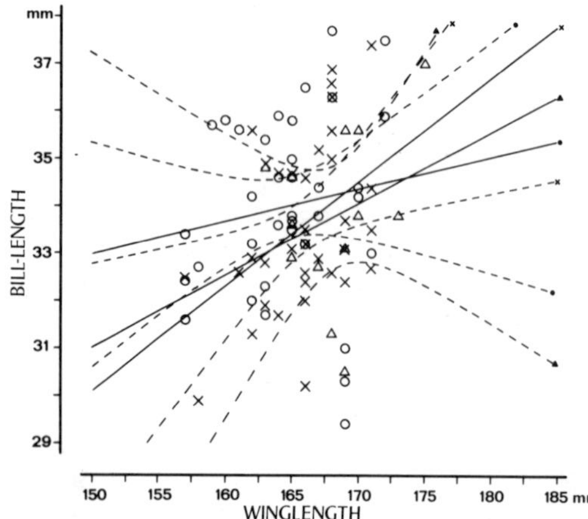

Fig. 95. Exposed bill- (in mm) and wing-lengths (mm) of first calendar-year Knots caught on the island of Vlieland from 1972-1975. Regression lines and 95% confidence intervals are indicated. △ denotes birds caught in August (n=13), ✕ September (n=35) and ○ October (n=37) (Boere, unpublished).

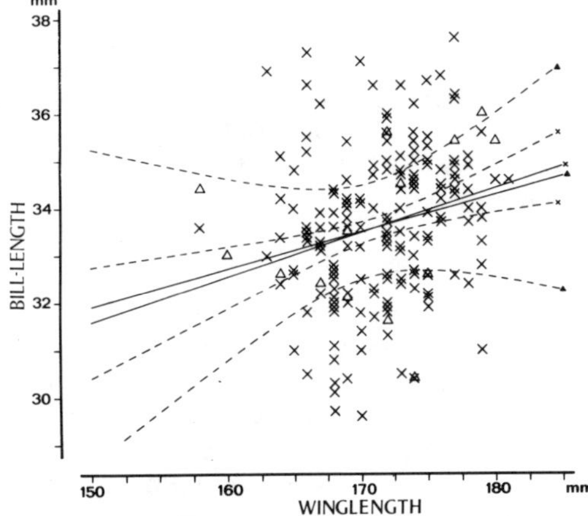

Fig. 96. Exposed bill- (in mm) and wing-lengths (mm) of adult Knots caught on the island of Vlieland from 1972-1975. Regression lines and 95% confidence intervals are indicated. △ denotes birds caught in August (n=14), **X** September (n=193). (Boere, unpublished).

3.18.2.3 Weight changes

The information available is restricted to the Dutch part of the Wadden Sea and only for a short period (Table 28). Information on morphometrics of birds caught on the island of Vlieland is shown in fig. 95 and 96. These data show especially the great variance in bill lengths according to:
- differences in sexes, males being slightly smaller than females (32.2 mm ± 0.2 (n=92); 34.0 mm ± 0.2 (n=71));
- differences between two populations. Birds from the Siberian populations have on the average a greater bill (34.2 mm ± 0.7) than those from Greenland and Eastern Canada (32.5 mm ± 0.3 (n=6))(Dick et al., 1976).

Table 28. Mean monthly weights of Knot, sexes combined, caught on the island of Vlieland, December 1971-December 1975 (Boere, unpubl.).

month	adult birds			first calendar-year birds		
	mean	s.d.	n	mean	s.d.	n
January	164	23	4			
February						
March	137	1	2			
April	165		1			
July	119		1			
August	142	26	10	124	16	9
September	137	11	200	122	17	41
October	161	22	4	137	17	38
November	168	42	2	143	16	3
December	149		1	144	26	4

3.18.3 Numbers

3.18.3.1 Population size

Prater (1976) estimates the size of the population wintering in W
Europe and NW Africa at 744,000. Of this number about 610,000 winter
in W Europe and 130,000 at the Banc d'Arguin. Piersma et al. (1980)
counted 366,000 at the Banc d'Arguin recently.

3.18.3.2 Numbers per area

Knots can be missed easily during a large scale wader count. They
have a strong preference for sandy high tide roosts without any vegetation. They can often be found in large compact flocks on uninhabitated islands and sandbanks which may be difficult to visit for counting purposes in winter and autumn. When the high tide is relatively
low sometimes considerable numbers will stay on the Wadden Sea sand
flats and not come to the high roosts. Because of their preference
for sandy places large high tide roosts are rarely found along the
mainland coast in Niedersachsen and the Netherlands.

As compared to other parts of the Wadden Sea Knots are rather
scarce in the Danish part. Autumn maxima occur in September but
amount to only about 1600. In the months following some hundreds
are present, in January and February numbers drop to about zero.
Peak numbers have been recorded in March with somewhat more than
5000. Until now only low numbers (of some tens) have been seen in
April, in May some hundreds are present again (Meltofte, 1980; reports
Vadefuglegruppen Dansk Ornithologisk Forening).

In the Wadden Sea area in Schleswig-Holstein first Knots arrive
by mid-July. Numbers increase considerably in August when adults arrive and again in September and October when juveniles arrive. All of
these birds start to moult (Drenckhahn et al., 1971). By this time
numbers may be as high as 400,000. Numbers drop very much in November
(Busche, 1980). As compared to autumn, peak numbers in winter are
small. According to Schlenker (1968) in normal winters up to 6000
are present, in very cold winters hardly any are seen (fig. 97).
Numbers in spring increase in March to peak in May. By this time
250,000 have been observed in the area. The number of summering Knots
is considerably smaller but may still amount to 10,000 (Heldt, 1968).
These are mostly second calendar-year birds starting wing moult by
the end of May and early June (Drenckhahn et al., 1971). Knots are
most common in the Western part of the area. Especially near Japsand,

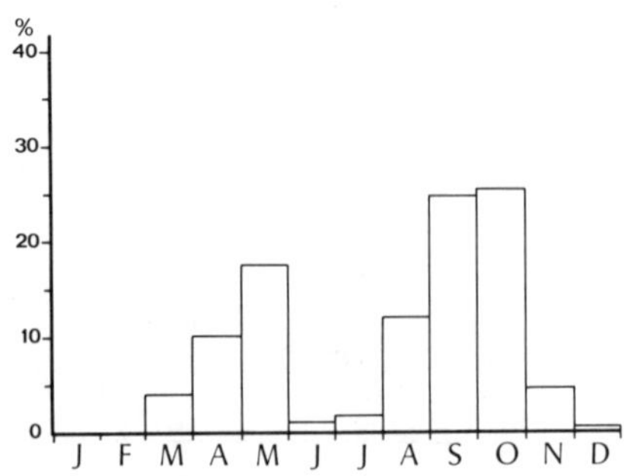

Fig. 97. Occurrence of Knots
in the Wadden Sea in Schleswig-
Holstein. Numbers for each month
are expressed as a percentage
of the total number observed in
all months (after Busche, 1980).

Fig. 98. Regular peak occurrences of Knots in the Wadden Sea in autumn per census area (after Smit, 1977; Busche, 1980; Meltofte, 1980 and data from Vadefuglegruppen Dansk Ornithologisk Forening and the Institut für Vogelforschung, Wilhelmshaven).

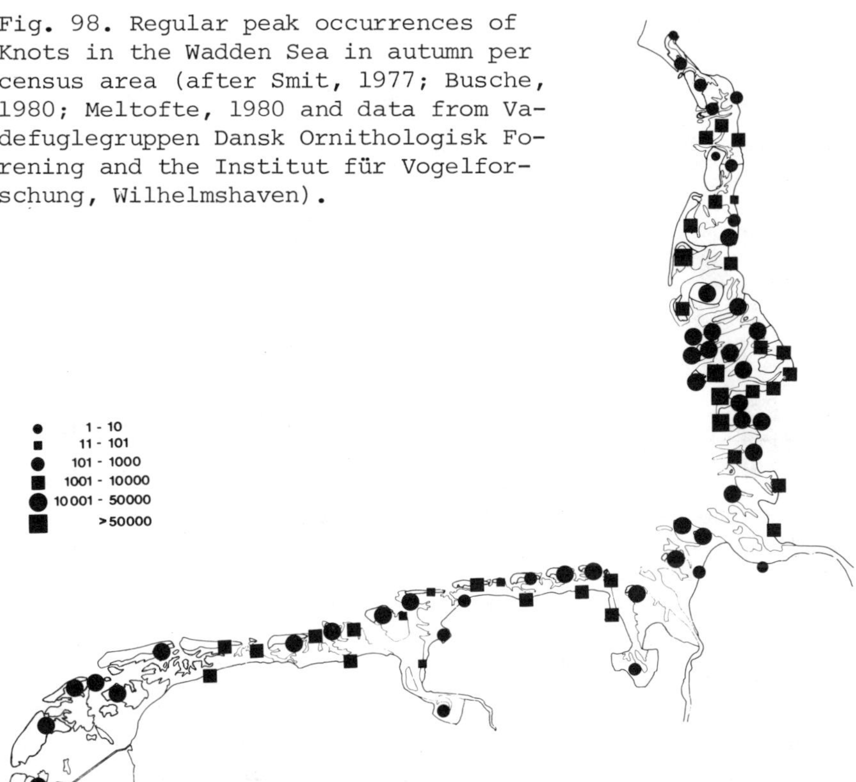

Norderoogsand, Süderoogsand and the Western part of the Eiderstedt peninsula enormous flocks may occur (fig. 98). Near St. Peter/Westerhever on September 5, 1971 180,000 have been counted (Busche, 1980).

In the Wadden Sea area in Niedersachsen Knots are not common along the mainland coast. They are more numerous around the islands where relatively large numbers may be observed in summer. Mean number per month for Wangerooge, Mellum, Scharhörn, Neuwerk and Grosser Knechtsand together for the period April-September are: April: 1128; May: 49,436; June: 31,948; July: 15,138; August: 10,113; September: 9,072 (data computed from Smit, 1977).

In June, when Knots are relatively scarce in other parts of the Wadden Sea area, these areas have the following maxima: Wangerooge (8000), Oldeoog (8000), Mellum (20,000), Scharhörn (25,000), Neuwerk (6500) and Grosser Knechtsand (60,000) (data: Institut für Vogelforschung, Wilhelmshaven). The only regions from which more or less comparable figures could be found are Norderoog (maximum 3000) and Trischen (1200), both in Schleswig-Holstein.

As in Niedersachsen Knots are relatively scarce along the mainland coast in the Dutch part of the Wadden Sea area. Fig. 99 shows mean numbers per month in the area. The non-hatched part in this figure denotes numbers in the Eastern part of the area (Terschelling, Ameland, Engelsmanplaat, Schiermonnikoog, Simonszand, Rottumerplaat, Rottumer-

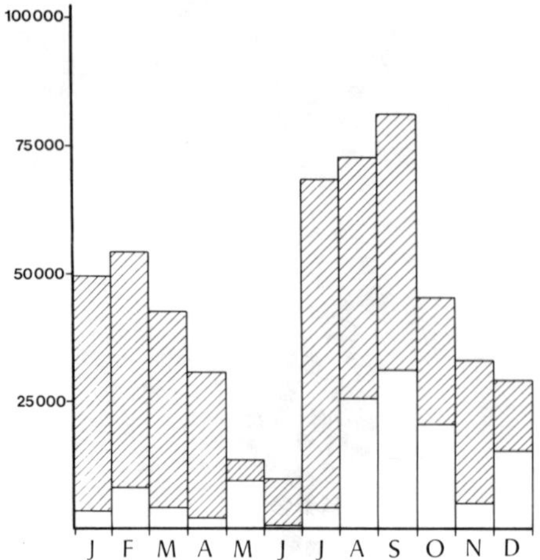

Fig. 99. Mean number of Knots per month in the Dutch part of the Wadden Sea. Hatched columns denote numbers in the Western part of the area (after data from Smit, 1977).

oog, Dollard and the mainland coast of Friesland and Groningen), showing relatively low numbers in winter there. Quite different from the situation in Schleswig-Holstein is that a spring peak in numbers is absent. Numbers in June are also very low. Counts of Knots in the area are listed in Table 29.

On September 1, 1973 the whole Wadden Sea was counted. The distribution was: Denmark 1,127; Schleswig-Holstein and Niedersachsen 219,000; the Netherlands 92,800, which yields a total number of 312,927. As mentioned above the population wintering in Europe and NW Africa is estimated to amount to about 750,000. At the days of the Wadden Sea count about 50% of this population was present in the area.

Table 29. Number of Knots in the Dutch part of the Wadden Sea (after various publications in Watervogels and Limosa and Zegers, in litt.).

Date	Year	Number
January 8	1977	19,620
January 12	1974	69,120
January 17	1976	75,500
January 18	1975	36,540
April 6	1973	> 18,400
April 19	1975	23,350
May 1	1976	3,150
May 11	1974	> 13,630
July 29	1972	24,225
August 22	1963	132,000
August 30	1975	23,900
September 1	1973	92,800
October 19	1974	55,725
November 13	1976	36,680
December 29	1966	58,500

3.18.4 Food

3.18.4.1 Food composition

According to the studies of Ehlert (1964) in the German Wadden Sea, Swennen (in prep.) and Piersma (in litt.) in the Dutch part the main prey items throughout the year are small molluscs of which *Macoma balthica* is most common. Small specimens of mussel and cockle are taken as well and sometimes large numbers of *Hydrobia*. If large numbers are present, crustaceans such as *Crangon crangon* and *Gammarus locusta* locally can be important as a food source, especially during summer and early autumn. Studies in British estuaries show the same food composition (Prater, 1972).

3.18.4.2 Feeding activities

Preferred feeding areas vary from place to place. In general however slightly muddy tidal flats and musselbeds are preferred. They feed in close groups, sometimes of several hundreds of birds.

Knots passing through in a certain area may build up very large flocks, the birds standing very close together. They often roost in flocks mixed with Dunlins of up to 50,000 birds or even more. Knots prefer open bare sandy areas where they roost close to or partly in the water. They also roost in coastal meadows with a short vegetation or, under certain circumstances, ploughed farm land.

3.18.4.3 Total food consumption

Little information is available from the area. Some observations show a food intake of 1.17 small cockles per minute per Knot. For flocks of several thousands of Knots this means a total consumption of 175,500 cockles per tidal cycle of 12 hours. The size of the cockles in this case was about 5.9 mm. Observations in other years show a lower intake of 0.79 prey items per minute per Knot, the size of the cockles in this case was about 5 mm (Rooth, unpublished reports of investigations on Vlieland in 1961 and 1962).

References

Andreassen, E.M. & O. Raad, 1977. Migration of the Knot, Calidris canutus based on Norwegian ringing results. Sterna 16: p. 31-45 (Norwegian, Engl. summ.).

Boere, G.C., 1976. The significance of the Dutch Waddenzee in the annual life cycle of arctic, subarctic and boreal waders. Part 1. The function as a moulting area. Ardea 64: p. 210-291.

Busche, G., 1980. Vogelbestände des Wattenmeeres von Schleswig-Holstein. Kilda, Greven (in press).

Dick, W.J.A., 1979. Results of the WSG project on the spring migration of Siberian Knot Calidris canutus 1979. Wader Study Group Bull 27: p. 8-13.

Dick, W.J.A., M.W. Pienkowski, M. Waltner & C.D.T. Minton, 1976. Distribution and geographical origins of Knots, Calidris canutus, wintering in Europe and Africa. Ardea 64: p. 22-47.

Dick, W.J.A., O. Fournier & P. Prokosch, 1980. W.S.G. project spring passage of Siberian Knot. Wader Study Group Bull. 28: p. 15.

Drenckhahn, D., R. Heldt Jun. & R. Heldt Sen., 1971. Die Bedeutung der Nordseeküste Schleswig-Holsteins für einige eurasische Wat- und Wasservögel mit besonderer Berücksichtigung des Nordfriesischen Wattenmeeres. Natur und Landschaft 46: p. 338-346.

Ehlert, W., 1964. Zur Oekologie und Biologie der Ernährung einiger Limikolen-arten. J. Orn. 105: p. 1-53.

Fournier, O. & F. Spitz, 1970. Etude biométrique des Limicoles III. Le Bécasseau maubêche (Calidris canutus). L'Oiseau 40: p. 69-81.

Glutz von Blotzheim, U.N., K.M. Bauer & E. Bezzel, 1975. Handbuch der Vögel Mitteleuropas, Vol. 6. Akademische Verlagsgesellschaft, Wiesbaden: 840 pp.

Harrington, B.A. & R.J.G. Morrison, 1980. Notes on the wintering areas of Red Knot Calidris canutus rufa in Argentina, South America. Wader Study Group Bull. 28: p. 40-42.

Haaland, A. & J.A. Kaalaas, 1980. Spring migration of the Siberian Knot Calidris canutus: additional information. Wader Study Group Bull. 28: p. 22-23.

Heldt, R., 1968. Uebersommernde Limikolen an der Westküste von Schleswig-Holstein. Corax 2: p. 108-130.

Meltofte, H., 1980. Fugle i Vadehavet. Vadefugletaellinger i Vadehavet 1974-1978. Miljøministeriet, Fredningsstyrelsen, København: 50 pp.

Meltofte, H., S. Pihl & B.M. Sørensen, 1972. Efteraarstraekket af vadefugle (Charadrii) ved Blaavandshuk 1963-1971. Dansk Orn. Foren. Tidsskr. 66: p. 63-69.

Morrison, M.J.G., 1975. Migration and morphometrics of European Knot and Turnstone on Ellesmere Island, Canada. Bird Banding 46: p. 290-301.

Morrison, R.J.G., B.A. Harrington & L.E. Leddy, 1980. Migration routes and stop over areas of North American Red Knot, Calidris c. rufa, wintering in South America. Wader Study Group Bull. 28: p. 35-39.

Netterstrøm, B., 1970. Efteraarstraekket af Islandks Ryle i Vestjylland. Dansk Orn. Foren. Tidsskr. 64: p. 223-228.

Nettleship, D.N., 1974. The breeding of the Knot at Hazen Camp, Ellesmere Island, N.W.T. Polarforschung 44: p. 8-26.

Oordt, G.J. van, 1928. Studies on the gonads of summering birds I and II. The Knot and the Turnstone. Tijdschrift Ned. Dierk. Ver. 3 ser. deel 1: p. 25-30.

Piersma, T., M. Engelmoer, W. Altenburg & R. Mes, 1980. A wader-expedition to Mauritania. Wader Study Group Bull. 29: p. 14.

Prater, A.J., 1972. The ecology of Morecambe Bay III. The food and feeding habits of Knot in Morecambe Bay. J. Appl. Ecol. 9: p. 179-194.

Prater, A.J., 1974. The population and migration of Knot in Europe. Proc. IWRB Wader Symposium, Warsaw 1973. Warsawa: p. 99-113.

Prater, A.J., 1976. The distribution of coastal waders in Europe and North Africa. Proc. Intern. Conference on Conservation of Wetlands and Waterfowl, Heiligenhafen 1974. IWRB, Slimbridge p. 255-271.

Schlenker, R., 1968. Ueber das Wintervorkommen von Limikolen an der Schleswig-Holsteinischen Westküste. Corax 2: p. 92-108.

Smit, C.J., 1977. On the occurrence of 32 bird species in the Danish,
    German and Dutch Wadden Sea. Unpubl. report International Wadden
    Sea Working Group, part 3: 174 pp.
Verwey, J., 1927. Maturity and breeding dress in birds III and IV. The
    plumages of Tringa canutus L. and Tringa crassirostris Temm. and
    Schlegel. Zoöl. Mededelingen (Leiden) 10: p. 158-183.
Verwey, J., 1930. Geschlechtsreife und Prachtkleid der Vögel V. Ge-
    schlechtsreife, Prachtkleid und Mauser, besonders bei einjährigen
    Calidris canutus (L). J. Orn. 78: p. 234-245.
Voous, K.H., 1960. Atlas of European birds. Nelson, London: 284 pp.

3.19 SANDERLING *(CALIDRIS ALBA* (PALLAS)*)*.
   G.C. Boere & C.J. Smit

Da: Sandløber; G: Sanderling; Du: Drieteenstrandloper

3.19.1 Distribution

3.19.1.1 Breeding area
   The Sanderling has a discontinuous circumpolar distribution, breeding only in the tundra climate zone (fig. 100; Voous, 1960). No subspecies have been described. Very few birds have been ringed in the
Wadden Sea which makes it difficult to determine the area of origin
exactly. According to the results of ringing programs in England, San-

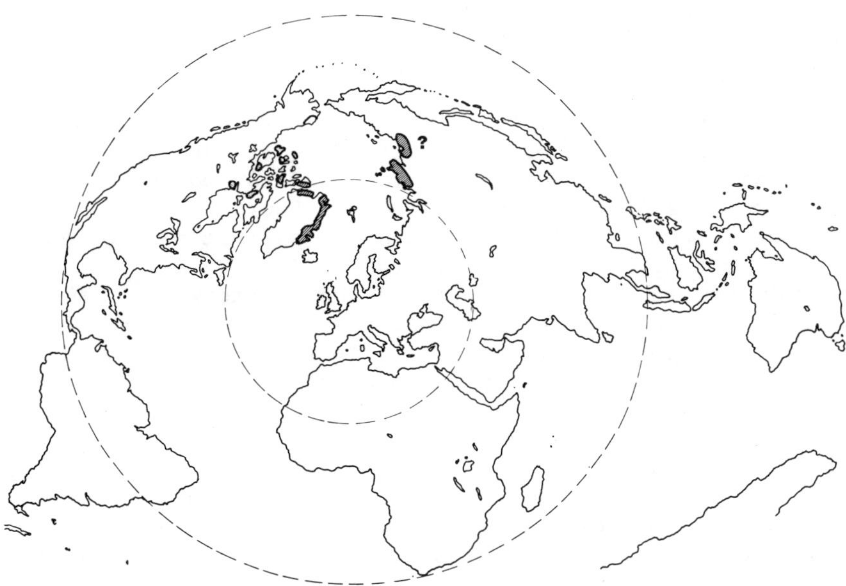

Fig. 100. Breeding area of the Sanderling (after Voous, 1960; modified). The dark shaded area represents approximately the region where
birds visiting the Wadden Sea originate from.

derlings which are present in the areas around the North Sea originate
from the Eurasian as well as the Greenland-NE Canadian populations (e.g.
Branson, 1979).

3.19.1.2 Migration routes

The species is almost strictly coastal which determines to a large
extent its migration routes in the Western Palaearctic. Like in many
other wader species observations of single birds inland mainly refer
to first calendar-year birds (Ferdinand, 1953; Meltofte et al., 1972;
Meltofte & Rabøl, 1977).

3.19.1.3 Wintering areas

Sanderlings winter widely spread and cosmopolitan on sandy, and
sometimes rocky shores along oceans and ocean islands south of the
frost line. Wintering flocks are small in the Wadden Sea area and generally do not exceed a few hundred birds. Recently flocks of tens or
even hundreds of thousands Sanderlings have been observed, e.g. along
the Namib coast (Underhill & Whitelaw, 1977) and Peru (Byskov, 1977;
Plenge, in Glutz et al., 1975).

3.19.4 Moulting areas

Though there is still insufficient detailed information it became
clear that generally moult is spread over a long period and at many
places, especially where a particular group winters or stays for a
longer period during migration. For the latter category the Wadden Sea
is a good example (see 3.19.2.2).

3.19.2 Annual cycle

3.19.2.1 Migration

Departure of adult birds from the breeding areas, at least those
in Greenland and NE Canada, takes place between early July and early
August (Pienkowski & Green, 1976). Migration of first calendar-year
birds starts some weeks later. Numbers of migrating Sanderlings at
Blaavandshuk peak in the last July decade and the first half of August (Meltofte et al., 1972). Numbers in the Wadden Sea are increasing from mid-July onwards to peak in late August. Afterwards numbers
decrease rapidly. Probably there is a continuous passage of birds to
areas in Southern Europe and Africa. Small flocks stay to winter.
Spring migration can be noticed from the end of March till late May.

3.19.2.2 Moult

The adult post-nuptial moult takes place outside the breeding area
and starts in the Wadden Sea by late July. Birds trapped about mid-
August were nearly all in full wing- and body moult. Specimens not
in moult tend to have an extremely high body weight and are supposed
to be transient birds (Boere, 1976). Resident birds finish moult about
early October.
First calendar-year birds trapped on August 22, 1974, have not started their body moult. Large numbers of Sanderlings undergo post-nuptial
moult in Africa including some first calendar-year birds which have a

Table 30; Weights (in g) of Sanderlings trapped on the island of Vlieland (Boere, unpubl.).

catching date			mean	s.d.	n
August 9, 1975	adult	moulting	58	5.8	38
		non-moulting	74	12.1	31
August 21, 1974	adult	moulting	54	4.7	129
		non-moulting	67	9.4	14
		first calendar-year	48		4

complete or partial wing moult (Underhill & Whitelaw, 1977).

Few information is available on the timing of pre-nuptial moult. At least partly it takes place in Europe just before spring migration towards breeding areas. On Süderoogsand (Schleswig-Holstein) the birds got into their brown coloured plumage about May, 20 (König, in litt.).

3.19.2.3 Weight changes

Very few information is available. Weights of birds trapped on the island of Vlieland are listed in Table 30.

3.19.3 Numbers

3.19.3.1 Population size

The number of Sanderlings wintering in NW Europe and W Africa can be estimated at 26,000 birds (Prater, 1976). This figure however can only be a rough estimation of the number actually present. In the Wadden Sea area the species may occur on sand banks which are often difficultly attainable. On the Wadden Sea islands it occurs on sandy beaches which are often not incorporated in wader counts. A part of the population passing the Wadden Sea winters along coasts of Central- and South Africa (Underhill & Whitelaw, 1977). These areas have been visited too infrequently to yield detailed information on the number of Sanderlings present. Recent observations at the Banc d'Arguin resulted in 34,000 wintering Sanderlings (Piersma et al., 1980). There are indications that the population breeding in Arctic Canada and wintering in South-America is larger than the population breeding in Greenland and the Arctic USSR. Plenge (in Glutz et al., 1975) even found 450,000 wintering Sanderlings in Peru. Whenever this is true this difference might be due to the differences in breeding densities in the Arctic. These have been found to differ considerably between areas (Parmelee, 1970; Meltofte, 1979; Pienkowski & Green, 1976).

3.19.3.2 Numbers per area

For many parts of the Wadden Sea insufficient information is available on numbers of this species because it has been overlooked too often. Therefore only a few data on numbers can be given.

Fig. 101. Regular peak occurrences of
Sanderlings in autumn or spring per
census area (after Smit, 1977;
Busche, 1980; Meltofte, 1980 and data
from Vadefuglegruppen Dansk Ornithologisk Forening and O.A.G. Schleswig-
Holstein und Hamburg.

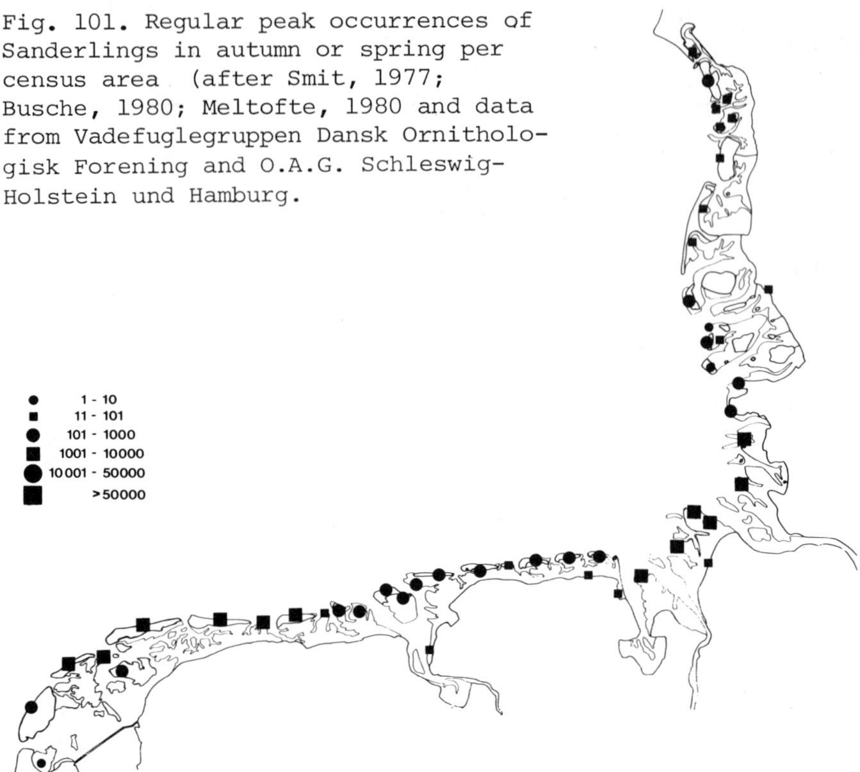

   In all parts of the Danish Wadden Sea the numbers are small (fig.
101). The maximum number recorded in autumn is 120, in spring only 30
(Meltofte, 1980; reports Vadefuglegruppen Dansk Ornithologisk Forening).
   Also in Schleswig-Holstein numbers are small. They are largest in
October and May when some thousands may be present. Larger flocks are
only known from the islands of Trischen, Amrum and Sylt and the mainland coast near St. Peter (Busche, 1980). In winter numbers drop to
60-120 (Schlenker, 1968).
   Along the mainland coast in Niedersachsen Sanderlings are absent.
Large numbers however occur, in the Mellum, Scharhörn, Neuwerk and
Grosser Knechtsand area. Table 31 gives an impression of the numbers
which may be present there.
   In the Dutch part of the Wadden Sea the species is absent along the
mainland coast. Fig. 101 and Table 31 show that sometimes on the islands
high numbers may occur. Fig. 102 and Table 32 show the fluctuation in
numbers present per month in the Dutch Wadden Sea. Generally at least
1000-2000 spend winter here but in very cold periods nearly all leave
the area. Numbers in Table 32 should be regarded as minimum figures.
The beaches of the islands were not always included in these counts.
Sometimes uninhabitated sandbanks could not be visited. Recent counts
on the sandbank Richel near the island of Vlieland show that some thousands occur regularly here (De Roos, pers. comm.).
   On September 1, 1973 the species was not observed in the Danish Wad-

Table 31. Maximum numbers of Sanderlings on some Wadden Sea islands in Niedersachsen and the Netherlands (data: Institut für Vogelforschung, Wilhelmshaven; Smit, 1977).

	April	May	July	August	September	October
Mellum	350	2,000	280	700	445	180
Scharhörn	510	2,000	700	3,000	3,000	2,000
Neuwerk	1,500	600	180	350	8,000	5,000
Grosser Knechtsand	?	5,000	100	280	450	?
Wangerooge	80	200	39	120	500	500
Schiermonnikoog	110	300	185	1,070	300	2,200
Engelsmanplaat	105	300	900	1,500	250	380
Ameland	25	130	140	240	570	1,030
Terschelling	900	375	530	1,000	1,890	1,900
Vlieland	600	?	250	3,500	250	500

Table 32. Numbers of Sanderlings in the Dutch part of the Wadden Sea. The figures should be regarded as minimum estimates (after various publications in Watervogels and Limosa and Zegers, in litt.).

Date	Year	Number
January 8	1977	1,490
January 12	1974	1,845
January 17	1976	2,260
January 18	1975	1,040
April 6	1973	170
April 19	1975	765
May 1	1976	500
May 11	1974	233
July 29	1972	1,750
August 22	1963	850
August 30	1975	2,430
September 1	1973	3,320
October 19	1974	3,800
November 13	1976	1,465
December 29	1966	1,200

den Sea, in Schleswig-Holstein and Niedersachsen 2860 were counted. Together with 3320 counted in the Netherlands 6180 were present in the whole Wadden Sea area.

3.19.4 Food

3.19.4.1 Food composition

No detailed studies are available, but polychaetes, especially *Scolelepis squamata*, are the most important prey items in the Wadden Sea

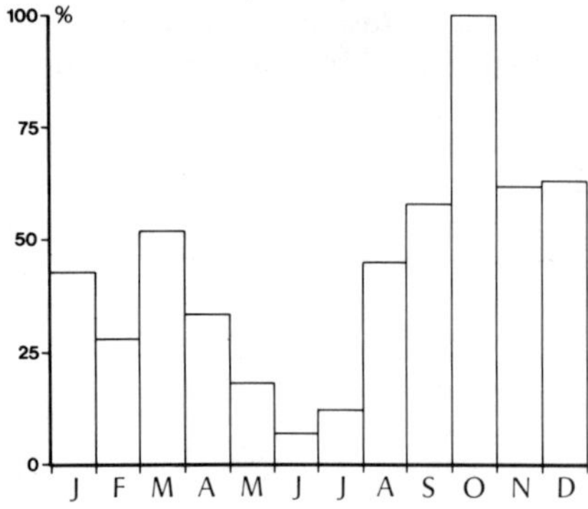

Fig. 102. Occurrence of Sanderlings per month in the Dutch
part of the Wadden Sea. The number per month is expressed as
a percentage of the peak recorded in October (after Smit,
1977).

area. Crustaceans have been noticed in small numbers too, as have been
small molluscs (Glutz et al., 1977).

3.19.4.2 Feeding activities

Feeding areas are almost restricted to the intertidal area of sandy
beaches along the coast of mainland and on the islands. Very characteristic is the feeding behaviour: fast running and preening close to
the waterline.

Favourite sites for the high tide roosts are sandy beaches just a-
bove the high tide level and sand ridges or shell banks close to the
high tide line. In the Dutch Wadden Sea, sometimes stone dikes are
used as well.

During heavy storms Sanderlings roost inland at various places or
on the beach sheltered between waste material and wreckage.

3.19.4.3 Total food consumption

No information available from the area.

References

Boere, G.C., 1976. The significance of the Dutch Wadden Sea in the an-
nual life cycle of arctic, subarctic and boreal waders. Part 1. The
function as a moulting area. Ardea 64: p. 210-291.
Branson, N.J.B.A. (ed.), 1979. Wash Wader Ringing Group. Report 1977-
1978. Cambridge: 68 pp.
Busche, G., 1980. Vogelbestände des Wattenmeeres von Schleswig-Holstein.
Kilda, Greven (in press).
Byskov, J.O., 1977. Peru & Ecuador, bird observations 1974. Unpublish-
ed expedition report: 45 pp.
Ferdinand, L., 1953. Sandløberens (Crocethia alba) traekforholdene i
Nord-europa. Dansk Orn. Foren. Tidsskr. 47: p. 69-95.
Glutz von Blotzheim, U.N., K.M. Bauer & E. Bezzel, 1975. Handbuch der
Vögel Mitteleuropas, Vol. 6. Akademische Verlagsgesellschaft, Wies-
baden: 840 pp.

Meltofte, H., 1979. The population of Waders, Charadriidae at Danmarks Havn, Northeast Greenland, 1975. Dansk Orn. Foren. Tidsskr. 73: p. 69-94.

Meltofte, H., 1980. Fugle: Vadehavet. Vadefugletaellinger i Vadehavet 1974-1978. Miljøministeriet, Fredningsstyrelsen, København: 50 pp.

Meltofte, H., Pihl, S. & B.M. Sørensen, 1972. Efteraarstraekket af Vadefugle (Charadrii) ved Blaavandshuk 1963-1971. Dansk Orn. Foren. Tidsskr. 66: p. 63-69.

Meltofte, H. & J. Rabøl, 1977. Vejrets indflydelse paa efteraarstraekket af Vadefugle ved Blaavandshuk, med et forsøg paa en analyse af traekkets geografiske oprindelse. Dansk Orn. Foren. Tidsskr. 71: p. 43-63.

Parmelee, D.F., 1970. Breeding behaviour of the Sanderling in the Canadian High arctic. Living Bird 9: p. 97-146.

Pienkowski, M.W. & G.H. Green, 1976. Breeding biology of Sanderlings in north-east Greenland. British Birds 69: p. 165-177.

Piersma, T., M. M. Engelmoer, W. Altenburg & R. Mes, 1980. A wader-expedition to Mauritania. Wader Study Group Bull. 29: p. 14.

Underhill, L.G. & D.A. Whitelaw, 1977. An ornithological expedition to the Namib coast, Summer 1976/1977. Report Western Cape Wader Group, Cape Town: 106 pp.

Schlenker, R., 1968. Ueber das Wintervorkommen von Limikolen an der Schleswig-Holsteinischen Westküste. Corax 2: p. 92-108.

Smit, C.J., 1977. On the occurrence of 32 bird species in the Danish, German and Dutch Wadden Sea. Unpubl. report Intern. Wadden Sea Working Group, part 3: 174 pp.

Voous, K.H., 1960. Atlas of European birds. Nelson, London: 284 pp.

3.20 CURLEW SANDPIPER *(CALIDRIS FERRUGINEA* (PONTOPPIDAN)*)*.
G.C. Boere & C.J. Smit

Da: Krumnaebbet Ryle; G: Sichelstrandläufer; Du: Krombekstrandloper

3.20.1 Distribution

3.20.1.1 Breeding area
The Curlew sandpiper has a rather limited distribution, breeding in the tundra climate zone on the Taymyr peninsula and some smaller spots in Eastern Siberia (fig. 103; Portenko, 1959). It incidentally breeds in Western Alaska. No subspecies are recognised. Curlew sandpipers migrating through the Wadden Sea originate from the Taymyr peninsula. Probably some birds originate from breeding areas even more eastward.

3.20.1.2 Migration routes
In Western Europe the Curlew sandpiper is mainly restricted to coastal habitats, but in some places it is also frequently observed at inland areas, adults as well as first calendar-year birds. This is particular true for large parts of Africa where the species is very com-

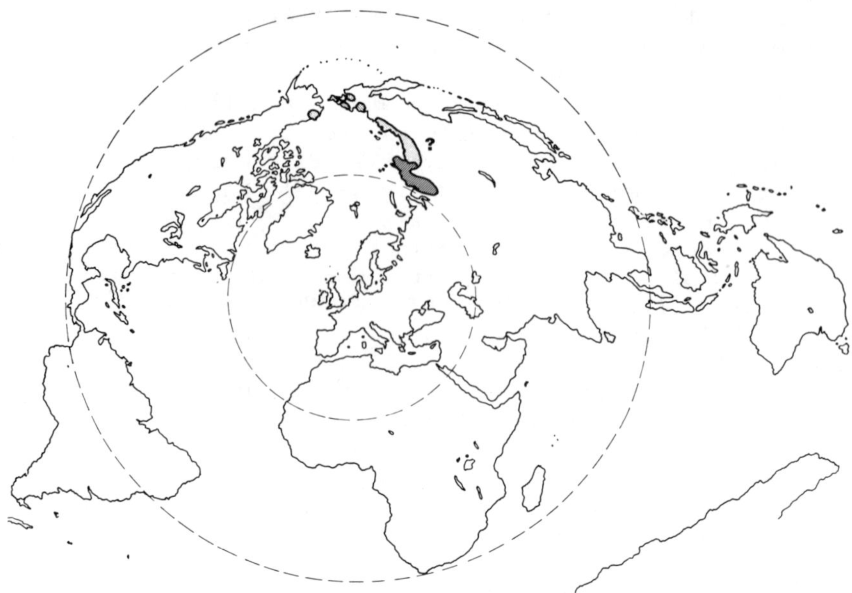

Fig. 103. Breeding area of the Curlew sandpiper (after Portenko, 1959). The dark shaded area represents approximately the region where birds visiting the Wadden Sea originate from.

mon along the shores of the lakes of the Rift Valley during spring migration (Fry et al., 1973). It is a common visitor of the Wadden Sea during a restricted period of autumn migration, but scarce during spring migration when the birds follow a more easterly route through Central- and Eastern Europe.

3.20.1.3 Wintering areas

Curlew sandpipers winter in Africa south of the Sahara, widely spread. Large concentrations occur in coastal areas in South-Africa where 30,000-50,000 birds have been counted (Elliott et al., 1976). Populations breeding in Eastern Siberia winter in Australia and New-Zealand. Ringing results from the Wadden Sea are scarce. There are several remarkable records of Curlew sandpipers ringed in South-Africa and controlled in the breeding area in Siberia (Glutz et al., 1975).

3.20.1.4 Moulting areas

The species moults in Africa at suitable areas along the migration route, particularly in Morocco, the Banc d'Arguin and in South-Africa. Very small numbers start their moult in Europe (Boere, 1976; Pienkowski et al., 1976). In Europe as well as Africa many birds arrest their moult after a certain period to finish it elsewhere after a long-distance migration.

3.20.2 Annual cycle

3.20.2.1 Migration

Non-breeding birds leave the breeding areas by late June. Breeding birds depart from July onwards, males being the first to leave. Numbers counted on migration at Blaavandshuk are limited. Most of them pass at the end of July and in August (Meltofte et al., 1972). There is a rapid increase of the numbers present in the Wadden Sea to a peak between late July and early August. First calendar-year birds arrive about one month later, from early August onwards and have left by the end of September. The birds stay probably for a very short time, although observations of colour-marked birds on Vlieland show that some may stay for at least three or four weeks. Spring observations are very rare and almost restricted to observations of single birds during May. Last birds have left NW Europe in October. Departure from the wintering quarters generally takes place in April. The spring peak around the Mediterranean and in Central Europe generally can be observed in May and some are even present here in early June (Glutz et al., 1975).

3.20.2.2 Moult

The adult post-nuptial moult starts with the body plumage which is in good progress during the period the birds stay in the coastal areas of NW Europe, including the Wadden Sea to moult their body-plumage completely as is shown by observations of colour marked birds on Vlieland in 1972 and 1973. The majority however arrests body-moult and leaves the area in an intermediate plumage while moult is completed in Africa (Boere, 1976; Elliott et al., 1976). Very few birds start their wing- and tail-moult in NW Europe and if they do, it is very soon arrested and finished in Africa from August till December. The difference in moult schedule of body feathers and wing- and tail feathers within the same bird is remarkable. Moult of first calendar-year birds involves the body plumage and part of the wing- and tail feathers and takes place during their stay in the African wintering areas. Pre-nuptial body moult into the breeding plumage takes place from mid-February onwards and is finished during migration before arriving in the breeding areas.

Table 33. Mean monthly weights, sexes combined, of adult and first calendar-year Curlew sandpipers, caught on the island of Vlieland, December 1971-December 1975 (Boere, unpubl.)

month	adult birds			first calendar-year birds		
	mean	s.d.	n	mean	s.d.	n
July	66	12	36			
August	73	11	71	52	7	16
September	82		1	84	9	17

3.20.2.3 Weight changes

During the very short time that Curlew sandpipers are present in the Wadden Sea, they increase considerable in weight (Table 33). No information is available from spring migration.

3.20.3 Numbers

3.20.3.1 Population size

Dick (1975) estimates the total number wintering at the Banc d'Arguin in Mauritania at 37,000. Part of the population winters further south (compare 3.20.1.3). The total size of the population passing the Wadden Sea is unknown.

3.20.3.2 Numbers per area

The Curlew sandpiper often occurs together with Dunlins. This species often occurs in large flocks and Curlew sandpipers may hide in these flocks. Regularly they occur in small flocks along creeks and little ponds on the salt marshes, places where they can be overlooked easily during a wader count. The numbers determined during large scale counts are therefore no accurate estimation of the numbers actually present and generally the real numbers present in a certain area will be higher. The picture obtained from its distribution will be incomplete as well. This is shown in study areas where large scale counts have been compared with detailed counts and results from ringing activities (Jukema, 1979).

In the Danish part of the Wadden Sea from July until September some tens of Curlew sandpipers are present. In the Danish part of the Wadden Sea area no Curlew sandpipers have been observed in spring yet (Meltofte, 1980; reports Vadefuglegruppen Dansk Ornithologisk Forening).

In Schleswig-Holstein the species never occurs in large numbers. From November-March they have not been observed at all. Maximum numbers per month from April-October are 2, 96, 16, 315, 520, 756 and 7, respectively (data: Ornithologische Arbeitsgemeinschaft Schleswig-Holstein und Hamburg).

Like in Schleswig-Holstein Curlew sandpipers are not recorded in very large numbers in Niedersachsen. Exceptions are Scharhörn, Neuwerk and Wangerooge where flocks of tens or hundreds have been seen (Smit,

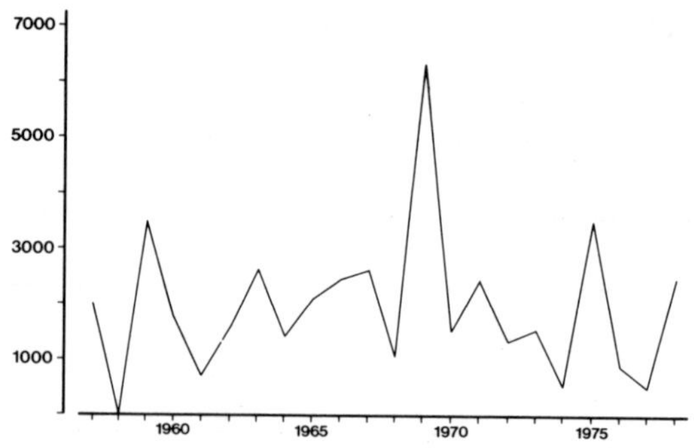

Fig. 104. Estimated number of Curlew sandpipers in the Dutch part of the Wadden Sea from 1957-1978 based on long-term counts, standardized observations in several areas and yearly ringing totals (after Roselaar, 1979).

Fig. 105. Regular peak occurrences of
Curlew sandpipers in autumn or spring
per census area (after Smit, 1977; Ju-
kema, 1979; Busche, 1980 and data from
Vadefuglegruppen Dansk Ornithologisk
Forening, O.A.G. Schleswig-Holstein
und Hamburg and the Institut für
Vogelforschung, Wilhemshaven).

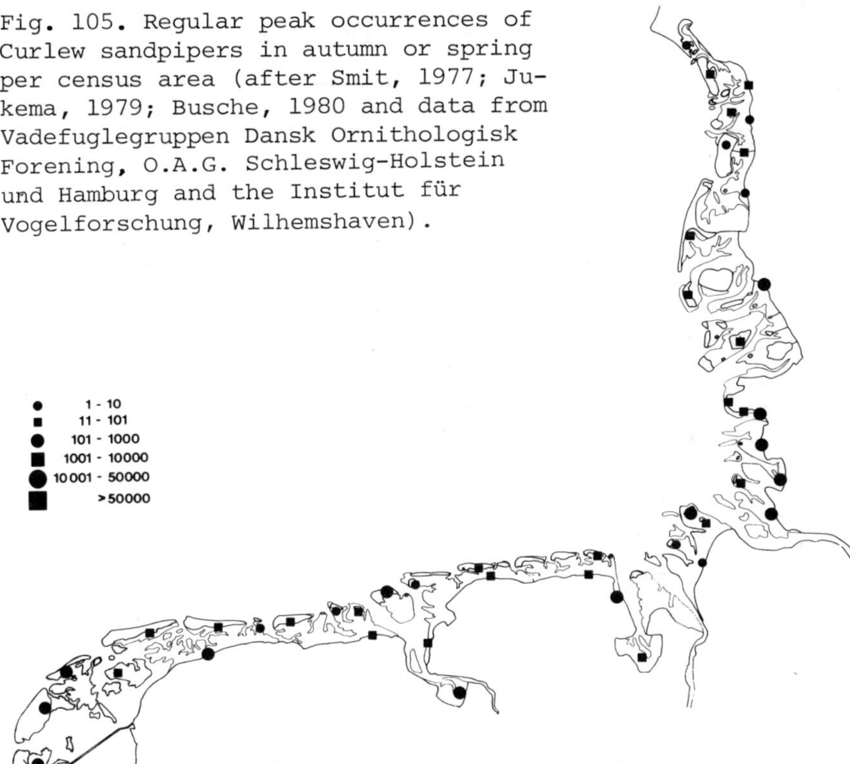

1977). They are only present during a short period in autumn and spring
and are hardly ever observed in winter and summer.

Also in the Dutch part of the Wadden Sea the species is a relatively rare guest. Roselaar (1979) estimates that annually about 1000-3000
Curlew sandpipers may be present in the area at the same time between
the end of July and early October. In some years however there seems
to be something like an invasion. In 1969 for instance several thousands occurred in the Dutch part of the Wadden Sea at the same time.
This phenomenon has been observed during autumn migration in other
countries as well and seems to be linked to population fluctuations
(fig. 104) (Kumerloeve, 1960; Roselaar, 1979; Gibb & Tucker, 1947;
Stanley & Minton, 1972; Glutz et al., 1975). On the island of Vlieland
and in the Balgzand area the species generally appears to be somewhat
more numerous than in other places as is shown in fig. 105. Numbers
counted in the Dutch part of the Wadden Sea are listed in Table 34.

On September 1, 1973 in the Danish and German part of the Wadden Sea
numbers were 35 and 1920 respectively. Because in the Dutch part of
the area 74 were present the number in the whole Wadden Sea therefore
was 2030.

Table 34. Number of Curlew sandpipers in the Dutch part of the Wadden Sea (after various publications in Watervogels and Limosa and Zegers (in litt.).

Date	Year	Number
January 8	1977	0
January 12	1974	0
January 17	1976	0
January 18	1975	0
April 6	1973	0
April 19	1975	0
May 1	1976	0
May 11	1974	50
July 29	1972	850
August 22	1963	110
August 30	1975	720
September 1	1973	74
October 19	1974	0
November 13	1976	0
December 29	1966	0

3.20.4 Food

3.20.4.1 Food composition

Unfortunately there is only very little information available. Food composition will be more or less comparable to that of Dunlin. Generally however prey items will be somewhat larger. Probably polychaetes will be somewhat more important.

3.20.4.2 Feeding activities

Curlew sandpipers are not bound to a certain kind of substrate but show a preference for somewhat muddy areas to feed on. Curlew sandpipers may use a great variety of places as a roost. They occur always in combination with other species, especially Dunlin, but also Knot, Redshank and Greenshank. During high tide they often continue feeding on shores of inland pools and ditches.

3.20.4.3 Total food consumption

No information available.

References

Boere, G.C., 1976. The significance of the Dutch Wadden Sea in the annual lifecycle of arctic, subarctic and borealwaders. Part 1. The function as a moulding area. Ardea 64: p. 210-291.

Busche, G., 1980. Vogelbestände des Wattenmeeres von Schleswig-Holstein. Kilda, Greven (in press).

Dick, W.J.A., 1975. Oxford and Cambridge Mauritanian expedition 1973. Report, Cambridge: 79 pp.

Elliott, C.C.H., M. Waltner, L.G. Underhill, J.S. Pringle & W.J.A. Dick,
  1976. The migration system of the Curlew sandpiper Calidris ferruginea in Africa. Ostrich 47: p. 191-213.
Fry, C.H., P.L. Britton & J.F.M. Horne, 1973. Lake Rudolf and the Palaearctic exodus from East-Africa. Ibis 116: p. 44-51.
Gibb, J.A. & B.W. Tucker, 1947. The exceptional passage of Curlew sandpipers and Little Stints in the autumn of 1946. Brit. Birds 40: p.
  354-359.
Glutz von Blotzheim, U.N., K.M. Bauer & E. Bezzel, 1975. Handbuch der
  Vögel Mitteleuropas, Vol. 6. Akademische Verlagsgesellschaft Wiesbaden: 840 pp.
Jukema, J., 1979. Krombekstrandlopers langs de Friese Waddenkust. Wa-
  tervogels 4: p. 3-6.
Kumerloeve, H., 1960. Zur Durchzugsfrequenz von Sichel- (Calidris fer-
  ruginea) und Zwergstrandläufer (Calidris minuta) auf Amrum. Beitr.
  Vogelk. 7: p. 33-37.
Meltofte, H., 1980. Fugle i Vadehavet. Vadefugletaellinger i Vadehavet
  1974-1978. Miljøministeriet, Fredningsstyrelsen, København: 50 pp.
Meltofte, H., S. Pihl & B.M. Sørensen, 1972. Efteraarstraekket af va-
  defugle (Charadrii) ved Blaavandshuk 1963-1971. Dansk Orn. Foren.
  Tidsskr. 66: p. 63-69.
Pienkowski, M.W., P.J. Knight, D.J. Stanyard & F.B. Argyle, 1976. The
  primary moult of waders on the Atlantic coast of Morocco. Ibis 118:
  p. 347-365.
Prater, A.J., 1976. The distribution of coastal waders in Europe and
  North Africa. In: M. Smart (ed.). Proceedings Intern. Conf. on Con-
  servation of Wetlands and Waterfowl, Heiligenhafen, 1974. IWRB,
  Slimbridge: p. 255-271.
Portenko, L.A., 1959. Studien an einigen seltenen Limicolen aus dem
  nördlichen und östlichen Siberien II. Der Sichelstrandläufer. J.
  Orn. 100: p. 141-172.
Roselaar, C.S., 1979. Fluctuaties in aantallen Krombekstrandlopers Ca-
  lidris ferruginea. Watervogels 4: p. 202-210.
Smit, C.J., 1977. On the occurrence of 32 bird species in the Danish,
  German and Dutch Wadden Sea. Unpubl. report Intern. Wadden Sea Work-
  ing Group, part 3: 174 pp.
Stanly, P.J. & C.D.T. Minton, 1972. The unprecedented westward migration of Curlew sandpipers in autumn 1969. Brit. Birds 65: p. 365-
  380.

3.21 DUNLIN *(CALIDRIS ALPINA* (L.)*)*
    G.C. Boere & C.J. Smit

Da: Almindelig Ryle; G: Alpenstrandläufer; Du: Bonte Strandloper

3.21.1 Distribution

3.21.1.1 Breeding area
   The Dunlin has an almost circumpolar holarctic distribution breeding predominantly in the tundra but also in the boreal and even in the

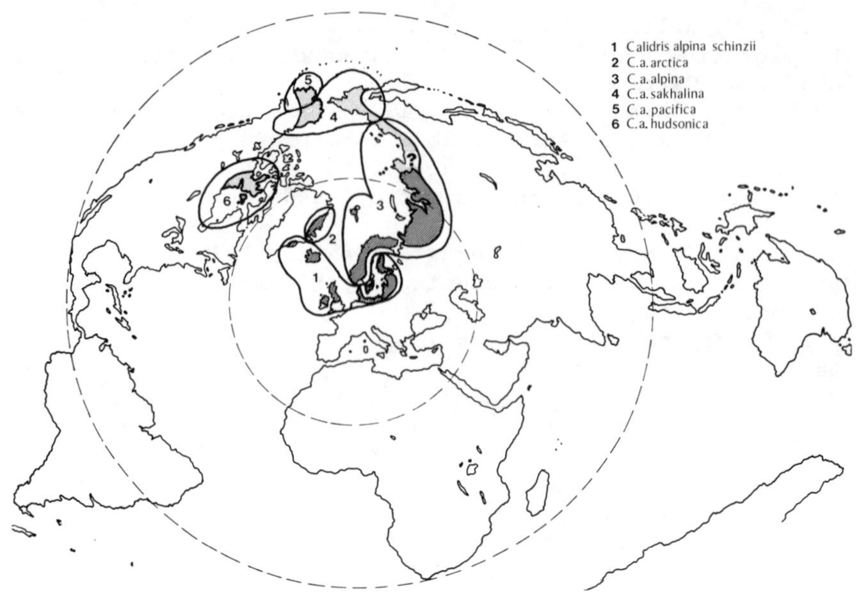

Fig. 106. Breeding area of the Dunlin (after Voous, 1960). The dark
shaded area represents the region where birds visiting the Wadden Sea
originate from. Question marks indicate a region from which is uncertain whether birds breeding there also visit the Wadden Sea.

temperate climate zones (Voous, 1960). The species shows much geographical variation and generally the following subspecies are distinguished: *C.a. schinzii*, *C.a. alpina*, *C.a. arctica*, *C.a. hudsonica*, *C.a. pa-
cifica* and *C.a. sakhalina* (Martin-Löf, 1954; Soikkeli, 1966, 1974;
Griffiths, 1970; McLean & Holmes, 1971). Their distribution is outlined
in fig. 106. There is a gradual increase in size from Western Europe
to Eastern Siberia. According to ringing- and biometrical studies (Nieboer, 1972; Boere, 1976; Boere et al., 1973) the majority of the Dunlins
in the Wadden Sea belongs to the subspecies *C.a. alpina*, but also small
numbers of *C.a. schinzii* and *C.a. arctica* are present (see also Table
35 and 36). According to the large size of some of the birds that have
been trapped in the Wadden Sea, it is likely that sometimes also *C.a.
sakhalina* is present. This means that the area of origin ranges at
least from Eastern Greenland through Scandinavia as far as the Ob, but
probably more to the east in Central Siberia.

3.21.1.2 Migration routes
   Migration takes place in a broad front across Europe. Large numbers
occur in coastal areas, although Dunlins are also frequently observed
at suitable inland localities all over Central Europe (Nørrevang, 1955;
Ogilvie, 1963; Mascher, 1966, 1971; Fuchs, 1973; O.A.G. Münster, 1976,
1980; Jansen, 1979; Leslie & Lessels, 1978).
   Greenland and Iceland breeding birds follow a route across the Atlantic Ocean, Iceland and England to the French Channel coast and then

to the south to Northern Africa and back in the opposite direction (Eades, 1974; Steventon, 1977; Pienkowski & Dick, 1975).

3.21.1.3 Wintering areas

The different subspecies of Dunlin to some extent show a differentiation in wintering areas. *C.a. alpina* winters mainly in the areas around the North Sea, the British Isles being most important, and in Southern Europe and North Africa. *C.a. schinzii* winters from Southern Europe to the Banc d'Arguin in Africa. Further south Dunlins are not common as a winter visitor (Vieillard, 1972; Pienkowski & Dick, 1975). Figs. 107 and 108 give some information on wintering areas and areas of origin of Dunlins passing the Netherlands.

3.21.1.4 Moulting areas

Moulting birds are present in the whole coastal area of Western Europe and North Africa, as far as the Banc d'Arguin in Mauritania. Larger concentrations gather at suitable areas within this total range.

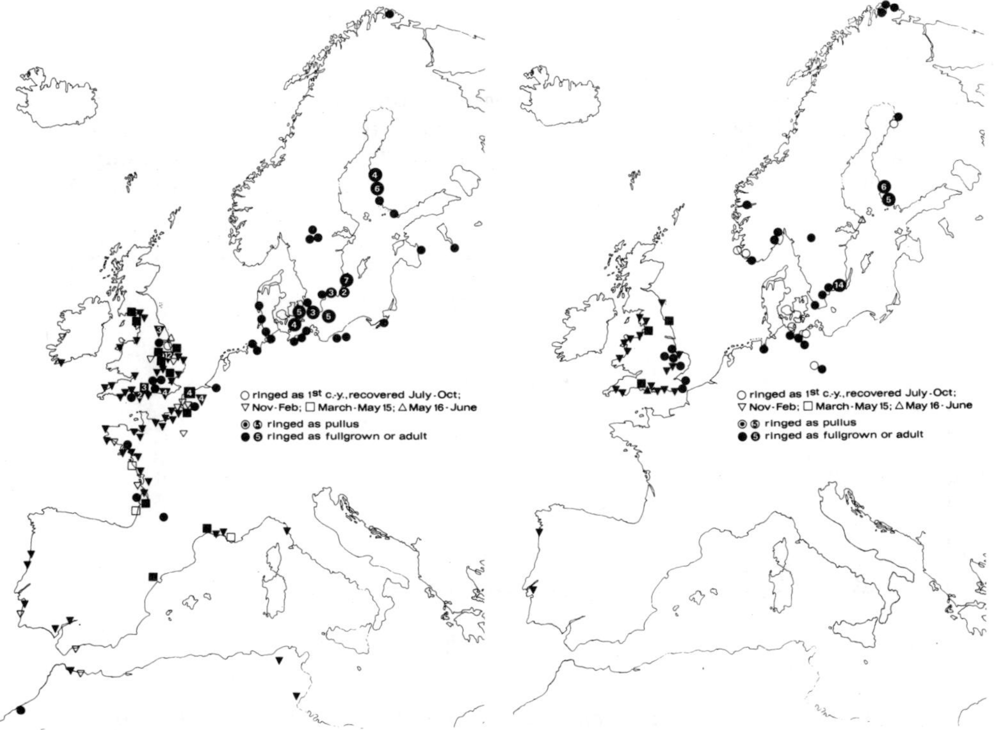

Fig. 107. Recoveries abroad of Dunlins ringed in the Dutch part of the Wadden Sea (records until 1979). The figures represent the number of recoveries in the specific area (Boere, unpubl.).

Fig. 108. Ringing places of Dunlins recovered in the Dutch part of the Wadden Sea (records from 1976-1979). The figures represent the number of recoveries in the specific area (Boere, unpubl.).

Some birds start their moult in the breeding area, but the majority migrates to special moulting areas after the breeding season. The Wadden Sea is one of the most important of these (Boere, 1976; Drenckhahn et al., 1971; Thelle & Netterstrøm, 1971) together with estuaries in Great-Britain. *C.a. schinzii* from the small Baltic population probably moults in the Wadden Sea, those breeding in Greenland and Iceland moult in Northern Africa (Pienkowski & Dick, 1975; Pienkowski et al., 1976).

3.21.2 Annual cycle

3.21.2.1 Migration

The first Dunlins arrive in the Wadden Sea by early July. These birds probably belong to *C.a. schinzii* (Boere, unpubl.) from the Baltic populations. Numbers increase rapidly and peak numbers occur during August and September. In the course of August flocks of first calendar-year birds arrive. At the same time adult birds, which have finished their moult, leave the Wadden Sea (fig. 109). From October onwards numbers in the Wadden Sea decrease to a midwinter minimum. Numbers present in winter vary from year to year due to the local weather conditions. From August on numbers on the British Isles increase considerably. From November till February a stable number of about 500,000 to 600,000 Dunlins is wintering. In March and the months following numbers drop fast (Prater, 1976a). Spring migration in the Wad-

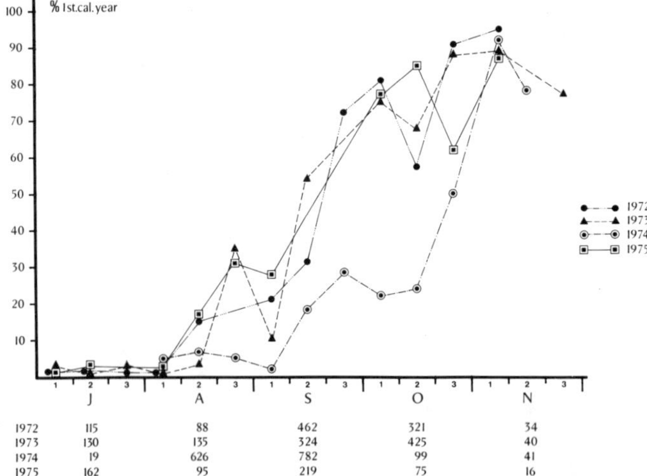

1972	115	88	462	321	34
1973	130	135	324	425	40
1974	19	626	782	99	41
1975	162	95	219	75	16

Fig. 109. Percentage of first calendar-year birds per decade in catches of Dunlins on the island of Vlieland. The figure illustrates the gradual increase of first calendar-year birds in the area from the end of July onwards and the departure of adult birds after the moulting period. Note the different schedule for 1974, which was a non-breeding year for arctic birds. A high percentage of first calendar-year birds was present only after departure of the adult birds. Absolute numbers of caught Dunlins for each month are listed below (Boere, unpublished).

den Sea becomes apparent by late February or early March with increasing numbers. Peak numbers occur from late March until early May, when numbers decrease rapidly. Small flocks stay to summer.

3.21.2.2 Moult

A complete post-nuptial moult of the whole *Calidris alpina* group takes place from early July till the end of September (Nieboer, 1972; Boere, 1976; Boere et al., 1973). There is some differentiation depending on the subspecies and geographical origin.

The majority of *C.a. sakhalina* in Eastern Siberia and Alaska has a complete moult in the breeding area starting already when still incubating (Holmes, 1966, 1971), the subspecies *C.a. alpina* starts moulting after the breeding season and after a short migration to the coastal areas around the North Sea. The Wadden Sea is the most important moulting area for these birds. Small numbers of *C.a. alpina* start to moult already in the breeding area (Lilja, 1969; Leslie & Lessels, 1978).

Between the end of July and the end of August nearly all adult Dunlins are in heavy moult of wing-, tail- and body feathers. The last birds in moult have been observed in early October. The total moult period for an individual bird takes 80-90 days as a maximum. Second calendar-year birds are supposed to have nearly the same moult schedule as adults. This is probably due to the fact that most Dunlins breed in their second year (at least go to the breeding area and mate).

Body moult of first calendar-year birds is spread over a large period from mid-August till the end of November. The moult schedule of the Icelandic and Greenland populations of *C.a. schinzii*, moulting in Northern Africa, is about two weeks behind that of *C.a. alpina* in the Wadden Sea. It is finished by early November. *C.a. schinzii*, breeding around the Baltic starts already by the end of June and is therefore completed about two weeks before the schedule of *C.a. alpina* (Glutz et al., 1975). Moult into the breeding plumage takes place from the beginning of March onwards. At the end of May many birds are still in heavy moult and probably finish it in the breeding area (Ferns & Green, 1979).

3.21.2.3 Weight changes

Weight studies are known from several localities in Europe and Northern Africa (Mascher, 1966; Mascher & Marcström, 1976; Eades & Okill, 1977; Dick & Pienkowski, 1979).

Information on weights from the Dutch part of the Wadden Sea is summarized in Table 37. It shows a weight increase towards the winter. During winter fat reserves are being used resulting in a low weight by the end of February and early March. Afterwards a steady increase occurs to heavy weights just before migration towards the breeding area. This weight schedule is only slightly different from that found by Pienkowski et al., (1979) in England and Branson (1979) in the Wash.

Tables 35 and 36 summarize bill- and wing lengths of first calendar-year and adult Dunlins caught in the Dutch Wadden Sea on the island of Vlieland.

Table 35. Mean exposed culmen- (in mm) and wing-lengths (mm) of first calendar-year Dunlins per decade, caught on the island Vlieland from 1972 up to and including 1975 (Boere, unpubl.).

month	decade	exposed culmen-length	wing-length	n
July	3	30.5	116.4	6
August	1	31.2	117.3	12
August	2	30.2	117.7	46
August	3	31.9	119.0	52
September	1	32.3	120.1	65
September	2	33.0	120.6	174
September	3	33.0	120.6	105
October	1	33.2	120.6	376
October	2	33.0	120.6	109
October	3	32.5	120.3	176
November	1	32.9	120.0	108
November	2	33.2	120.7	18
November	3	32.8	120.8	8

Table 36. Mean exposed culmen- (in mm) and wing-lengths (mm) of adult Dunlins per decade, caught on the island Vlieland from 1972 up to and including 1975. o.p. = old primary; n.p. = new primary (Boere, unpubl.).

month	decade	exposed culmen-length	wing-length		n
July	1	32.5	117.7	(o.p)	86
July	2	32.4	118.2	"	64
July	3	32.2	119.1	"	193
August	1	33.0	119.5	"	164
August	2	32.7	118.8		159
August	3	33.0	118.8		191
September	1	33.9	119.0	"	68
September	2	32.6	119.9	"	151
September	3	31.9	120.6	"	22
October	1	32.7	120.5	(n.p)	79
October	2	32.0	120.3	"	50
October	3	32.2	120.7	"	47

Table 37. Mean weights per decade, sexes and subspecies combined, of adult and first calendar-year Dunlins, caught on the island of Vlieland, December 1971-December 1975 (Boere, unpubl.).

month	decade	adult birds			first calendar-year birds		
		mean	s.d.	n	mean	s.d.	n
January	1	57.6	5.6	42			
	3	53.0		1			
February	3	44.5	2.1	2			
March	1	50.8	2.6	16			
	2	49.5	4.4	49			
	3	51.8	3.9	26			
April	1	51.1	4.3	9			
	2	52.2	4.4	26			
	3	51.3	3.7	7			
May	1	53.3	5.1	8			
	2	54.8	6.3	70			
	3	62.2	7.2	70			
June	1	33		1			
	3	53.7	2.3	3			
July	1	49.4	4.7	88			
	2	50.7	4.4	64	56.0		1
	3	52.4	4.4	199	46.6	7.2	5
August	1	52.8	4.4	166	42.7	6.1	10
	2	51.6	4.2	169	43.7	4.4	43
	3	52.6	3.7	211	46.2	6.1	40
September	1	53.0	4.5	184	53.3	7.1	63
	2	52.0	4.7	520	49.4	5.7	172
	3	54.4	4.8	67	50.6	5.1	105
October	1	53.8	4.9	142	51.0	4.6	374
	2	54.4	5.9	62	50.6	4.7	109
	3	55.7	6.3	54	53.3	4.5	177
November	1	61.8	5.3	11	56.5	6.1	109
	2	57.5	4.7	12	57.4	5.6	18
	3	49.0	1.4	2	54.4	4.2	8
December	1	64.0	4.4	3	58.7	7.5	3
	2				58.5	0.7	2
	3	64.0		1	61.0		1

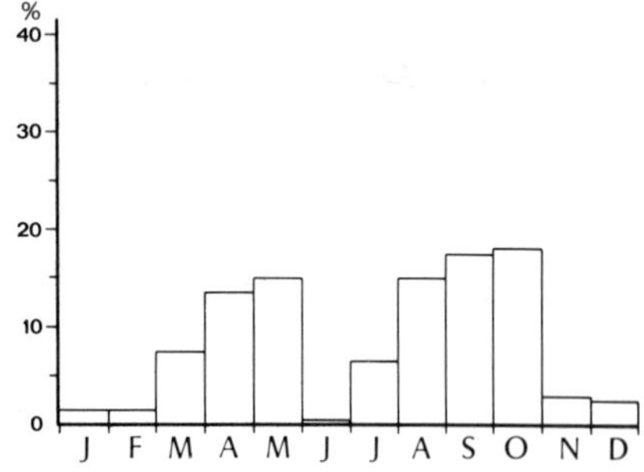

Fig. 110. Occurrence of Dunlins in the Wadden Sea area in Schleswig-Holstein. Numbers for each month are expressed as a percentage of the total number observed in all months (after Busche, 1980).

3.21.3 Numbers

3.21.3.1 Population size

Prater (1976b) estimates the size of the population wintering along the coasts in W Europe and NW Africa at about 1,410,000. Most important wintering areas are the coasts of Britain (550,000) and France (300,000) and the Banc d'Arguin (180,000). Probably the Banc d'Arguin is much more important as Piersma et al. (1980) counted 818,000 Dunlins in this area recently.

In the 19th century and partly still in the 20th Dunlins were breeding on many Wadden Sea islands. Observations of this kind were made at Texel, Terschelling, Borkum, Langeoog, Wangerooge, Süderoog, Pellworm, Nordstrand and the Grüne Insel in the Eider estuary. Incidentally they bred on Juist, Baltrum, Neuwerk, Nordstrandischmoor and Hooge. Dunlins nowadays occur more or less regularly as a breeding bird on Ameland (1968, 1970), Spiekeroog, Nordmarsch-Langeness, Amrum, Sylt, Rømø, Mandø, Langli and Fanø. The total breeding population amounts at present to about 25-40 pairs.

3.21.3.2 Numbers per area

Like in other parts of the area, Dunlins belong to the most abundant wader species in the Danish part of the Wadden Sea. Numbers increase in July, when 100,000 may be present in the area already, and August (with maximum numbers of about 240,000). Dunlins are most numerous how-

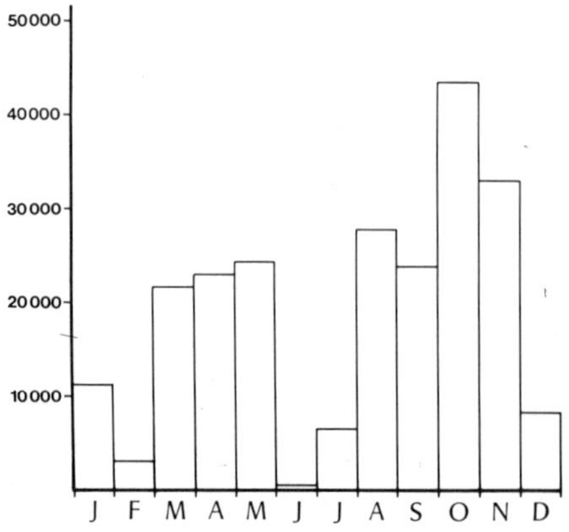

Fig. 111. Mean number of Dunlins per month in a 50 km long stretch along the mainland coast in Niedersachsen (after data from Smit, 1977 and the Institut für Vogelforschung, Wilhelmshaven).

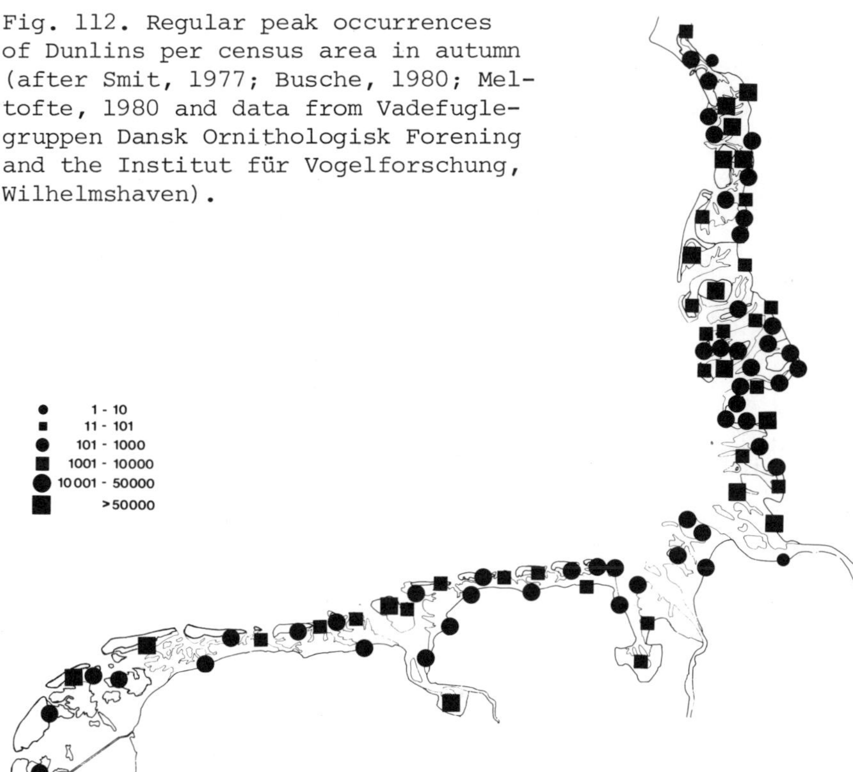

Fig. 112. Regular peak occurrences of Dunlins per census area in autumn (after Smit, 1977; Busche, 1980; Meltofte, 1980 and data from Vadefuglegruppen Dansk Ornithologisk Forening and the Institut für Vogelforschung, Wilhelmshaven).

ever in September with a peak number of 358,000. A gradual decrease follows in the months afterwards. From December till February numbers may drop to several hundreds, in mild winters however some tens of thousands may still be present. In March, April and May 100,000-200,000 may be present again (Meltofte, 1980; Dansk Ornithologisk Forening).

In the Wadden Sea area in Schleswig-Holstein numbers increase from July on (fig. 110). Peak numbers occur in the months following. Probably in October about 300,000 are present. In mild winters about 20,000 remain in the area, in cold winters numbers drop to a few thousands (Busche, 1980). Peak numbers in spring are of the same magnitude as those in autumn. In May up to 250,000 have been counted (Drenckhahn et al., 1971). Dunlins summer in relatively large numbers in this part of the Wadden Sea. Heldt (1968) and Busche (1980) both estimate numbers summering in June at about 5000.

Migration along the mainland coast in Niedersachsen seems to occur in two waves (fig. 111). Numbers from the Wadden Sea islands however give no indication that this phenomenon occurs all over the area in Niedersachsen. Information on total numbers in the area is far from complete but peak numbers in autumn will exceed 150,000. In the area around Wangerooge, Oldeoog, Mellum, Scharhörn, Knechtsand and Neuwerk relatively high numbers occur in June. During this month a mean number of 10,000 has been registered in this area (after data from Institut für Vogelforschung, Wilhelmshaven). Fig. 112 shows that Dunlins may

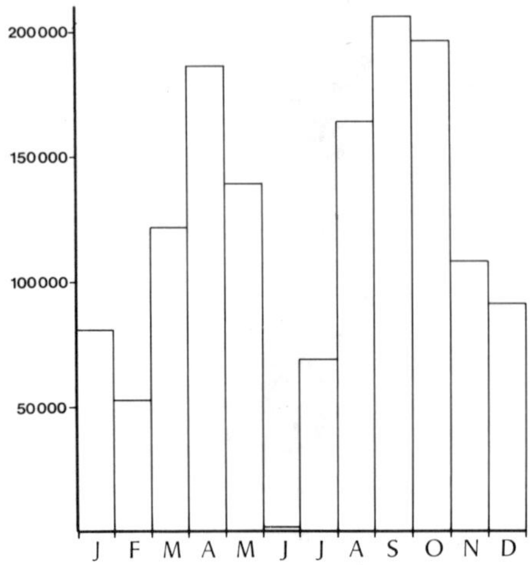

Fig. 113. Mean number of Dunlins per month in the Dutch part of the Wadden Sea (after data from Smit, 1977).

be very numerous on some of the islands in Niedersachsen in autumn.

Also in the Dutch part of the Wadden Sea Dunlins occur in large numbers reaching maxima in August and September though also in April high numbers have been registered (fig. 113). Numbers actually counted in the Dutch part are listed in Table 38. On the islands numbers in spring are considerably lower than in autumn. Along the mainland coast however they are of the same magnitude or even higher. This difference is demonstrated in fig. 114.

On September 1, 1973 in the Wadden Sea in Denmark 45,000 Dunlins have been counted, in Schleswig-Holstein and Niedersachsen 409,000 and in the Netherlands 274,000, which brings the Wadden Sea total at 728,000. The population of the species wintering along the shores of Europe and NW Africa is estimated to amount to 1,410,000. This means that at the day of the count more than 50% of the population was present in the Wadden Sea. Probably this was even much more because especially in the Danish part of the Wadden Sea the count was very incomplete. The weather was bad, there was a high tide and there was much disturbance by hunters. By this time usually more than 200,000 are present in that part of the area (Meltofte, in litt.).

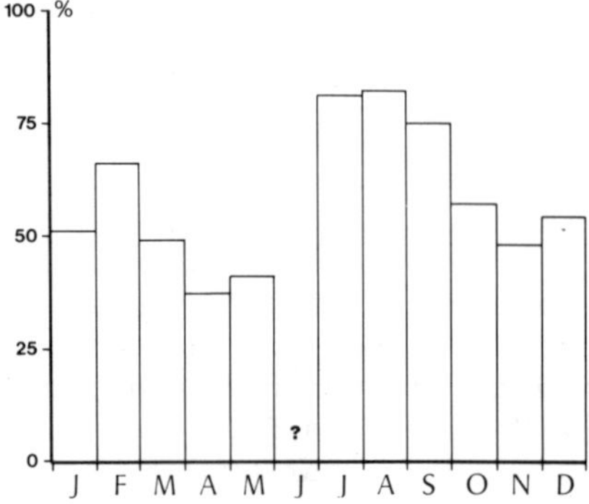

Fig. 114. Numbers of Dunlins present on the Dutch Wadden Sea islands, expressed as a percentage of the number present in the whole area in that particular month (after data from Smit, 1977). The percentage for June is obscure due to the very small number of Dunlins in the area in that month.

Table 38. Number of Dunlins in the Dutch part of the Wadden Sea (after various publications in Watervogels and Limosa and Zegers, in litt.).

Date	Year	Number
January 8	1977	71,410
January 12	1974	106,140
January 17	1976	126,550
January 18	1975	109,470
April 6	1973	197,700
April 19	1975	197,170
May 1	1976	157,750
May 11	1974	151,950
July 29	1972	109,400
August 30	1975	268,700
September 1	1973	274,130
October 19	1974	187,600
November 13	1976	141,630
December 29	1966	78,000

3.21.4 Food

3.21.4.1 Food composition

In the breeding area and during migration insects and insect larvae are important food sources. In the Wadden Sea a great variety of prey items are eaten. Most common are *Nereis*, *Corophium*, *Hydrobia* and small specimens of other crustaceans and molluscs (Ehlert, 1964; Höfmann & Hoerschelmann, 1969; Glutz et al., 1975). Near the island of Ameland Kersten & Piersma (in litt.) saw Dunlins feed on *Nereis*, siphons of small bivalves (probably *Macoma*), *Carcinus* and several small prey items. At inland sites chironomid larvae are taken as main food source.

3.21.4.2 Feeding activities

A great variety of substrates is chosen as feeding areas. Dunlins however mainly feed on muddy sandflats covering large areas. Peck- and success rate of foraging Dunlins near Ameland amounted to 24.0 pecks and 1.6 preys per minute (Kersten & Piersma, in litt.). On high tide roosts Dunlins may gather in very large flocks of over 30,000 birds. These flocks are mainly situated in open sandy areas close to the water. Short grassed salt marshes and meadows are also favoured as roost sites.

3.21.4.3 Total food intake

No information available.

References

Boere, G.C., 1976. The significance of the Dutch Waddenzee in the annual life cycle of arctic, subarctic and boreal waders. Part. I. The function as a moulting area. Ardea 64: p. 210-291.

Boere, G.C., J.W.A. de Bruyne & E. Nieboer, 1973. Onderzoek naar de betekenis van het Nederlandse Waddenzeegebied voor Bonte Strandlopers in nazomer en herfst. Limosa 46: p. 205-227.

Branson, N.J.B.A. (ed.), 1979. Wash Wader Ringing Group Report 1977-1978. Cambridge, 68 pp.

Busche, G., 1980. Vogelbestände des Wattenmeeres von Schleswig-Holstein. Kilda, Greven (in press).

Dick, W.J.A. & M.K. Pienkowski, 1979. Autumn and early winter weights of waders in North-West Africa. Ornis Scand. 10: p. 117-123.

Drenckhahn, D., R. Heldt jun. & R. Heldt sen., 1971. Die Bedeutung der Nordseeküste Schleswig-Holsteins für einige Wat- und Wasservögel mit besondere Berücksichtigung des Nordfriesischen Wattenmeeres. Natur und Landschaft 46: p. 338-346.

Eades, R.A., 1974. Monthly variation in foreignringed Dunlins in the Dee-estuary. Bird Study 21: p. 155-157.

Eades, R.A. & J.D. Okill, 1977. Weight changes of Dunlins on the Dee-estuary in May. Bird Study 24: p. 62-63.

Ehlert, W., 1964. Zur Ökologie und Biologie der Ernährung einiger Limikolen-Arten. J. Orn. 105: p. 1-53.

Ferns, P.N. & G.H. Green, 1979. Observations on the breeding plumage and prenuptial moult of Dunlins Calidris alpina, captured in Britain. Gerfaut 69: p. 287-303.

Fuchs, E., 1973. Durchzug und Überwinterung des Alpenstrandläufers in der Camargue. Orn. Beob. 70: p. 113-134.

Glutz von Blotzheim, U.N., K.M. Bauer & E. Bezzel, 1975. Handbuch der Vögel Mitteleuropas, Vol. 6. Akademische Verlagsgesellschaft, Wiesbaden: 840 pp.

Griffiths, J., 1970. The bill-length of Dunlins. Bird Study 17: p. 42-44.

Heldt, R., 1968. Übersommernde Limikolen an der Westküste von Schleswig-Holstein. Corax 2: p. 108-130.

Höfmann, H. & H. Hoerschelmann, 1969. Nahrungsuntersuchungen bei Limikolen durch Mageninhaltanalysen. Corax 3: p. 7-22.

Holmes, R.T., 1966. Molt cycle of the Red-backed Sandpiper, Calidris alpina, in Northern Alaska. Condor 68: p. 3-46.

Holmes, R.T., 1971. Latitudinal differences in the breeding and molt schedules of Alaskan Red-backed Sandpipers, Calidris alpina. Condor 73: p. 93-99.

Jansen, F.H., 1979. Enige aspecten van de trek van de Bonte Strandloper Calidris alpina voor Scheveningen. Limosa 52: p. 34-52.

Leslie, R. & C.M. Lessels, 1978. The migration of Dunlin Calidris alpina through northern Scandinavia. Ornis Scand. 9: p. 84-86.

Lilja, S., 1969. Muuttavien Suosirrien (Calidris alpina) post nuptiaalisesta siipisulkasadostra. Ann. Report Orn. Soc. Pori 1968: p. 84-86.

Martin-Löf, P., 1958. Storleksskillnader hos genomsträckande Kärrsnäppor (Calidris alpina (L.)) vid Ottenby. Vår Fågelvärld 17: p. 287-301.

Mascher, J.W., 1966. Weight variations in resting Dunlins (Calidris a. alpina) on autumn migration in Sweden. Bird Banding 37: p. 1-34.

Mascher, J.W., 1971. A plumage painting study of autumn Dunlin, Calidris alpina, migration in the Baltic and North Sea Areas. Ornis Scand. 2: p. 27-33.

Mascher, J.W. & V. Marcström, 1976. Measures, weights and lipid levels in migrating Dunlins, Calidris a. alpina L., at the Ottenby Bird Observatory, South Sweden. Ornis Scand. 7: p. 49-59.

McLean, S.F., jr. & R.T. Holmes, 1971. Bill length, wintering areas and taxonomy of North-American Dunlins, Calidris alpina. Auk 88: p. 893-901.

Meltofte, H., 1980. Fugle i Vadehavet. Vadefugletaellinger i Vadehavet 1974-1978. Miljøministeriet, Fredningsstyrelsen, Kobenhavn: 50 pp.

Nieboer, E., 1972. Preliminary notes on the primary moult in Dunlins Calidris alpina. Ardea 60: p. 112-119.

Nørrevang, A., 1955. Rylens (Calidris alpina (L.)) traek over Nord-Europa. Dansk Orn. Foren. Tidsskr. 49: p. 18-49.

O.A.G. Münster, 1976. Zur Biometrie des Alpenstrandläufers Calidris alpina in den Rieselfeldern Münster. Vogelwarte 28: p. 278-293.

O.A.G. Münster, 1980. Wader ringing on the sewage farm of Münster, Federal Republic of Germany. Wader Study Group Bull. 28: p. 17-21.

Ogilvie, M.A., 1963. The migration of European Redshank and Dunlin. Wildfowl 14: p. 141-149.

Pienkowski, M.W. & W.J.A. Dick, 1975. The migration and wintering of Dunlin Calidris alpina in North West Africa. Ornis Scand. 6: p. 151-167.

Pienkowski, M.W., C.S. Lloyd & C.D.T. Minton, 1979. Seasonal and migrational weight changes in Dunlins. Bird Study 26: p. 134-148.

Piersma, T., M. Engelmoer, W. Altenburg & R. Mes, 1980. A wader-expedition to Mauritania. Wader Study Group Bull. 29: p. 14.

Prater, A.J., 1976a. Birds of estuaries enquiry 1973-74. British Trust for Ornithology, Royal Soc. Prot. Birds, Wildfowl. Trust: 48 pp.

Prater, A.J., 1976b. The distribution of coastal waders in Europe and North Africa. In: M. Smart (ed.). Proceedings International Conference on Conservation of Wetlands and Waterfowl, Heiligenhafen, 1974. IWRB, Slimbridge: p. 255-271.

Smit, C.J., 1977. On the occurrence of 32 bird species in the Danish, German and Dutch Wadden Sea. Unpubl. report Intern. Wadden Sea Working Group, part. 3: 174 pp.

Soikkeli, M., 1966. On the variation in bill- and wing length of the Dunlin in Europe. Bird Study 13: p. 256-269.

Soikkeli, M., 1974. Size variation of breeding Dunlins in Finland. Bird Study 21: p. 151-154.

Steventon, D.J., 1977. Dunlin in Porthsmouth, Langstone and Chichester harbours. Ringing and migration 1: p. 141-147.

Thelle, T. & B.O. Netterstrøm, 1971. Vadefugle optaellinger i Vadehavet i Juli og August 1969. Dansk Orn. Foren. Tidsskr. 65: p. 164-172.

Vieillard, J., 1972. Définition du Bécasseau variable. Alauda 40: p. 321-342.

Voous, K.H., 1960. Atlas of European birds. Nelson, London: 284 pp.

3.22 BAR-TAILED GODWIT (*LIMOSA LAPPONICA* L.)
G.C. Boere & C.J. Smit

Da: Lille Kobbersneppe; G: Pfuhlschnepfe; Du: Rosse Grutto

3.22.1 Distribution

3.22.1.1 Breeding area

The Bar-tailed godwit has a trans-palaearctic and Northwest nearctic distribution, breeding in the arctic, subarctic and boreal climate zones of Eurasia and Alaska. Within the Eurasian breeding range there is a clinal increase of body length from west to east. Based on these measurements generally two or three subspecies are recognized: *Limosa l. lapponica*, (*L. l. menzbieri*) and *L. l. baueri* (fig. 115) (Voous, 1960; Kozlova, 1962; Glutz et al., 1977). According to the ringing results of several European countries, including birds ringed in the Wadden Sea (fig. 116a), Bar-tailed godwits breeding on Taymyr peninsula still migrate to Europe and Africa. There is no good evidence that also more eastern populations migrate to Europe although very large birds have been caught in the Dutch Wadden Sea area.

3.22.1.2 Migration routes

The species is a long-distance migrant. Probably the main migration routes are along the coasts of Western Europe and Africa. After the breeding season flocks of adult birds concentrate along the Arctic coast and migrate through the White- and Baltic Sea area to Western

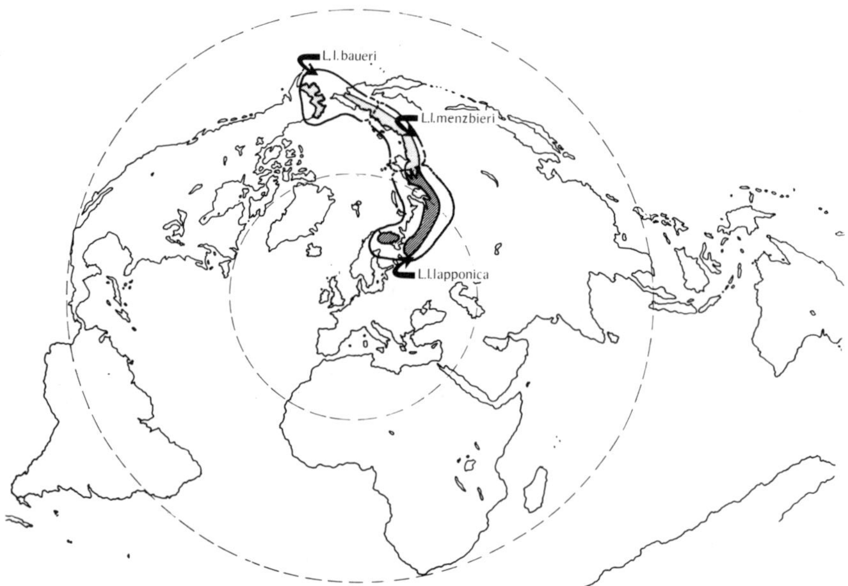

Fig. 115. Breeding area of the Bar-tailed godwit (after Voous 1960). The dark shaded area represents approximately the region where the birds visiting the Wadden Sea originate from.

Europe. Small numbers are observed inland, mainly first calendar-year birds. Few birds migrate through Central Russia to the Caspian lowlands (Glutz et al., 1977). The Eastern breeding population migrates to the Malaysian and Indonesian archipelago, e.g. Australia, the islands in the Pacific Ocean and New-Zealand (Edgar et al., 1970).

3.22.1.3 Wintering areas

Although widely spread over a large area from NW Europe till South-Africa there are two areas where about 80-90% of the total population is wintering: the shallows and estuaries along the Southern part of the North Sea (with about 55,000 wintering birds) (Prater, 1976a, 1976b; Smit, 1977) and Banc d'Arguin in Mauritania, where 200,000 to 543,000 birds have been counted (Dick; 1975; Piersma et al., 1980) (figs. 116ab). South of Mauritania it is a rare species. Some smaller flocks of first calendar-year and immature birds occur in the coastal lagoons of South-Africa. The Eastern palaeartic (including Alaska) breeding population has its main wintering areas in New-Zealand (De Long & Thompsom, 1968; Edgar et al., 1970).

3.22.1.4 Moulting areas

Generally moult of flight feathers takes place in the wintering areas, i.c. the North Sea area and Banc d'Arguin (Mauritania) (with interchange between wintering and moulting areas in the North Sea region, according to ringing results). Banc d'Arguin must be considered

Fig. 116a. Recoveries of Bar-tailed godwits ringed in the Dutch part of the Wadden Sea and in Denmark (based on published records from 1911-1976).

Fig. 116b. Ringing places of Bar-tailed godwits recovered in the Dutch part of the Wadden Sea (based on published records from 1911-1976).

as the most important moulting area. Further south only small numbers of Bar-tailed godwits moult. These are mainly groups of first calendar-year birds which also spend the Northern summer in tropical Africa. The Wadden Sea is the most important moulting area in spring when adult birds moult body feathers. During this partial moult from mid-April till the end of May, the birds come into their breeding plumage.

3.22.2 Annual cycle

3.22.2.1 Migration

Birds are leaving their Eurasian breeding areas from early July (non-breeders) onwards till early August. The number of adult birds passing Blaavandshuk has a maximum in the last July decade. Most juveniles pass in the second September pentade (Meltofte et al., 1972). In the Wadden Sea peak numbers occur about mid-August. Very large numbers must have passed the area rapidly to NW Africa because several hundreds of thousands have been counted there, already by late September, especially at Banc d'Arguin (Dick, 1975). Many birds leave the Wadden Sea after the moulting period to winter elsewhere. Exact data on spring migration in Southern Europe and Africa are scarce. Wintering birds leave these areas from March onwards, corresponding with an increase of the numbers in the Wadden Sea. Peak numbers occur throughout May when numbers are twice as high as in autumn. By the end of May numbers decrease rapidly, sometimes during one night. Numbers of summering immature birds vary from year to year. Generally several hundreds or thousands are present in the whole Wadden Sea area.

3.22.2.2 Moult

Adults carry out a complete post-nuptial moult, starting late July or early August. It is finished during the first half of November. Birds moulting in Africa are about 10-15 days behind on this schedule. The pre-nuptial moult into the breeding plumage, involving the body feathers, starts already early February and is finished by late May just before the birds leave for the breeding places. A small percentage of Bar-tailed godwits moult some of the tail feathers in spring.

First calendar-year birds undergo a partial moult involving the body feathers from late September till the end of November. However little information is available about this group. Summering immatures start a complete moult in late June. Their moulting schedule is about three weeks in advance of that of the adult Bar-tailed godwits (Boere, 1976).

3.22.2.3 Weight changes

Table 39 summarizes the mean monthly weights of male and female Bar-tailed godwits. Separation between the sexes has been made according to bill lengths (Glutz et al., 1977): 89 mm or less: males and 92 mm or more: females.

Noteworthy is the considerable weight increase of both sexes during May. This suggests a long non-stop flight after the departure from the Wadden Sea and the disposal of fat reserves for bad conditions at the beginning of the breeding season. Fig. 117 shows a significant sexual dimorphism of bill- and wing length.

3.22.3 Numbers

3.22.3.1 Population size

Prater (1976b) estimates the size of the population wintering in NW Europe and W Africa at about 300,000 birds.

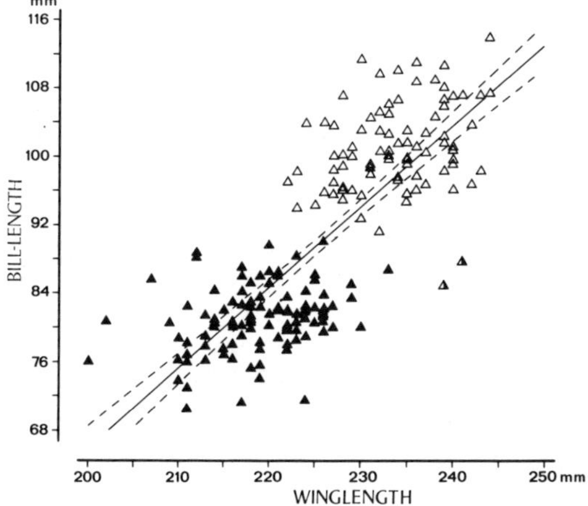

Fig. 117. Diagram showing the individual variation in biometrics of Bar-tailed godwits and the considerable sexual dimorphism (regression line on 95 % confidence interval). △ denote females, ▲ males. Data of birds caught from 1972-1975 at the island of Vlieland (Boere, unpublished).

Table 39. Mean monthly weights of male and female Bar-tailed godwits, caught on Vlieland, December 1971-December 1975. Possible subspecies have been combined (Boere, unpubl.).

		adult birds			first calendar-year birds		
		mean	s.d.	n	mean	s.d.	n
January	male	323	28	17			
	female	404	44	13			
February	male	286	17	10			
	female	350	15	4			
March	male	286	26	112			
	female	355	31	85			
April	male	255	47	8			
	female	326	60	8			
May	male	338	49	18			
	female	420	61	9			
July	male	287	23	5			
	female	301	47	2			
August	male	300	48	27			
	female	356	37	13			
September	male	285	28	29	249	31	6
	female	351	31	17	389	10	2
October	male	286	27	15	303	28	6
	female	329	28	29	339		1
November	male	309	25	75			
	female	366	30	32			
December	female	360	26	3			

3.22.3.2 Numbers per area

Maximum numbers in the Danish part of the Wadden Sea occur in April and May with maximum numbers of 56,300 and 43,400 respectively. In autumn numbers are 10,000 to 25,000, in winter only a few still are present in the area. Fig. 118 shows that the largest numbers occur on the tidal flats near Mandø (with up to 38,000 birds) and Jordsand (23,000) and along the coast between the Danish-German border and the Rømø-dam. In the latter two areas a maximum number of 37,000 has been counted (Meltofte, 1980; reports Vadefuglegruppen Dansk Ornithologisk Forening).

Fig. 118. Regular peak occurrences of
Bar-tailed godwits in autumn and spring
per census area (based on Smit, 1977;
Busche, 1980; Meltofte, 1980; reports
from Vadefuglegruppen Dansk Ornithologisk Forening and unpublished data from
the Institut für Vogelforschung, Wilhelmshaven).

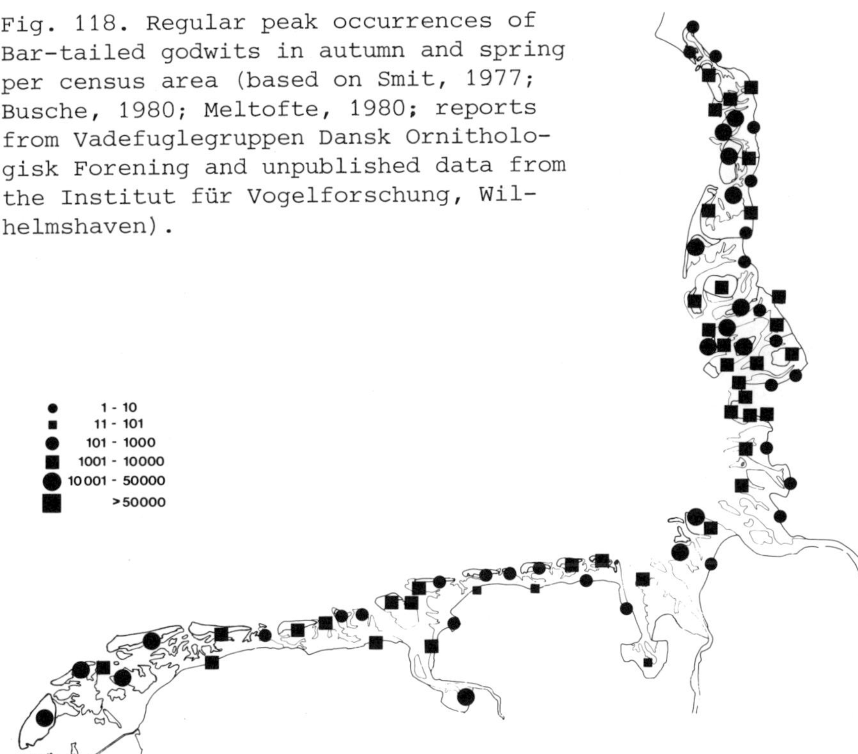

In Schleswig-Holstein numbers increase from early July on but especially in the first part of August. By this time in some years up
to 60,000 Bar-tailed godwits may occur in the area, generally however
numbers will be lower. In October and November numbers decrease and
in winter relatively few occur. Only some thousands are left, mainly
gathered in the Sylt-Amrum-Föhr-Pellworm-area. Fluctuations in numbers are demonstrated in fig. 119, showing that numbers in April and
May increase consiberably. In May probably up to 60,000 are present
in the area, the number of summering Bar-tailed godwits amounts to
2000-4000. In summer the species is most numerous around the Rantum-
Becken (Sylt) (maximum number 600), Norderoog (500), Hooge (460),
Hauke Haienkoog (900) and Südfall (500) (Busche, 1980).
    In Niedersachsen peak numbers occur in August, but already in July
large numbers can be observed. Generally numbers decrease considerably
in September, though in September 1971 up to 30,000 were seen at Grosser Knechtsand. Mean numbers per month in a 50 km long stretch along
the coast between the Dutch-German border and the Jadebusen were:
April 1840; May 5600; June 20; July 810; August 1250; September 340;
October 55 and November 20. From December-March Bar-tailed godwits
have not been observed. Generally somewhat larger numbers occur on the
islands. Several thousands can be observed in the German part of the
Dollart and in the area between Knock and Rysum along the Ems estuary
(after data from several working groups; source Institut für Vogel-

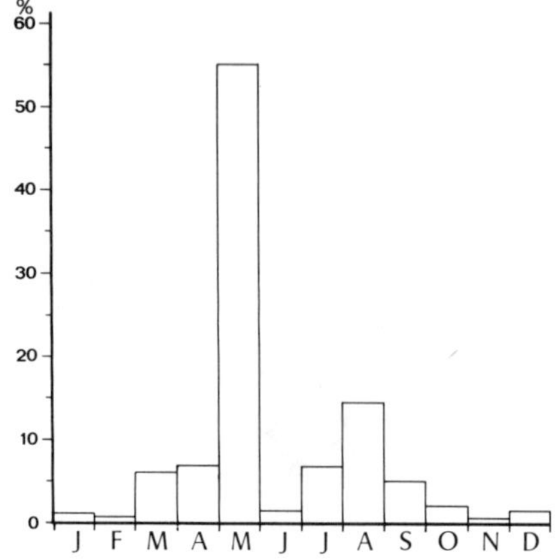

Fig. 119. Occurrence of Bar-
tailed godwits in the Wadden
Sea area in Schleswig-Holstein.
Numbers for each month are expressed as a percentage of the
total number observed in all
months (after Busche, 1980).

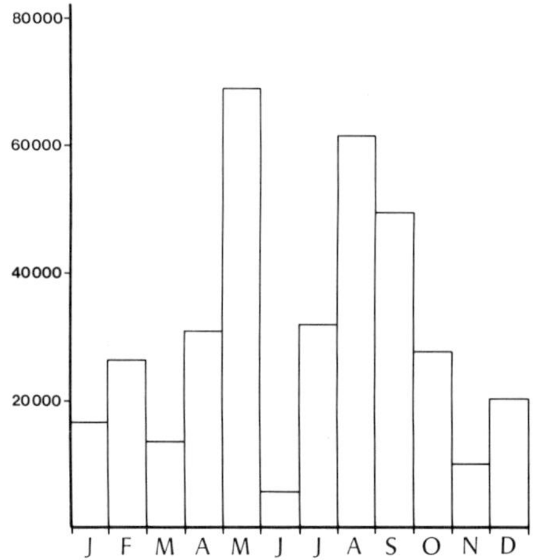

Fig. 120. Mean number of Bar-
tailed godwits per month in the
Dutch part of the Wadden Sea
(after data from Smit, 1977).

forschung, Wilhelmshaven).

In the Western part of the Dutch Wadden Sea area generally numbers
are much higher as compared to the Eastern part, especially during winter. In the area around the islands of Texel, Vlieland, Terschelling
and Griend and in the Balgzand area in winter more than 20,000 Bar-
tailed godwits are present. Fig. 120 shows mean numbers per month for
the Dutch part of the Wadden Sea. Numbers actually counted in the area
are listed in Table 40. The Western part of the Dutch Wadden Sea is
the only area in the Wadden Sea where Bar-tailed godwits winter in
rather large numbers. In the Eastern part peak numbers are reached
from July to August, in the Western part from August to October.
From December till February only some thousands are present in the
Eastern part. In the Netherlands as well as in Germany numbers in
May exceed those of the autumn months.

Table 40. Numbers of Bar-tailed godwits in the Dutch part of the Wadden Sea (after various publications in Watervogels and Limosa and Zegers, in litt.).

Date	Year	Number
January 8	1977	15,280
January 12	1974	18,300
January 17	1976	31,160
January 18	1975	16,470
April 6	1973	28,250
April 19	1975	30,100
May 1	1976	61,300
May 11	1974	78,130
July 29	1972	42,400
August 22	1963	23,000
August 30	1975	38,250
September 1	1973	51,700
October 19	1974	18,280
November 13	1976	13,320
December 29	1966	21,000

3.22.4 Food

3.22.4.1 Food composition

Few material is available from the Wadden Sea area. Main prey items are large polychaetes. Besides small crustaceans, like *Corophium*, small molluscs are eaten (Pauw, 1970; Höfmann & Hoerschelmann, 1969). Kersten & Piersma (in litt.) saw Bar-tailed godwits near Ameland feeding on *Nereis*, *Macoma*, *Carcinus* and small unidentified prey species. Piersma et al. (in litt.) saw Bar-tailed godwits along the mainland coast of Friesland taking *Nereis*, *Arenicola* and small prey items (partly consisting of *Heteromastus*) and incidentally *Crangon*, *Carcinus*, *Ammodytes* and bivalves (probably *Macoma*). *Nereis* dominated strongly. It was striking that female Bar-tailed godwits took especially large specimens. More detailed studies are available for British estuaries (Smith & Evans, 1973; Evans & Smith, 1975) showing differences in prey size taken by males or females, with *Arenicola* as most important food item.

3.22.4.2 Feeding activities

Feeding areas are large, more or less sandy tidal flats with a small percentage of clay, but also musselbeds. Bar-tailed godwits frequently feed just in or in front of the water edge when the tide comes up.

In August 1980 Piersma et al. (in litt.) found that on the muddy sandflats near Moddergat (along the mainland coast of Friesland) on the average 85% of the males was foraging and 77% of the females. Kersten & Piersma (in litt.) found in September 1978 and May 1979 on the island of Ameland probe rates of foraging Bar-tailed godwits of 17.2 and 20.3 pecks per minute, success rates of 2.8 and 2.1 per minute, which yields success percentages of 16% and 10% respectively.

Like Oystercatchers and Sanderlings, Bar-tailed godwits may be observed feeding on the beaches along the North Sea coast of the islands, especially during peak migration. During high tide they concentrate in large flocks of several thousands of birds. They prefer large open sandy areas close to or even in the water. They rather often join Curlews, during the moulting period flocks of Bar-tailed godwits may even join Curlew roosts in the dunes. Occasionally they roost on salt marshes and meadows sometimes even at 2 km from the coast.

3.22.4.3 Total food consumption

Near the island of Ameland Kersten & Piersma (in litt.) in September found a food intake of 9.2 g ash-free dry weight, in May of 16 g per emersion period. Sight records of birds in August on Schiermonnikoog show an intake of about three prey items per minute. This corresponds with 189-216 gram wet weight intake during one low tide feeding period (Pauw, 1970). Daily intake must be somewhat higher.

References

Boere, G.C., 1976. The significance of the Dutch Wadden Sea in the annual life cycle of arctic, subarctic and boreal waders. Part. 1. The function as a moulting area. Ardea 64: p. 210-291.

Busche, G., 1980. Vogelbestände des Wattenmeeres von Schleswig-Holstein Kilda, Greven (in press).

Dick, W.J.A. (ed.), 1975. Oxford and Cambridge Mauritanian Expedition 1973. Cambridge: 79 pp.

Edgar, A.T., H.R. McKenzie & R.B. Sibson, 1970. Arctic waders in northern New Zealand 1968, 1969. Notornis 26: p. 285-287.

Evans, P.R. & P.C. Smith, 1975. Studies of shorebirds at Lindisfarne, Northumberland. 2. Fat and pectoral muscle as indicators of bodycondition in the Bar-tailed Godwit. Wildfowl 26: p. 64-74.

Glutz von Blotzheim, U.N., K.M. Bauer & E. Bezzel, 1977. Handbuch der Vögel Mitteleuropas, Vol. 7. Akademische Verlaggesellschaft, Wiesbaden: 895 pp.

Höfmann, H. & H. Hoerschelmann, 1969. Nahrungsuntersuchungen bei Limikolen durch Mageninhaltanalysen. Corax 3: p. 7-22.

Kozlova, E.V., 1962. Charadriiformes, suborder Charadrii. Fauna SSSR, Pticy, Vol. 2. Akad. Nauk, Moscow: 434 pp.

Long, R.L. de & M. Thompson, 1968. Bar-tailed Godwit from Alaska, recovered in New-Zealand. Wilson Bull. 80: p. 490-491.

Meltofte, H., 1980. Fugle i Vadehavet. Vadefugletaellinger i Vadehavet 1974-1978. Miljøministeriet, Fredningsstyrelsen, København: 50 pp.

Meltofte, H., S. Pihl & B.M. Sørensen, 1972. Efteraarstraekket af Vadefugle (Charadrii) ved Blaavandshuk 1963-1971. Dansk Orn. Foren. Tidsskr. 66: p. 63-69.

Pauw, P., 1970. De aktiviteit van de Rosse Grutto. Schierboek 4: p. 89-101.

Piersma, T., M. Engelmoer, W. Altenburg & R. Mes, 1980. A wader expedition to Mauritania, Wader Study Group Bull. 29: p. 14.

Prater, A.J., 1976a. Birds of estuaries enquiry 1973-1974. British
    Trust for Ornithology, Royal Soc. Prot. Birds, Wildfowl Trust:
    48 pp.
Prater, A.J., 1976b. The distribution of coastal waders in Europe and
    North Africa. In M. Smart (ed.): Proceedings Intern. Conf. on Con-
    serv. of Wetlands and Waterfowl, Heiligenhafen, 1974. IWRB, Slim-
    bridge: p. 255-271.
Smit, C.J., 1977. On the occurrence of 32 bird species in the Danish,
    Dutch and German Wadden Sea. Unpubl. report Intern. Wadden Sea Work-
    ing Group, part 3: 174 pp.
Smith, P.C. & P.R. Evans, 1973. Studies of shorebirds at Lindisfarne,
    Northumberland. 1. Feeding ecology and behaviour of the Bar-tailed
    Godwit. Wildfowl 24: 135-139.
Voous, K.H., 1960. Atlas of European birds. Nelson, London: 284 pp.

3.23 CURLEW *(NUMENIUS ARQUATA L.)*
G.C. Boere & C.J. Smit

Da: Storspove; G: Grosser Brachvogel; Du: Wulp

3.23.1 Distribution

3.23.1.1 Breeding area
    The Curlew has an almost trans-palaearctic distribution breeding
in the boreal, temperate and steppe climate zones (fig. 121; Voous,
1960). Two subspecies have been described: *Numenius arquata arquata*
and *Numenius arquata orientalis*. The Eastern subspecies is a very
rare guest in Western Europe (Voous 1960; Glutz et al., 1977).
    The large numbers of Curlews in the Wadden Sea area originate from
the West European breeding population, according to ringing results
probably as far as from Western Russia (fig. 122; Sach, 1969;
Bainbridge & Minton, 1978).

3.23.1.2. Migration routes
    Probably no specific migration routes are taken. Large numbers
migrate through Central Europe. At suitable places in Central Europe,
such as the Caspian lowlands and the Hungarian plains, roosts of
several thousands birds are kwown (Beretz et al., 1958). Tens of
thousands migrate along coastal Western Europe, where they concentrate in certain areas.

3.23.1.3 Wintering areas
    The breeding populations of different parts of Europe have more or
less separated wintering areas (Bainbridge & Minton, 1978), but there
is a great overlap. Large number winter in the Wadden Sea, especially in the Dutch part, and on the British Isles. Small concentrations
occur in France and Northern Africa. More to the south, for instance
on the Banc d'Arguin (Mauritania) fairly large numbers are present.

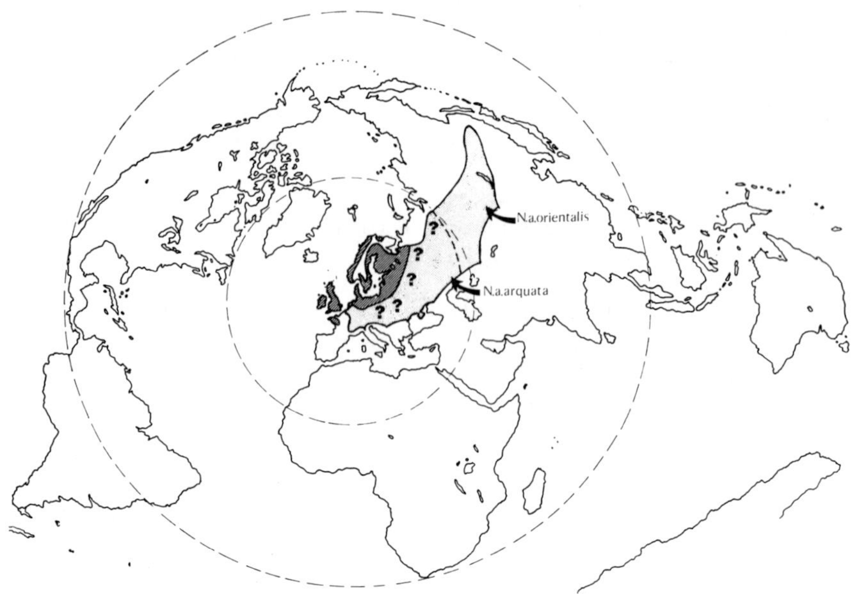

Fig. 121. Breeding area of the Curlew (after Voous, 1960). The dark
shaded area represents the region where the birds visiting the Wadden Sea originate from. Question marks indicate a region from which
it is uncertain whether birds breeding there also visit the Wadden
Sea.

These birds however belong to the Eastern subspecies *Numenius arquata
orientalis*.

3.23.1.4 Moulting areas

Moulting birds are scattered over a large area in Europe and Northern Africa. The most important concentrations of moulting birds can
be found in the Wadden Sea and on the British Isles (Sach, 1968, 1970;
Boere, 1976; Bainbridge & Minton, 1978). Also on the inland roosts in
the Hungarian plains and the Caspian lowlands concentrations of moulting Curlews are present (Poslawski, 1968, 1969; Beretz et al., 1958).

3.23.2 Annual cycle

3.23.2.1 Migration

Departure from the breeding areas starts already at the end of May
and by mid-June at several places in the Wadden Sea roosts of hundreds
of birds are present, non-breeding birds (second calendar-year as
well as adult birds). The breeding population arrives from the end of
June onwards and numbers increase very rapidly during July due to the
arrival of birds of the Scandinavian breeding populations.

Migration along Blaavandshuk peaks early July but decreases considerably by mid-July. A second wave of migrating birds however is observed during the whole month of August (Meltofte et al., 1972). In

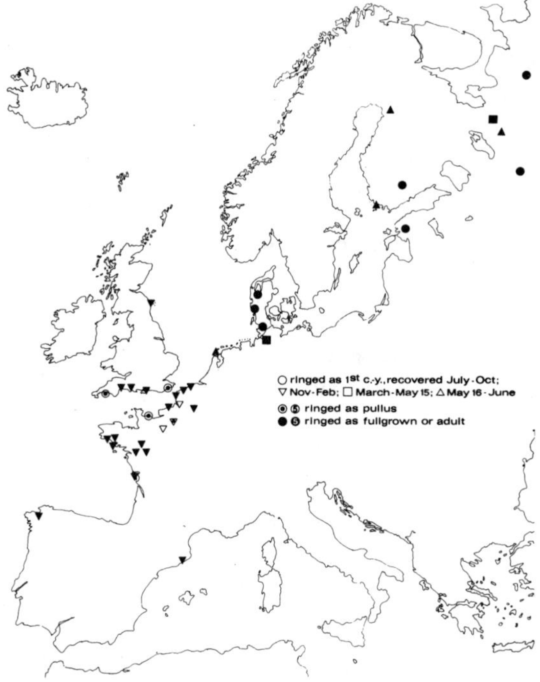

Fig. 122. Recoveries of Curlews ringed in the Dutch part of the Wadden Sea (based on published records from 1911-1976).

the Danish part of the Wadden Sea peak numbers of Curlews are present in August, a rather stable and somewhat lower number is present in the months following. In the German and Dutch parts of the Wadden Sea numbers peak somewhat later in the season.

After moult, migration towards the wintering areas starts from the end of August onwards. It ends by the end of October or early November. Large numbers stay in the Wadden Sea to winter. Autumn- and winter migratory movements are partly depending on the local weather conditions.

Spring migration, as indicated by increasing and/or strong fluctuating numbers, starts in February (compare 3.23.3.2) or March (Prater, 1976a), in inland Europe by late February or early March. It is finished by early May.

Groups of several hundreds of birds summer in the Wadden Sea and have their roosts at isolated places on the islands.

3.23.2.2 Moult

Post-nuptial moult of second calendar-year birds and non-breeding individuals starts already by the end of May and is in full progress during June and early July. The majority finishes moult at the end of August. Post-nuptial moult of adult breeding birds starts after migration from the breeding area to the moulting areas. Breeding birds from Central Europe sometimes start their moult on the breeding grounds. In early July the first adult birds can be observed in primary moult. Adult post-nuptial moult is finished at the end of October. Some birds however still are in moult till mid-November.

First calendar-year birds have a partial moult of body feathers from September until October.

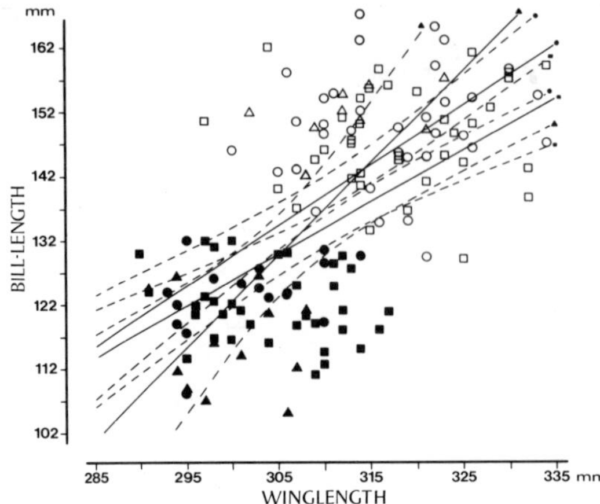

Fig. 123. Exposed bill- (in mm) and wing-lengths (in mm) of adult Curlews caught on the island of Vlieland from 1972-1975. Results are shown for January (▲) (n=21), February (■) (n=71) and March (●) (n=50) with regression lines and 95% confidence intervals. Open symbols denote females, dark symbols males (Boere, unpubl.).

Pre-nuptial moult starts early February and involves the body feather. Incidentally moult of tail feathers has been observed. Pre-nuptial moult is finished in April (Boere, 1976).

3.23.2.3 Weight changes

Annual weight changes in the Dutch Wadden Sea are summarized in Table 41. There is no information available from the German or Danish part of the area.

Fig. 123 shows morphometrics of Curlews caught in the Dutch Wadden sea. No significant differences between the three months appear to exist in bill- and wing lengths. This in combination with many retraps within the same winter or in the same period after one or more years, suggests a stable wintering population and high site fidelity.

Table 41. Mean monthly weights of adult male and female Curlews, caught on the island of Vlieland from December 1971-December 1975 (Boere, in Glutz et al., 1977).

month	males			females		
	mean	s.d.	n	mean	s.d.	n
January	864	71	11	1050	100	10
February	773	48	25	907	89	37
March	771	58	11	961	106	37
April	841	83	4	1127	119	7
August	674	86	9			
September	728	57	51	876	57	35
October	761	62	38	865	78	41
November				1006	159	6
December	839	69	5	979	59	7

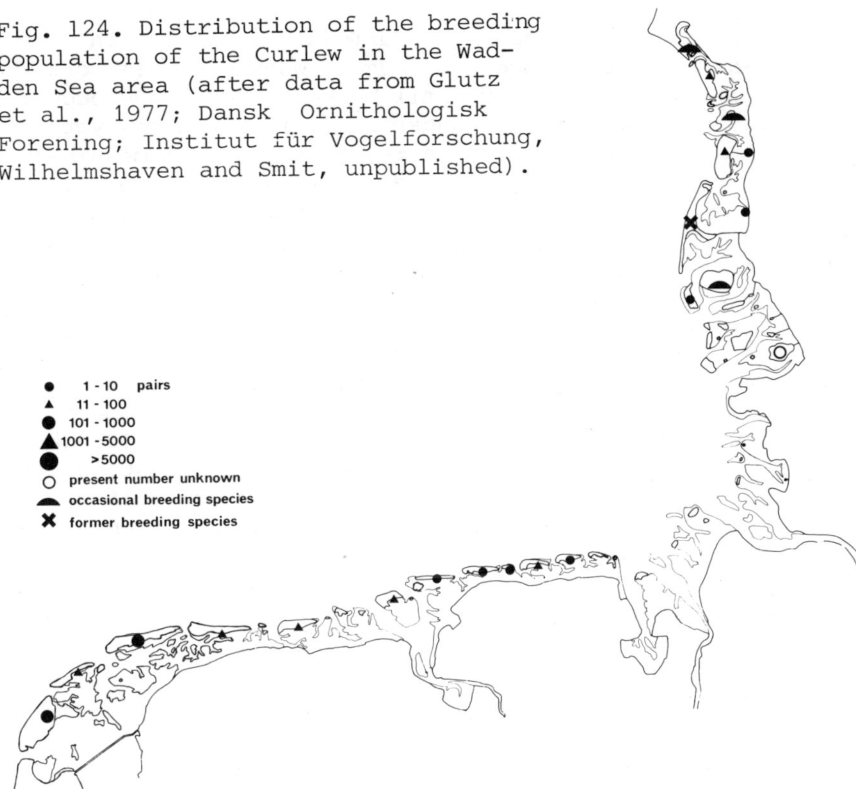

Fig. 124. Distribution of the breeding population of the Curlew in the Wadden Sea area (after data from Glutz et al., 1977; Dansk Ornithologisk Forening; Institut für Vogelforschung, Wilhelmshaven and Smit, unpublished).

3.23.3 Numbers

3.23.3.1 Population size

Prater (1976b) estimates the population wintering along the coasts in W Europe and NW Africa at 151,000. Of this number about 145,000 have been recorded to winter in W Europe. According to Prater this number probably is too low as autumn censuses in Britain (70,000), Ireland (15,000), the Wadden Sea (102,000) and the Delta region in the Netherlands (11,000) yielded a total of 198,000). The true figure for the number wintering in W Europe and NW Africa is likely to amount to about 225,000 as Curlews are also present in autumn along the coasts of France, the Iberian peninsula and NW Africa.

The size of the population breeding in the Wadden Sea area can be estimated at about 600-650 pairs of which about 450-500 pairs breed on the Dutch Wadden Sea islands. Except for some pairs breeding along the mainland coast in the Danish part of the Wadden Sea area the distribution is limited to the islands as is shown in fig. 124 (after data from Dansk Ornitologisk Forening; Institut für Vogelforschung, Wilhelmshaven and Smit).

3.23.3.2 Numbers per area

In the Danish part of the Wadden Sea Curlews never occur in very large numbers. Maximum numbers occur in August when somewhat more than 5000 may be present. From September until December 2000-4500 occur. Numbers in January and February are under 1500, in March however already a maximum number of 5600 could be registered. Maximum numbers in the months following drop from 1500 in April to less than 100 in June. In July again more than 5000 may be present. Curlews are most numerous along the Rømø-dam, along the coast near Emmerlev, in the Ho Bugt and on the tidal flats between Rømø and Fanø (Meltofte, 1980; reports Vadefuglegruppen Dansk Ornithologisk Forening.

Numbers in Schleswig-Holstein increase in June (fig. 125) when second calendar-year birds arrive in the area to moult (Sach, 1968). In July numbers increase again due to the arrival of adult and first calender-year birds from Finland and the Baltic area. Peak numbers occur somewhere between August (Drenckhahn et al., 1971) and October (Busche, 1980). Numbers in August-October amount to about 40,000 birds In mild winters at least 12,000 are still present (Busche, 1980). According to Heldt (1968) considerable numbers stay over summer, generally at least 2000. The distribution is shown in fig. 126.

On most of the Wadden Sea islands in Niedersachsen peak numbers occur in August and September, along the mainland coast however in September and October (fig. 127 and Bub, 1967). In the whole area in Sep-

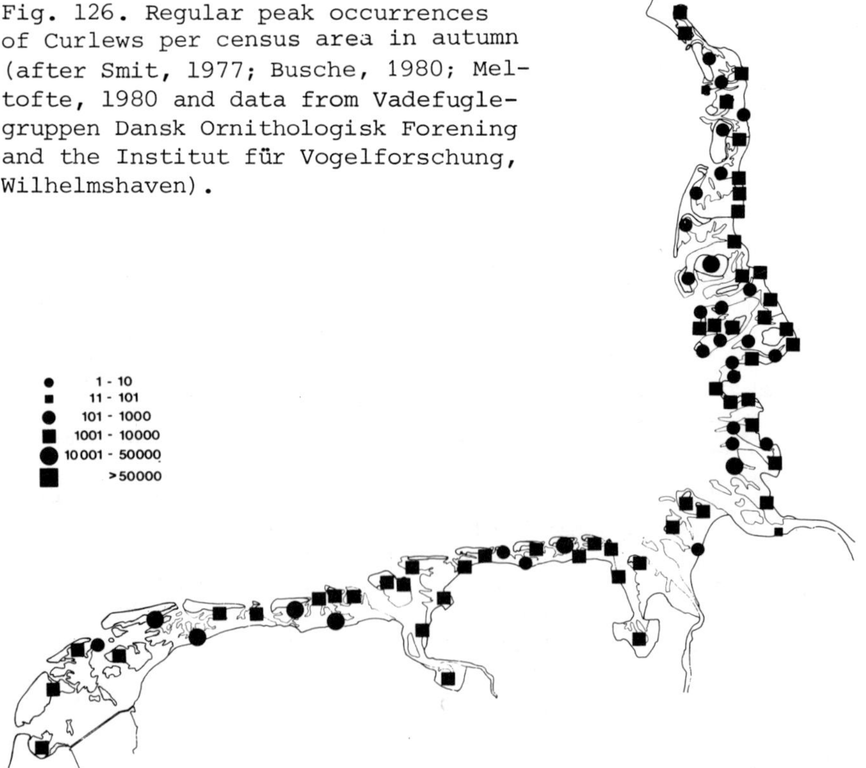

Fig. 126. Regular peak occurrences of Curlews per census area in autumn (after Smit, 1977; Busche, 1980; Meltofte, 1980 and data from Vadefuglegruppen Dansk Ornithologisk Forening and the Institut für Vogelforschung, Wilhelmshaven).

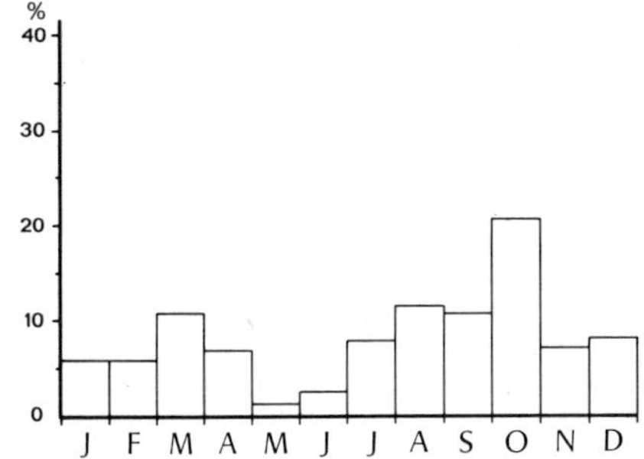

Fig. 125. Occurrence of Curlews in the Wadden Sea area in Schleswig-Holstein. Numbers for each month are expressed as a percentage of the total number observed in all months (after Busche, 1980).

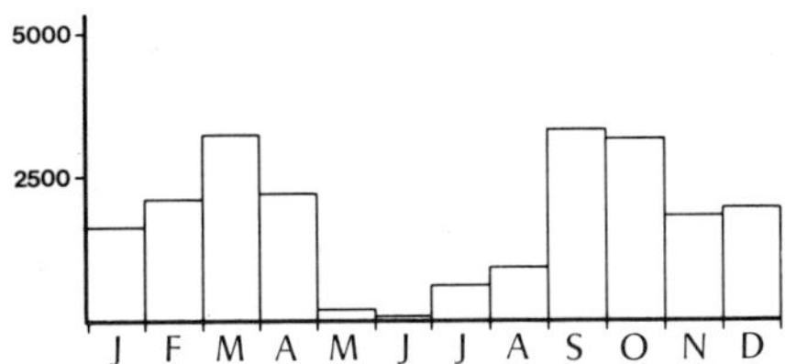

Fig. 127. Mean number of Curlews per month in a 50 km long stretch along the mainland coast of Niedersachsen (after Smit, 1977 and data from the Institut für Vogelforschung, Wilhelmshaven).

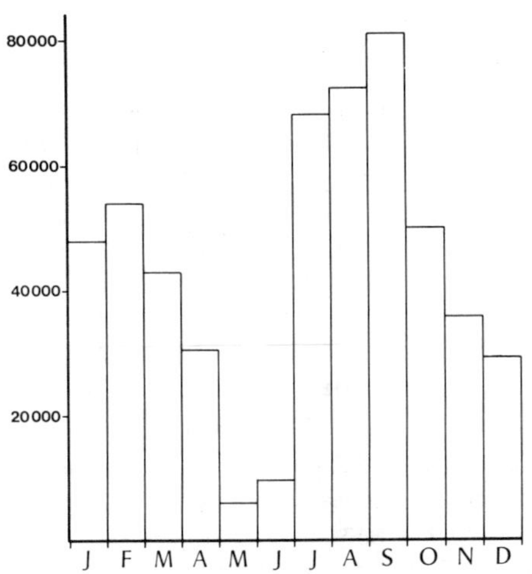

Fig. 128. Mean number of Curlews per month in the Dutch part of the Wadden Sea (after data from Smit, 1977).

tember about 30,000 will be present, in the winter of 1973-1974 along the mainland coast about 20,000 occurred (after data from Institut für Vogelforschung, Wilhelmshaven; Smit, 1977). Bub (1956) however, counted only 6800 along 180 km of mainland coast in September.

As compared to numbers in other parts of the Wadden Sea Curlews are numerous in the Dutch part. Fig. 128 shows mean numbers per month for the area. Numbers counted in the area are listed in Table 42. Curlews are most numerous along the mainland coast of Friesland though on most of the islands high numbers are counted as well (fig. 126).

Table 42. Numbers of Curlews in the Dutch part of the Wadden Sea (after various publications in Watervogels and Limosa and Zegers, in litt.).

Date	Year	Number
January 8	1977	33,000
January 12	1974	51,610
January 17	1976	65,200
January 18	1975	53,840
April 6	1973	35,600
April 19	1975	34,070
May 1	1976	6,140
May 11	1974	5,500
July 29	1972	89,920
August 30	1975	96,450
September 13	1976	101,650
October 19	1974	56,400
November 13	1976	35,810
December 29	1966	26,000

3.23.4 Food

3.23.4.1 Food composition

From June until late September small specimens of *Carcinus maenas* are the most important food source. Later on polychaetes, such as *Nereis, Arenicola, Lanice* and small molluscs or siphons of these become important. Along the mainland coast of Friesland some even specialize in eating large specimens of *Mya* which may be buried 13 cm into the sediment (De Vries in Zwarts, 1979). The diet of Curlews foraging on mussel beds near the island of Schiermonnikoog in late summer consisted of *Macoma, Cerastoderma, Carcinus* and *Mytilus* (Voss & Koolhaas, 1969). Kersten & Piersma (in litt.) saw Curlews near the island of Ameland in September taking polychaetes (47% of all prey specimens) and *Carcinus* (43%), to a lesser extent small unidentified prey items and bivalves. During high tides when Curlews feed in meadows inland large numbers of *Lumbricus* are taken. During winter this may happen during low tides as well (Goss-Custard & Jones, 1976; Ens & Zwarts, 1980a). In England Elphi (1979) and Townshend (1979) found that in such a case relatively many males were foraging in meadows inland. The same applies for Curlews along the mainland coast of Friesland. In this latter area especially shortbilled males were involved (Ens & Zwarts, 1980a).

3.23.4.2 Feeding activities

Curlews feed on muddy tidal flats. In particular areas with mussel beds are frequently visited. Prey choise and feeding activities appear to vary much between individual birds (Ens & Zwarts, 1980a). On the mudflats along the mainland coast of Friesland part of the Curlews defend territories during a part of the year (Ens & Zwarts, 1980b).

In September near the island of Ameland Kersten & Piersma (in
litt.) found peck- and success rates of 22.8 and 3.8 (17%) respectively, per minute.

The birds are very shy in comparison to other wader species and
therefore high tide roosts are situated on isolated places on the islands and the mainland coasts, often in or very close to the edge of
the water. Sometimes high tide roosts lie in open polder areas far
from the water. They often share the roost with Bar-tailed godwits
and sometimes with small numbers of Redshanks.

3.23.4.3 Total food consumption.

No information available from the area.

References

Bainbridge, J.P. & C.D.T. Minton, 1978. The migration and mortality
    of the Curlew in Britain and Ireland. Bird Study 25: p. 39-50.
Beretz, P., A. Keve, B. Nagy & J. Szij, 1958. Economic importance
    of the Curlews. Aquila 65: p. 89-126 (Hungarian, Engl. summ.).
Boere, G.C., 1976. The significance of the Dutch Wadden Sea in the an-
    nual life cycle of arctic, subarctic and boreal waders. Part 1.
    The function as a moulting area. Ardea 64: p. 210-291.
Bub, H., 1956. Eine Seevogel-Bestandsaufnahme an der ostfriesisch-
    oldenburgischen Küste. Ornithol. Mitt. 8: p. 49-50.
Bub, H., 1967. Ueber den Säbelschnäbler (Recurvirostra avosetta) und
    den Grossen Brachvogel (Numenius arquata) im Jadebusen bei Hochwas-
    ser. Vogelwarte 24: p. 135-142.
Busche, G., 1980. Vogelbestände des Wattenmeeres von Schleswig-Holstein.
    Kilda, Greven (in press).
Drenckhahn, D., R. Heldt jun. & R. Heldt sen., 1971. Die Bedeutung der
    Nordseeküste Schleswig-Holsteins für einige eurasische Wat- und Was-
    servögel mit besonderer Berücksichtigung des Nordfriesischen Watten-
    meeres. Natur und Landschaft 46: p. 338-346.
Elphick, D., 1979. An island flock of Curlews in Mid-Cheshire, Eng-
    land. Wader Study Group Bull. 26: p. 31-35.
Ens, B., & L. Zwarts, 1980a. Wulpen op het wad van Moddergat. Water-
    vogels 5: p. 108-120.
Ens, B., & L. Zwarts, 1980b. Territoriaal gedrag bij wulpen buiten het
    broedgebied. Watervogels 5: p. 155-169.
Glutz von Blotzheim, U.N., K.M. Bauer & E. Bezzel, 1977. Handbuch der
    Vögel Mitteleuropas, Vol. 7. Akademische Verlagsgesellschaft, Wies-
    baden: 894 pp.
Goss-Custard, J.D. & R.E. Jones, 1976. The diets of Redshank and Cur-
    lew. Bird Study 23: p. 233-243.
Heldt, R., 1968. Uebersommernde Limikolen an der Westküste von Schles-
    wig-Holstein. Corax 2: p. 108-130.
Meltofte, H., 1980. Fugle i Vadehavet. Vadefugletaellinger i Vadehavet
    1974-1978. Miljøministeriet, Fredningsstyrelsen, København: 50 pp.
Meltofte, H., S. Pihl & B.M. Sørensen, 1972. Efteraarstraekket af va-
    defugle (Charadrii) ved Blaavandshuk 1963-1971. Dansk Orn. Foren.
    Tidsskr. 66: p. 63-69.

Prater, A.J., 1976a. Birds of estuaries enquiry 1973-1974. British
    Trust for Ornithology, Royal Soc. Protection of birds, Wildfowl
    Trust: 48 pp.
Prater, A.J., 1976b. The distribution of coastal waders in Europe and
    North Africa. In: M. Smart (ed.) Proceedings International Confe-
    rence on Conservation of Wetlands and Waterfowl, Heiligenhafen,
    1974a. IWRB, Slimbridge: p. 255-271.
Poslawski, A.N., 1968. Durchzug und übersommern von Limikolen im nörd-
    lichen Vorland des Kaspis. J. Orn. 109: p. 1-10.
Poslawski, A.N., 1969. Zug und Mauser des Grossen Brachvogels. Falke
    16: p. 184-188.
Sach, G., 1968. Die Mauser des Grossen Brachvogels, Numenius arquata.
    J. Orn. 109: p. 486-511.
Sach, G., 1969. Ringfunde des Grossen Brachvogels. Auspicium 3: p. 153-
    158.
Sach, G., 1970. Zur Handschwingenmauser von Numenius arquata. J. Orn.
    111: p. 105-106.
Smit, C.J., 1977. On the occurrence of 32 bird species in the Danish,
    German and Dutch Wadden Sea. Unpubl. report International Wadden
    Sea Working Group, part 3: 174 pp.
Townshend, D.J., 1979. The use of space by individual Grey Plovers
    (Pluvialis squatarola) and Curlews (Numenius arquata) on their
    winter feeding grouds. Wader Study Group Bull. 26: p. 29.
Voous, K.H., 1960. Atlas of European birds. Nelson, London: 284 pp.
Voss, A. & J. Koolhaas, 1969. Verslag van het Wulpenonderzoek op Schier
    - 3, 1966. Schierboek 3: p. 173-178.
Zwarts, L., 1979. Feeding ecology of Curlew. Wader Study Group Bull.
    26: p. 28.

3.24 SPOTTED REDSHANK *(TRINGA ERYTHROPUS* (PALLAS)*)*
G.C. Boere & C.J. Smit

Da: Sortklire; G: Dunkler Wasserläufer; Du: Zwarte Ruiter

3.24.1 Distribution

3.24.1.1 Breeding area
   The Spotted redshank has a trans-palaearctic distribution breeding
only in the lower arctic and the boreal climate zones. The breeding
area generally lies north of that of the Redshank, and ranges from
NE Norway as far as Siberia (fig. 129). Very few ringing results are
available from the Wadden Sea area, which makes it difficult to determine the exact area of origin. Birds passing through the Wadden Sea
probably originate from Scandinavia and Western Russia. No subspecies
are distinguished (Voous, 1960; Glutz et al., 1977).

3.24.1.2 Migration routes
   The Spotted redshank is a long distance migrant with a typical
broad-front migration across Europe and no special routes. Locally

Fig. 129. Breeding area of the Spotted redshank. The dark shaded area represents a region where birds passing through the Wadden Sea originate from. Question marks indicate a region from which it is uncertain whether also from there birds migrate towards the Wadden Sea.

larger concentrations occur, sometimes along the coast but more often at inland wetland areas like the Hungarian plains. Peak numbers of up to 6000 in August have been counted there (Sterbetz in Glutz et al., 1977).

3.24.1.3 Wintering areas

The species winters in a large area ranging from NW Europe to Central Africa where it winters at suitable areas widely spread over the continent, generally however north of the equator. Concentrations are present in delta regions in Senegal and Ghana.

3.24.1.4 Moulting areas

Moulting areas are not known in detail but the Wadden Sea, especially the areas with very muddy tidal flats, acts as a moulting station for several thousands of birds. In the Dutch part, especially the Dollard, about 2000-3000 moulting birds may be present, in the German part about 4500. All of these birds probably did not breed. Many birds moult in the Mediterranean area, South Asia and tropical Africa.

3.24.2 Annual cycle

3.24.2.1 Migration

First groups arrive in the Wadden Sea during the second half of June. The numbers are rather stable till the end of July and beginning of August, when new birds arrive. These are mainly first calendar-year birds. As a result of this the total number increases. In the course of September many birds depart. At Blaavandshuk peak numbers of Spotted redshanks pass by during August. Migration continues until mid-September (Meltofte et al., 1972). During winter only very few birds are observed in the Wadden Sea area, numbers however vary from year to year. During spring migration numbers are smaller than in autumn and concentrations occur less often. Peak numbers occur in early May.

3.24.2.2 Moult

Very little information is known from the Wadden Sea. Holthuyzen (in litt.) collected moulted primaries on the high tide roosts in the Dollard during one summer period. It appeared that birds present there are in moult by late June. These birds are probably all immatures or non-breeding birds. Breeding birds start their moult also very early, even in the breeding area. The post-nuptial adult moult continues through July and August for birds moulting in NW and Southern Europe and till November-December for birds moulting in tropical areas (Glutz et al., 1977). Pre-nuptial body moult takes place in the wintering area and is not recorded in the Wadden Sea.

3.24.2.3 Weight changes

Very little information is available from the area. Weights of two first calendar-year birds trapped on Vlieland in early August 1975 were 118 and 113 gram (Boere, unpubl.).

3.24.3 Numbers

3.24.3.1 Population size

Small numbers winter along the coasts of W Europe and N Africa. Prater (1976) estimates only about 500. Because Spotted redshanks winter in a huge area, partly inland, which has not been studied sufficiently, no statements can be made on the size of the population passing through the Wadden Sea area.

3.24.3.2 Numbers per area

In the Wadden Sea the species is mostly rather difficult to count. It often occurs along creeks, mostly in relatively small numbers, often mixed with Redshanks which occur in larger numbers. During many counts they will have been missed partly. The numbers shown here will therefore represent minima.

In the Danish part of the Wadden Sea Spotted redshanks are never numerous. The autumn maximum amounts to 144 birds, the spring maximum to only 24 (Meltofte, 1980). Fig. 130 shows the distribution of the small numbers.

In Schleswig-Holstein numbers increase already from the mid-June on. By the end of the month some hundreds may be present. The first to arrive are females leaving the breeding grounds soon after laying the eggs. They join the birds present in the area for the whole sum-

Fig. 130. Regular peak occurrences of
Spotted redshanks in autumn in the
Wadden Sea area (data: Holthuyzen,
1975; Smit, 1977; Busche, 1980; Meltofte, 1980; Vadefuglegruppen Dansk
Ornithologisk Forening and Institut
für Vogelforschung, Wilhelmshaven).

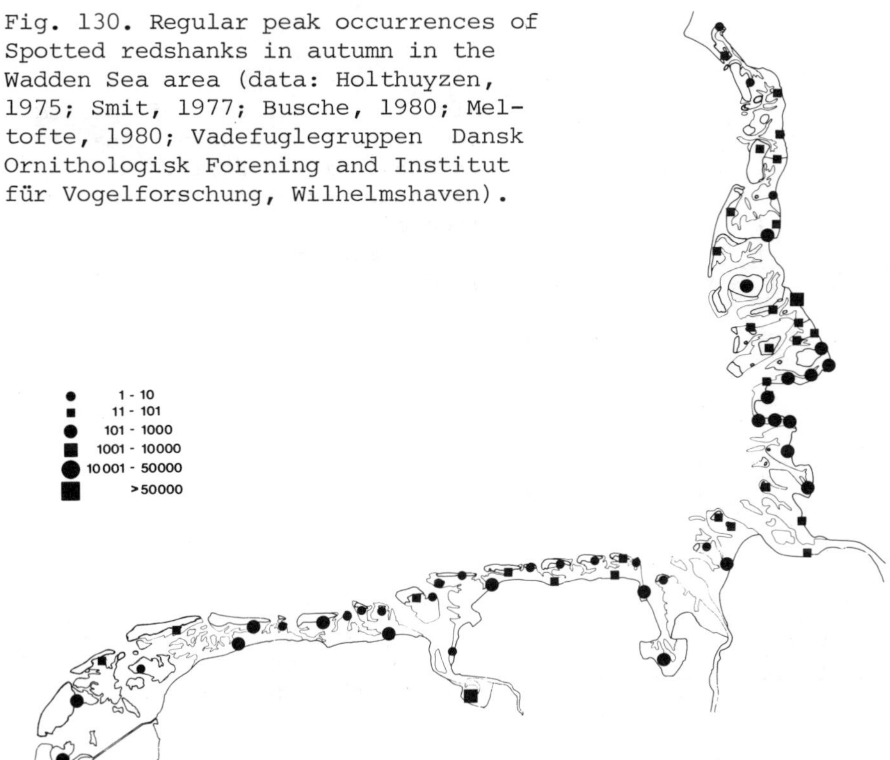

mer, generally a few hundreds. In the beginning of July the species
becomes more numerous, by the end of the month and the beginning of
August however peak numbers are reached when males and juveniles arrive. They concentrate in the following regions: Meldorfer Bucht, Grüne Insel, St. Peter-Süderhöft, Nordstrander Damm and Hauke Haienkoog.
In some of these regions sometimes more than 1000 can be present. Numbers decrease in September and October, in November and December only
a few are left. Spring migration is less marked. It lasts from mid-
March to April and May (Drenckhahn et al., 1971; Smit, 1977). The number present in autumn in the area generally amounts to about 3000, in
some years probably up to 4000 birds (Busche, 1980). Fig. 131 shows
something of the phenology of the species in the area.

Apart from the Dollart and the Jadebusen the species does not occur
in large numbers in Niedersachsen. Though data are scarce the phenology pattern in several areas seems to be quite different from that in
Schleswig-Holstein and the Netherlands. In Niedersachsen numbers in
spring dominate much more, especially in the Dollart (fig. 132) (source:
Institut für Vogelforschung, Wilhelmshaven; Holthuyzen, 1975; Smit,
1977).

Fig. 133 shows that in the Dutch part of the Wadden Sea Spotted redshanks are most numerous in July and August. Numbers may amount to 3000
birds in these months.

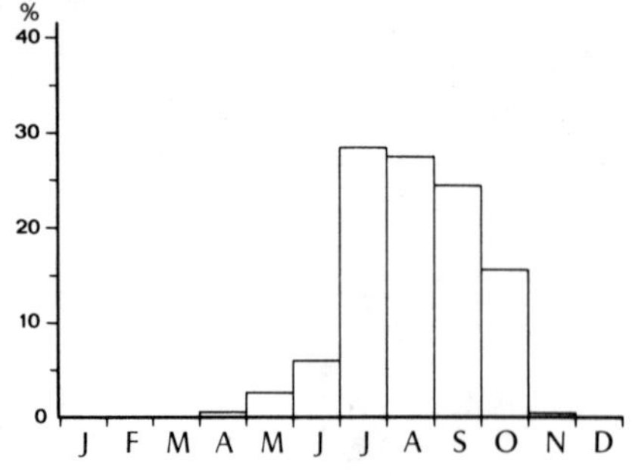

Fig. 131. Occurrence of Spotted redshanks in the Wadden Sea area in Schleswig-Holstein. Data per month are expressed as a percentage of the total number observed in all months (after Busche, 1980).

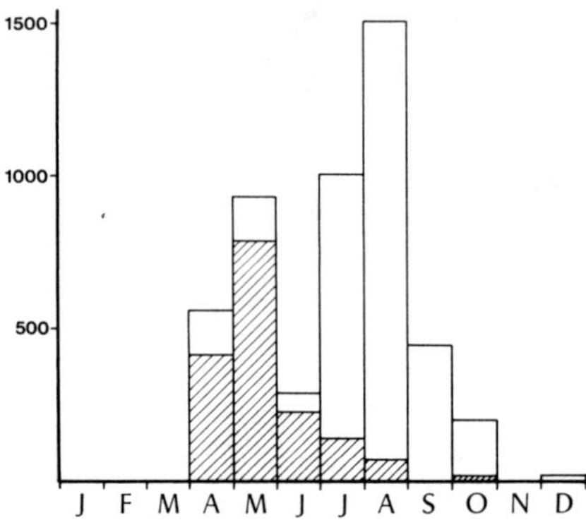

Fig. 132. Numbers of Spotted redshanks per month in the Dollard. Columns indicate total numbers, hatched parts indicate numbers in the German part of the area (after data from Holthuyzen, 1975 and Smit, 1977).

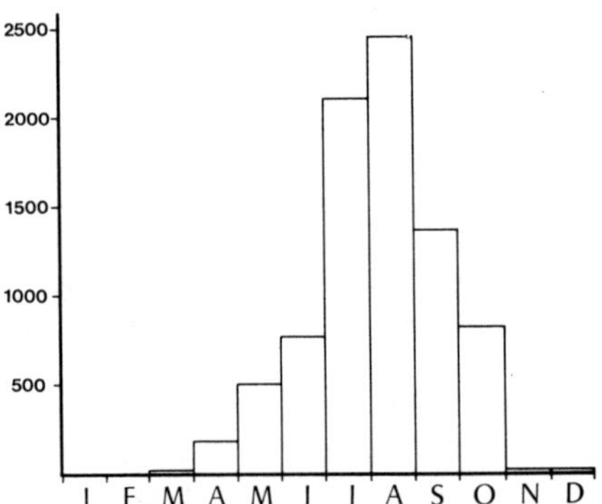

Fig. 133. Mean number of Spotted redshanks per month in the Dutch part of the Wadden Sea (after data from Smit, 1977).

From Table 44 can be seen that the species prefers the Dollard and the mainland coast of Friesland though it can also be numerous along creeks in the salt marshes of the islands. On the muddy borders of the Balgzand flats relatively high numbers may be present. In the Dutch part of the Wadden Sea two migration waves can be distinguished, the first from the end of June to the beginning of August, the second at the end of August and the beginning of September. In spring, migration proceeds fast. Numbers are low nearly everywhere (Smit, 1977).

Table 43. Number of Spotted redshanks in the Dutch part of the Wadden Sea (after various publications in Watervogels and Limosa and Zegers in litt.).

Date	Year	Number
January 8	1977	1
January 12	1974	4
January 18	1975	2
April 6	1973	440
April 19	1975	285
May 11	1974	920
July 29	1972	2,200
August 30	1975	1,150
September 13	1976	2,930
October 19	1974	560
November 13	1976	65
December 29	1966	56

Table 44. Maximum numbers of Spotted redshanks in the Dollard and several parts of the Dutch Wadden Sea (sources: Holthuyzen, 1975; Smit, 1977; Wadvogelwerkgroep Groningen, unpublished).

	Jan	April	May	July	Aug	Sept	Oct
Dollart (German part)	0	952	1,760	410	120	11	22
Dollard (Dutch part)	5	430	477	1,530	1,920	1,429	650
Groningen	17	58	166	24	230	245	84
Friesland	0	35	80	885	605	1,200	92
Balgzand	0	120	240	400	480	900	80
Schiermonnikoog	0	25	160	144	300	55	350
Engelsmanplaat	0	0	4	3	0	0	0
Ameland	0	19	42	28	185	90	150
Terschelling	3	43	90	12	70	82	69
Griend	0	0	6	2	3	1	0
Vlieland	0	0	0	1	30	2	0
Texel	2	20	50	3	150	12	7

3.24.4 Food

3.24.4.1 Food composition

The way of feeding in shallow water results in a substantial percentage of small fish in the diet, which is more or less comparable to the diet of the Greenshank. Prey items however are generally smaller. *Pomatoschistus microps*, but also *Crangon crangon, Nereis diversicolor* and *Carcinus maenas* are important prey items (Holthuyzen, 1979).

3.24.4.2 Feeding activities

In the Wadden Sea Spotted redshanks prefer very muddy shallows along the coast and in the neighbourhood of some islands. Generally feeding takes place in shallow water, with a water level up above the tarsus or even higher (Daanje, 1933; Holthuyzen, 1979). Birds have often been observed feeding in the creeks of the salt marshes or ditches in polders close to the Wadden Sea. High tide roosts vary from place to place but are mainly situated on salt marshes and in shallow pools. They often share roosts with Redshanks. The remarkable difference in numbers between the German and the Dutch part of the Dollard as is shown in fig. 125 probably has several reasons. Holthuyzen (1975, 1979) suggests that the distribution of *Crangon crangon* and the more intensive disturbance by man of the high tide roosts on the German part in late summer might be the reason for this difference.

3.24.4.3 Total food consumption

No information available.

References

Busche, G., 1980. Vogelbestände des Wattenmeeres von Schleswig-Holstein. Kilda, Greven (in press).

Daanje, A., 1933. Vischetende Zwarte Ruiters. Ardea 22: p. 183-184.

Drenckhahn, D., R. Heldt jun. R. Heldt sen., 1971. Die Bedeutung der Nordseekuste Schleswig-Holsteins für einige eurasische Wat- und Wasservögel mit besonderer Berücksichtigung des Nordfriesischen Wattenmeeres. Natur und Landschaft 46: p. 338-346.

Glutz von Blotzheim, U.N., K.M. Bauer & E. Bezzel, 1977. Handbuch der Vögel Mitteleuropas, Vol. 7. Akademische Verlagsgesellschaft, Wiesbaden: 895 pp.

Holthuyzen, Y.A., 1975. Voedsel en voedseloecologie van de Zwarte Ruiter (Tringa erythropus) in de Dollard. Report Zool. Lab. University of Groningen: 81 pp.

Holthuyzen, Y.A., 1979. Het voedsel van de Zwarte Ruiter, Tringa erythropus, in de Dollard. Limosa 52: p. 22-33.

Meltofte, H., 1980. Fugle i Vadehavet. Vadefugletaellinger i Vadehavet 1974-1978. Miljøministeriet, Fredningsstyrelsen, København: 50 pp.

Meltofte, H., S. Pihl & B.M. Sørensen, 1972. Efteraarstraekket af vadefugle (Charadrii) ved Blaavandshuk 1963-1971. Dansk Orn. Foren. Tidsskr. 66: p. 63-69.

Prater, A.J., 1976. The distribution of coastal waders in Europe and North Africa. Proceedings Intern. Conference on the Conservation of Wetlands and Waterfowl, Heiligenhafen 1974. IWRB, Slimbridge: p. 255-271.

Smit, C.J., 1977. On the occurrence of 32 bird species in the Danish, German and Dutch Wadden Sea. Unpubl. report Intern. Wadden Sea Working Group, Part 3: 174 pp.

Voous, K.H., 1960. Atlas of European birds. Nelson, London: 284 pp.

3.25 REDSHANK *(TRINGA TOTANUS* L.*)*
G.C. Boere & C.J. Smit

Da: RØdben; G: Rotschenkel; Du: Tureluur

3.25.1 Distribution

3.25.1.1 Breeding area

The Redshank has a trans-palaearctic distribution, breeding in the temperate, boreal, steppe, desert and to a limited extent also in the Mediterranean climate zone (Voous, 1960). The breeding area ranges from Iceland to the Chinese coast throughout the Palaearctic with a disjunct breeding distribution in the Mediterranean area and the Scandinavian countries (fig. 134).

The considerable geographical variation in size and colour makes it difficult to distinguish clear subspecies and the systematic position of these has been and still is subject of discussion (Harrison, 1944; Verheyen, 1947; Salomonsen, 1954; Hale, 1971). Generally six subspecies are distinguished, based on measurements as well as plumage characters, of which three are mentioned on the map. According to biometrical studies and ringing results (fig. 135) two groups of birds make use of the Wadden Sea:

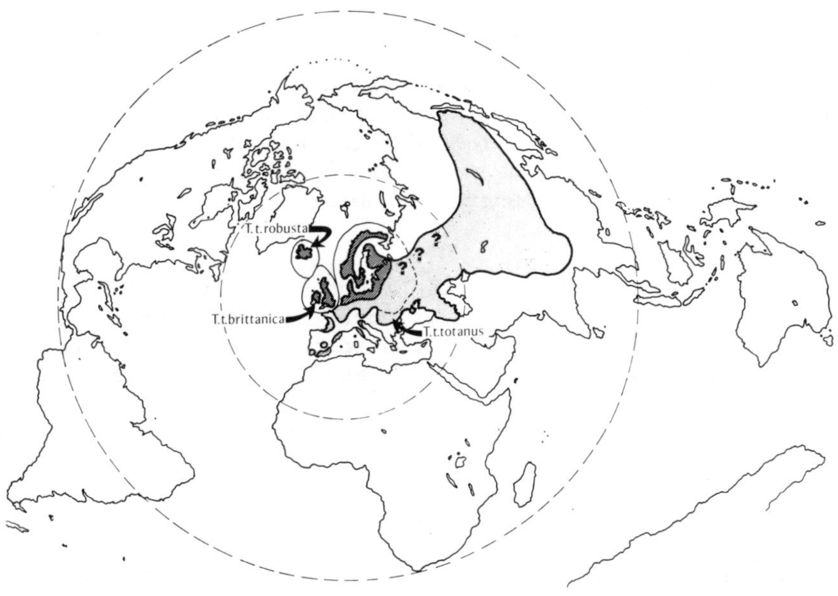

Fig. 134. Breeding area of the Redshank (after Voous, 1960). The dark shaded area represents the region where birds visiting the Wadden Sea originate from. Question marks indicate an area from which is uncertain whether birds breeding there also visit the Wadden Sea. Because subspecies separation still is in discussion only those subspecies breeding in Europe have been indicated.

1) The small birds of the continental breeding populations belonging
to the subspecies *T.t. totanus*, breeding in NW Europe, Scandinavia and
Western Russia.
2) The larger subspecies: *T.t. robusta*, breeding on Iceland, (Hale,
1971; Glutz et al., 1977).

3.25.1.2 Migration routes

The Redshank is a short, medium as well as a long distance migrant,
while the breeding population of the British Isles does not migrate
at all. No particular migration routes can be recognized. The species
shows a broad front migration across Europe and Central Asia with local concentrations at suitable places inland, such as river valleys,
marshes, large ponds and wet meadows. Relatively large concentrations
occur along the coast. Ringing recoveries show that birds ringed along
the North Sea coast do not necessarily follow the coast, but cross
Central Europe to, for instance the Camargue or other wetland areas
in the Mediterranean, (Grosskopf, 1971; Hale, 1973; Verheyen & Le
Grelle, 1950; Fournier & Spitz, 1969; Glutz et al., 1977).

There is much evidence that at least in spring Icelandic Redshanks
make a non-stop flight across the North Sea and Atlantic Ocean to Iceland.

3.25.1.3 Wintering areas

The wintering area ranges from the North Sea coast to tropical Africa. In Western Europe wintering areas are mainly coastal shallows and
estuaries. This is also true for the Mediterranean and Africa. In Africa, south of the Sahara, however, also fresh water marshes, river valleys and flooded areas act as important wintering areas. Redshanks wintering in the Wadden Sea belong to the Icelandic subspecies. The largest numbers are present in the Dutch part of the area. According to
investigations on the island of Vlieland, wintering Redshanks have a
great site fidelity: several birds have been controlled at the same
spot 10-15 years after ringing (Boere, unpubl.).

3.25.1.4 Moulting areas

Moulting takes place within the whole range of the wintering area.
Concentrations of moulting birds however occur in the coastal areas
of Western Europe, along the African Atlantic coast, for instance in
Morocco, at the Banc d'Arguin (Mauritania) and in the Delta areas of
the Niger and Senegal rivers.

There is a separation in moulting areas between the two groups mentioned before:
1) Continental Redshanks start moulting at several places along their
migration routes, i.e. in the Wadden Sea. They finish moult elsewhere,
mainly in Africa.
2) Icelandic Redshanks moult in the wintering areas along the North
Sea coast (Boere, 1976).

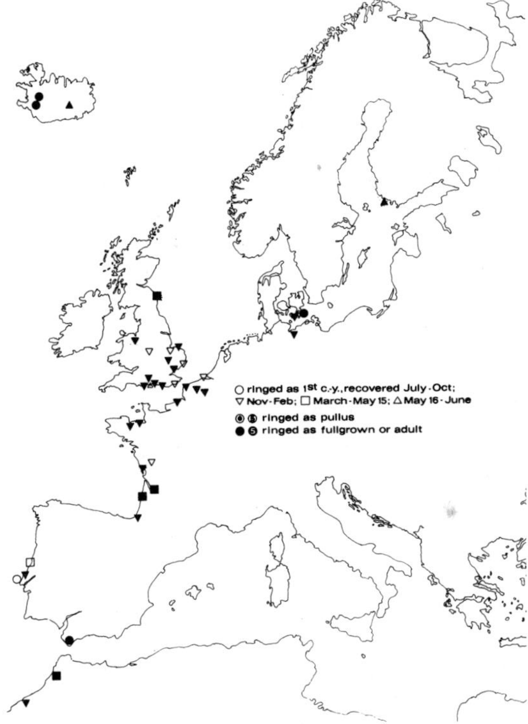

Fig. 135a. Recoveries of Redshanks ringed in the Dutch part of the Wadden Sea (based on published records from 1911 to 1979).

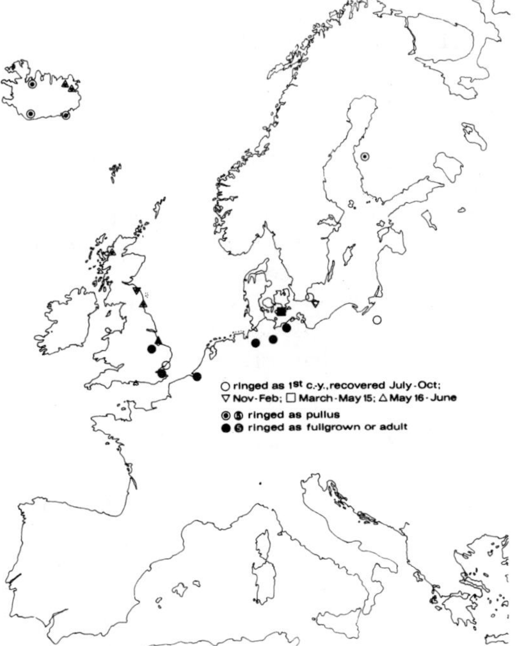

Fig. 135b. Ringing places of Redshanks recovered in the Dutch part of the Wadden Sea (based on published records from 1911 to 1979).

3.25.2 Annual cycle

3.25.2.1 Migration

Departure from the breeding areas starts immediately after the breeding season. The first arrivals of small flocks in the Wadden Sea take place at the end of June. These probably include the local breeding birds and those of the Northern part of the Netherlands. Visible migration at Blaavandshuk has peak values in the last July decade and the first August pentade (Meltofte et al., 1972). Numbers in the Wadden Sea area increase rapidly to peak numbers at the end of July. At the same time many birds pass through the area. In the course of August the numbers decrease.

The leaving birds are partly replaced by Icelandic Redshanks (Boer, 1966; Goethe, 1972; Boere, 1976). These birds stay to winter and leave during April, at the same time when total numbers increase because of an influx of Continental Redshanks from their wintering areas in · Southern Europe and Africa (see fig. 136). During May the numbers decrease rapidly and by the end of May very few birds are left, mainly belonging to the local breeding population.

3.25.2.2 Moult

The adult post-nuptial moult of *T.t. totanus* starts in the last June decade soon after the breeding season when the birds concentrate at the moulting areas mainly along the coast. From investigations in the Dutch Wadden Sea (Boere, 1976) it became clear that Continental Redshanks leave the area before their moult is completed. They arrest moult or even migrate in full wing moult and finish it at other localities in Southern Europe and NW Africa. The moult is completed there from mid-September (SW Europe) until early November (Banc d'Arguin) (Johnson, 1974; Dick, 1975; Pienkowski et al., 1976). Icelandic Redshanks *(T.t. robusta)* start their post-nuptial moult somewhat later in the Wadden Sea and other areas around the North Sea (Mackie, 1976) and have completed it by mid-October.

First calendar-year birds of both subspecies moult large parts of their body feathers and have been observed in active moult between the end of July until November. There is no evidence of a separate moult schedule of second year birds. Pre-nuptial moult of the body feathers of *T.t. robusta* starts early March and is completed when the birds leave the wintering area by the end of April. Continental Redshanks trapped at the island of Vlieland during May did not show any sign of moult and had a complete breeding plumage (Boere, 1976).

3.25.2.3 Weight changes

The mean monthly weights of Continental and Icelandic Redshanks trapped in the Dutch Wadden Sea are shown in Table 45. Information from other areas in the Wadden Sea is restricted to the breeding population of Wangerooge. One hundred males had a mean weight of 123 grams (range 107-142), 100 females of 140 g (range 121-152) (Grosskopf, 1968).

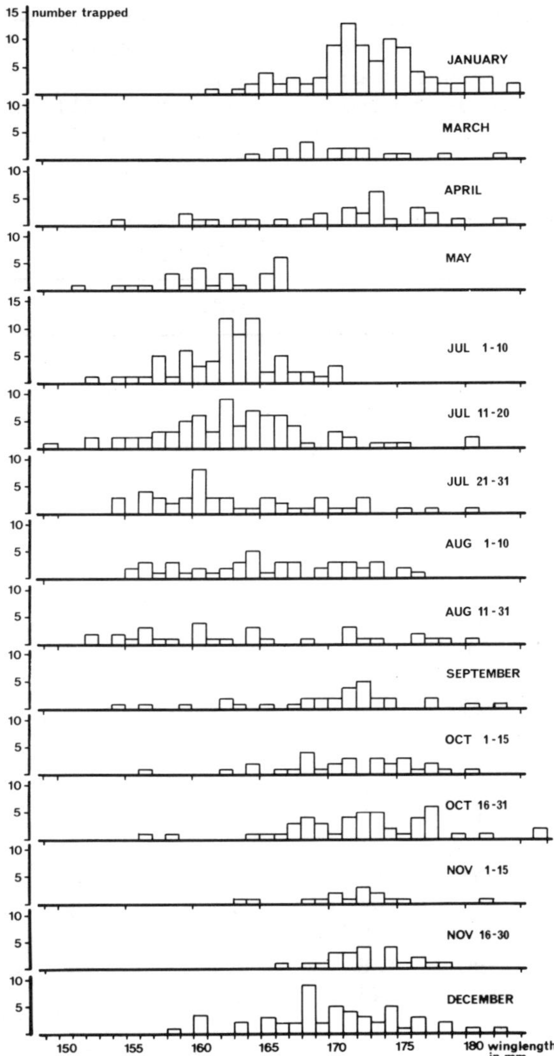

Fig. 136. Wing-length frequency distribution for Redshanks caught at the island of Vlieland in 1972 and 1973. The graph shows the seasonal replace of Tringa t. totanus (short-winged) by T. t. robusta (long-winged) (Boere, unpublished).

3.25.3 Numbers

3.25.3.1 Population size

Prater (1976) estimates the size of the population wintering in NW Europe and W Africa at 235,000, about 100,000 of these being present at the Banc d'Arguin and 125,000 in W Europe. The Dutch Ornithological Mauritanean expedition counted 70,000 at the Banc d'Arguin in winter and early spring 1980 (Piersma et al., 1980). The total size of the population passing the Wadden Sea is unknown because of insufficient information on numbers present in wintering areas inland in Africa.

Table 45. Mean monthly weights in grams of *Tringa t. totanus* (Continental Redshanks, wing length less than 168 mm) and *T.t. robusta* (Icelandic Redshanks, wing length larger than 168 mm), sexes are combined, adults and first calendar-year birds are separated. All birds were caught on the island of Vlieland from December 1971-December 1975 (Boere, unpubl.).

		adult birds			first calendar-year birds		
		mean	s.d.	n	mean	s.d.	n
January	T.t. totanus	147	14	14			
	T.t. robusta	176	17	85			
February	T.t. totanus	136		1			
	T.t. robusta	158	15	3			
March	T.t. totanus	132	26	3			
	T.t. robusta	154	18	40			
April	T.t. totanus	131	27	12			
	T.t. robusta	194	24	26			
May	T.t. totanus	127	14	27			
	T.t. robusta						
June	T.t. totanus	125	14	41	118		1
	T.t. robusta	139	8	12			
July	T.t. totanus	137	20	231	123	21	36
	T.t. robusta	151	15	81	165	20	3
August	T.t. totanus	138	21	90	128	15	76
	T.t. robusta	152	9	149	143	11	50
September	T.t. totanus	149	23	14	147	13	10
	T.t. robusta	152	9	18	156	16	12
October	T.t. totanus	157	17	10	159	13	9
	T.t. robusta	179	22	53	168	17	33
November	T.t. totanus	168	16	6	175		2
	T.t. robusta	184	16	85	168	11	19
December	T.t. totanus	166	17	3	159	15	10
	T.t. robusta	192	22	26	172	13	14

The distribution of the population breeding in the Wadden Sea area
is shown in fig. 137. As a rule only Redshanks breeding on the islands
and in the salt marshes along the coast have been taken into account.
Exceptions are the populations breeding in the Hauke Haienkoog and the
Lauwersmeer. Insufficient information was available on size and distribution of the breeding population along the mainland coast in Schleswig-Holstein and Niedersachsen. Redshanks will be more numerous in this
area than indicated on the map. The size of the population amounts to:

Denmark   : 1100-1250 pairs
Schleswig-Holstein   : at least 3000 pairs
Niedersachsen   : at least 4000 pairs
The Netherlands   : 2300-2700 pairs

The total size of the population breeding in the Wadden Sea area
amounts to at least 11,400 pairs (sources: Glutz et al., 1977; Dansk
Ornithologisk Forening; Institut für Vogelforschung, Wilhelmshaven;
Smit, unpubl.).

3.25.3.2 Numbers per area

In the Danish part of the Wadden Sea Redshanks are most numerous
in August when up to 10,000 may be present. In July and in September
and October 2000-4000 have been observed, in winter and early spring
numbers drop somewhat. Spring maxima occur in April and May with 2000-
3000 (Meltofte, 1980; Vadefuglegruppen Dansk Ornithologisk Forening).

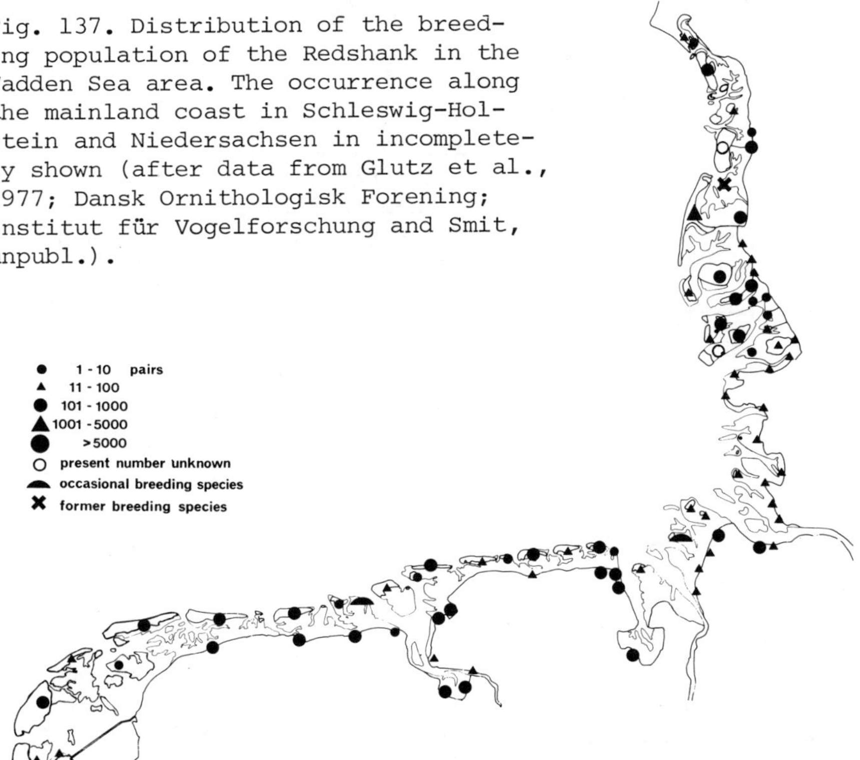

Fig. 137. Distribution of the breeding population of the Redshank in the
Wadden Sea area. The occurrence along
the mainland coast in Schleswig-Holstein and Niedersachsen in incomplete-
ly shown (after data from Glutz et al.,
1977; Dansk Ornithologisk Forening;
Institut für Vogelforschung and Smit,
unpubl.).

    In Schleswig-Holstein especially the relatively large numbers in May
and June are remarkable. Numbers in April and May may include breeding birds. Numbers in June are unsuccessful breeding birds from coastal areas and probably also from further inland, which gather along the
Wadden Sea coast. Whether summering Redshanks contribute to the numbers present in June is unknown. Numbers increase in July when adult
and juvenile birds from the breeding population in Schleswig-Holstein
arrive and probably also the first birds of the Fenno-Scandinavian population. Numbers in July amount to about 16,000 birds. Fig. 138 shows
that in August and afterwards numbers decrease gradually. In mild winters about 1500 Redshanks are present in the area (Busche, 1980).

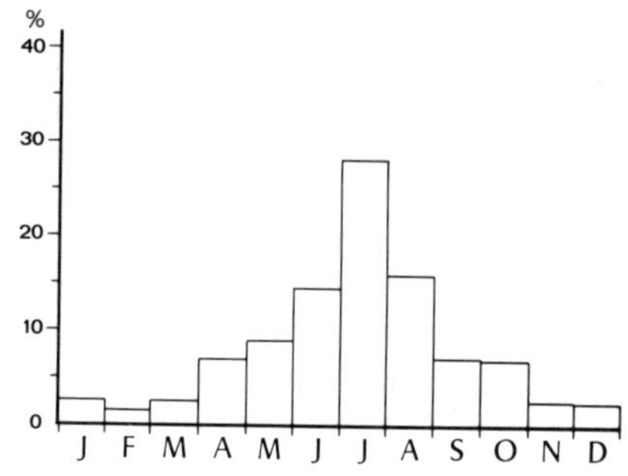

Fig. 138. Occurrence of Redshanks
in the Wadden Sea area in Schleswig-Holstein. Numbers for each
month are expressed as a percentage of the total number observed in all months (after Busche,
1980).

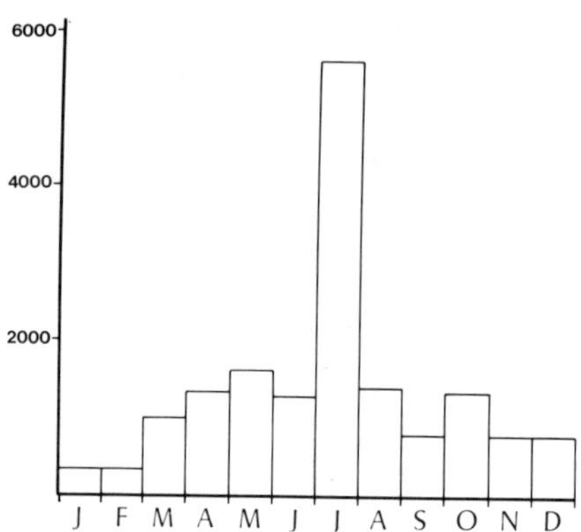

Fig. 139. Mean number of Redshanks per month present in a
50 km long stretch along the
mainland coast in Niedersachsen
(after Smit, 1977 and data from
the Institut für Vogelforschung,
Wilhelmshaven).

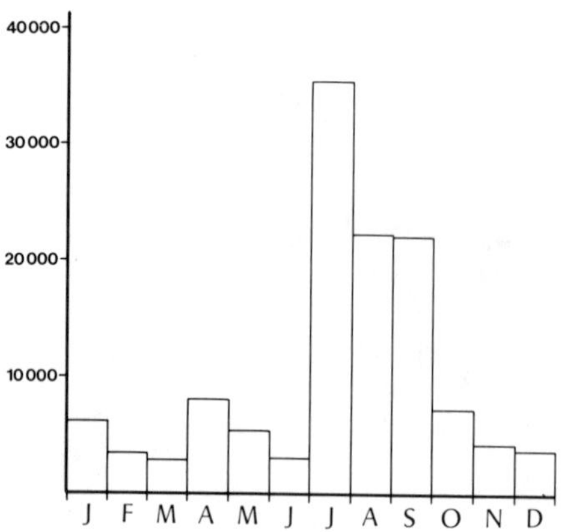

Fig. 140. Mean number of Redshanks
per month in the Dutch part of
the Wadden Sea (after data from
Smit, 1977).

Phenology along the mainland coast in Niedersachsen is somewhat comparable to the Netherlands with relatively high numbers in July, a limited number in winter and a relatively low peak in spring (fig. 139). Mean numbers for six months from the islands of Wangerooge, Mellum, Scharhörn and Neuwerk show a quite different phenology, more or less comparable to that of Schleswig-Holstein. Mean numbers were for April 152, for May 328; June 430; July 636; August 1048 and September 520 (data: Institut für Vogelforschung, Wilhelmshaven; Smit, 1977).

Fig. 140 shows that in the Dutch part of the Wadden Sea Redshanks are most numerous in July when a mean number of about 35,000 is present in the area. Numbers drop considerably afterwards. From October until the following June numbers do not vary considerably. During winter 3000-6000 Redshanks are present (after Smit, 1977).

Actually counted numbers for the whole Dutch part of the Wadden Sea are listed in Table 46. Because *T. t. totanus* and *T. t. robusta* are difficult to distinguish in the field, Wadden Sea counts give no information on their occurrence. Fig. 141 shows that in July, August and September large numbers of Redshanks occur along the coast of Friesland. Timmerman (1974) gives a maximum of 32,000 for July. Maximum numbers in the area drop to 21,500 in August and 16,750 in September. In autumn relatively large numbers are also present along the mainland coast of Groningen and in the Ballumerbocht-area south of the island of Ameland. All these areas have a muddy soil.

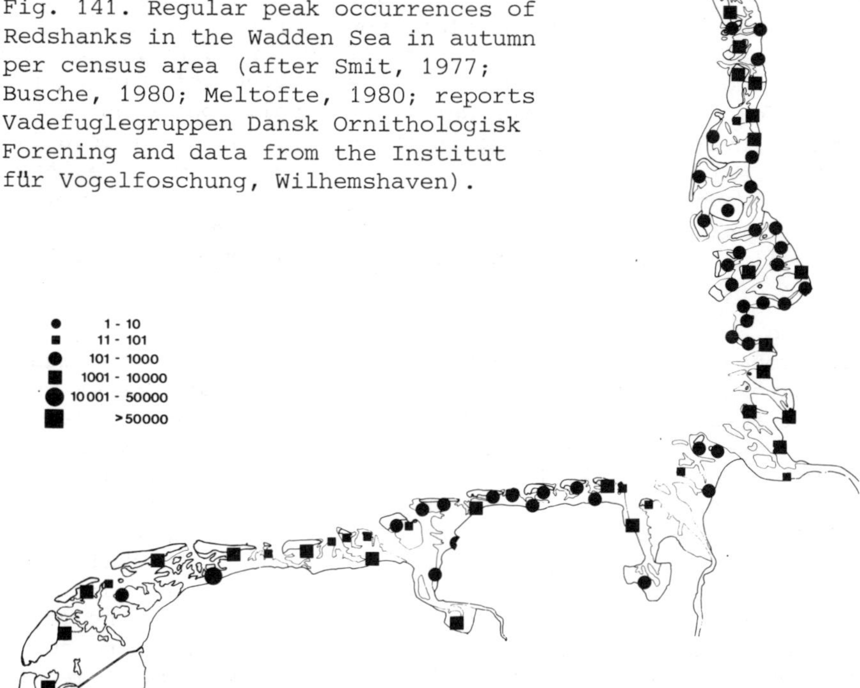

Fig. 141. Regular peak occurrences of Redshanks in the Wadden Sea in autumn per census area (after Smit, 1977; Busche, 1980; Meltofte, 1980; reports Vadefuglegruppen Dansk Ornithologisk Forening and data from the Institut für Vogelfoschung, Wilhemshaven).

Table 46. Numbers of Redshanks in the Dutch part of the Wadden Sea (after various publications in Watervogels and Limosa and Zegers, in litt.).

Date	Year	Number
January 8	1977	8,680
January 12	1974	7,730
January 17	1976	11,610
January 18	1975	9,410
April 6	1973	17,400
April 19	1975	10,715
May 1	1976	5,830
May 11	1974	4,780
July 29	1972	37,760
August 30	1975	28,950
September 1	1973	34,750
October 19	1974	10,950
November 13	1976	12,775
December 29	1966	4,200

3.25.4 Food

3.25.4.1 Food composition

Food composition of the Redshank has been studied in detail in England (Goss-Custard, 1969, 1970, 1976, 1977; Goss-Custard & Jones, 1976). Information from the Wadden Sea area is less detailed. Prey items found by Zwarts (1974) on the flats south of the island of Schiermonnikoog were *Corophium, Nereis*, small molluscs, amphipods and some insects. Along the mainland coast in the province of Groningen De Vlas (1970) found that in winter *Nereis* and small specimens of *Carcinus* and small invertebrates living in the salt marshes (especially spiders and beetles) were taken. In summer main prey items were *Carcinus, Nereis*, and *Crangon*. Apart from *Nereis* in all seasons polychaetes were eaten to a smaller extent *(Harmothoe, Nephtys, Lanice* and *Pectinaria)*. Some molluscs were found in summer too *(Hydrobia, Macoma)*.

3.25.4.2 Feeding activities

Very large concentrations of Redshanks are present along the muddy coast of the mainland. Also birds roosting on the islands prefer muddy shallows for feeding. The species is very scarce in sandy areas and absent on beaches. A detailed account of foraging activity and consumption of *Corophium* on the tidal flats near Schiermonnikoog is given by Zwarts (1974). Foraging occurs by pecking as well as by bill swaying. High tide roosts vary from place to place and from tide to tide, but are mainly situated on salt marshes, coastal meadows or in shallow pools. During a spring high tide the roosts may be situated far from the water on the salt marshes, during a neap low tide along the edges of the salt marshes (Zwarts & Zwarts, 1969). Along the mainland coast, large roosts can be present on wet meadows several hundreds of meters from the water. They often share the roost with Greenshanks and Dunlins and sometimes Curlew sandpipers.

3.25.4.3 Total food consumption

Zwarts (1974) determined that Redshanks foraging on *Corophium* took 10-50 preys per minute. He calculated that one Redshank may take 7500-12,000 specimens of *Corophium* per tidal cycle. *Corophium* densities in the area amounted to 5000-20,000.m^{-2}.

References

Boer, P., 1966. Enkele wintergegevens van de Tureluur. Limosa 39: p. 214-216.

Boere, G.C., 1976. The significance of the Dutch Wadden Sea in the annual life cycle of arctic, subarctic and boreal waders. Part 1. The function as a moulting area. Ardea 64: p. 210-291.

Busche, G., 1980. Vogelbestände des Wattenmeeres von Schleswig-Holstein. Kilda, Greven (in press).

Dick, W.J.A., (ed.) 1975. Oxford and Cambridge Mauritanian Expedition 1973. Cambridge: 79 pp.

Fournier, O. & F. Spitz, 1969. Etude biométrique et cycle de présence des populations de Tringa totanus stationnant dans le Sud de la Vendée. Oiseau 39: p. 242-251.

Glutz von Blotzheim, U.N., K.M. Bauer & E. Bezzel, 1977. Handbuch der Vögel Mitteleuropas, Vol. 7. Akademische Verlagsgesellschaft Wiesbaden: 894 pp.

Goethe, F., 1972. Isländische Rotschenkel an der winterlichen Jade. Vogelkdl. Ber. Niedersachsen 4: p. 34-38.

Goss-Custard, J.D., 1969. The winter feeding ecology of the Redshank (Tringa totanus). Ibis 111: p. 338-356.

Goss-Custard, J.D., 1970. The respons of Redshank (Tringa totanus L.) to spatial variation in the density of their prey. J. Anim. Ecol. 39: p. 91-113.

Goss-Custard, J.D., 1976. Variation in the dispersion of Redshank Tringa totanus on their winter feeding grounds. Ibis 118: p. 257-263.

Goss-Custard, J.D., 1977. Optimal foraging and the size selection of worms by Redshank, Tringa totanus, in the field. Anim. Behav. 29: p. 10-29.

Goss-Custard, J.D. & R.E. Jones, 1976. The diets of Redshank and Curlew. Bird Study 23: p. 233-243.

Grosskopf, G., 1971. Ringfunde des Rotschenkels. Auspicium 4: p. 311-323.

Hale, W.G., 1971. A revision of the taxonomy of the Redshank. Zool. J. Linn. Soc. 50: p. 199-268.

Hale, W.G., 1973. The distribution of the Redshank in the winter range. Zool. J. Linn. Soc. 53: p. 177-236.

Harrison, J.M., 1944. Some remarks upon the Western Palaearctic races of Tringa totanus (Linnaeus). Ibis 86: p. 493-503.

Johnson, A.R., 1974. Wader Research in the Camargue. Proc. IWRB Wader Symposium Warsaw 1973, Warsawa: p. 63-82.

Mackie, P., 1976. A short note on Redshank in the Upper Clyde Estuary.
    Wader Study Group Bull. 17: 5-10.
Meltofte, H., 1980. Fugle i Vadehavet. Vadefugletaellinger i Vadehavet
    1974-1978. Miljøministeriet, Fredningsstyrelsen, København: 50 pp.
Meltofte, H., S. Pihl & B.M. Sørensen, 1972. Efterraarstraekket af Va-
    defugle (Charadrii) ved Blaavandshuk 1963-1971. Dansk Ornithologisk
    Foren. Tidsskr. 66: p. 63-69.
Ogilvie, M.A., 1963. The migration of the European Redshank and Dunlin.
    Wildfowl 14: p. 141-149.
Pienkowski, M.W., P.J. Knight, D.J. Stanyard & F.B. Argyle, 1976. The
    primary moult of waders on the Atlantic coast of Morocco. Ibis 118:
    347-365.
Piersma, T., M. Engelmoer, W. Altenburg & R. Mes, 1980. A wader-expe-
    dition to Mauritania. Wader Study Group Bulletin 29: p. 14.
Prater, A.J., 1976. The distribution of coastal waders in Europe and
    North Africa. Proceedings Intern. Conference of Conservation of Wet-
    lands and Waterfowl, Heiligenhafen 1974. IWRB, Slimbridge: p. 255-
    271.
Salomonsen, F., 1954. The migration of the European Redshank. Dansk
    Orn. Foren. Tidskr. 48: p. 94-122.
Smit, C.J., 1977. On the occurrence of 32 bird species in the Danish,
    German and Dutch Wadden Sea. Unpubl. report Intern. Wadden Sea
    Working Group, part 3: 174 pp.
Timmerman, A., 1974. Over de betekenis van de Friese Waddenkust tussen
    Harlingen en Lauwersoog voor wad- en watervogels. Oenkerk: 54 pp.
Verheyen, R., 1947. La question des races geographiques du Chevalier
    gambette en Belgique. Gerfaut 37: p. 113-116.
Verheyen, R. & G. Le Grelle, 1950. La migration de la Gambette d'Europe
    d'après les résultats du bagage. Gerfaut 40: 201-206.
Vlas, J. de, 1970. Een inleidend onderzoek naar de aantallen vogels,
    hun potentiële prooidieren, en het dieet van de Tureluur in een ge-
    bied ten noorden van de Groninger kust, nabij een toekomstig lozings-
    punt voor afvalwater. Doctoraalscriptie University of Groningen:
    78 pp.
Voous, K.H., 1960. Atlas of European birds. Nelson, London: 284 pp.
Zwarts, L., 1974. Voedselopname en fourageergedrag van Tureluurs op
    het wad onder Schiermonnikoog. Schierboek 5: p. 183-227.
Zwarts, L. & J. Zwarts, 1969. Tureluurs in de Oosterkwelder 1966.
    Schierboek 3: p. 191-200.

3.26 GREENSHANK (*TRINGA NEBULARIA* (GUNNERUS))
G.C. Boere & C.J. Smit

Da: Hvidklire; G: Grünschenkel; Du: Groenpootruiter

3.26.1 Distribution

3.26.1.1 Breeding area
The Greenshank has a trans-palaearctic distribution, breeding mainly in the boreal, but also in the tundra climate zone. The breeding

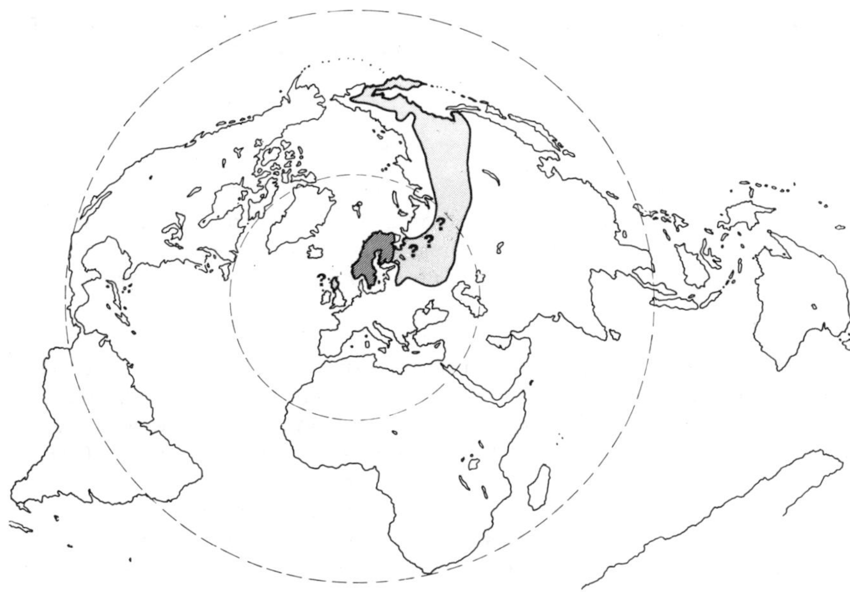

Fig. 142. Breeding area of the Greenshank (afer Voous, 1960). The dark
shaded area represents the region where birds visiting the Wadden
Sea originate from. Question marks indicate a region for which is
uncertain whether birds breeding there pass through the Wadden Sea.

area ranges from Scotland as far as Eastern Siberia (fig. 142). The
number of ringing recoveries is small and gives only a slight indication of the origin of the birds present in the Wadden Sea. Probably
they come from Scandinavia and as far as Western Russia.

No geographical variation or subspecies have been described (Voous,
1960; Glutz et al., 1977; Nethersole-Thompson, 1979).

3.26.1.2 Migration routes

As the Spotted redshank no particular migration routes can be distinguished. Migration takes place in a broad front across Europe and
Asia. Greenshanks can be observed at all suitable places along the
coast and inland. Larger concentrations, as in the Wadden Sea, occur
only locally.

3.26.1.3 Wintering areas

The wintering area ranges from NW Europe (incidentally) to tropical
Africa, far south of the Sahara, for instance along river borders and
lakes and in swamps. Probably Greenshanks winter further south in Africa than Spotted redshanks (Glutz et al., 1977). Some birds occasionally winter in the Wadden Sea. Local concentrations are rare and the species is widely spread over the whole wintering area.

3.26.1.4 Moulting areas

There are no particular moulting areas known, as moult is spread
over a long period and takes place in the whole wintering area as

well as during migration. One particular bird probably may moult at several places after periods of arrested moult (Boere, 1976).

3.26.2 Annual cycle

3.26.2.1 Migration
   Departure from the breeding area starts in the end of June and continues till early September. Maximum numbers passing Blaavandshuk are seen in the last July pentade and the first August decade (Meltofte et al., 1972).
   During the first half of July the number of Greenshanks in the Wadden Sea increases rapidly and peak numbers are present from late July until early September. During this period probably many birds have passed without staying for a long period. Depending on the weather conditions hundreds of birds can still be present by the end of October or even early November. Observations of single birds are known from the winter period. Spring migration is observed from April until the end of May with largest numbers being present during May. Total figures however are very much beneath those present during autumn migration.

3.26.2.2 Moult
   The timing of moult within the species and between individual birds shows great variation. Only a small percentage of the adult birds present in the Wadden Sea in July and August starts post-nuptial moult. Whenever this happens it is arrested to be finished elsewhere, mainly in the wintering areas in Africa. There is some evidence from collected primaries on the high tide roosts at Vlieland (Boere, 1976) that few birds probably undergo a complete moult in the Wadden Sea. First calendar-year birds probably undergo body moult, but good evidence is not available.
   There is no information about the pre-nuptial moult, which probably takes place in the wintering area as birds observed during spring migration are in breeding plumage.

3.26.2.3 Weight changes
   There is only information from the Dutch part of the Wadden Sea and from a restricted period. It is listed in Table 47.

3.26.3 Numbers

3.26.3.1 Population size
   Prater (1976) estimated the number of Greenshanks wintering in NW Europe and W Africa at 2000, 800 of these wintering on the Banc d'Arguin (Mauritania). In the winter of 1980 1450 have been counted there (Piersma et al., 1980). Numbers wintering further south in Africa are largely unknown but will exceed by far the numbers counted.

Table 47. Mean monthly weights, sexes combined, of adult and first calendar-year Greenshanks, caught on the island of Vlieland, December 1971-December 1977 (Boere, unpubl.).

month	adult birds			first calendar-year birds		
	mean	s.d.	n	mean	s.d.	n
May	200		1			
June	180	15	49			
July	188	21	68	131	8	4
August	201	38	6	140	12	5
September	241	20	3	246	55	2
October	221	19	2			

3.26.3.2 Numbers per area

In the Danish Wadden Sea Greenshanks are somewhat more numerous along the mainland coast (fig. 143). Peak numbers in autumn are about 1750, in spring about 900. The species is most numerous in July, August and May. Numbers in winter and summer are very low, rarely exceeding some tens (Meltofte, 1980; Vadefuglegruppen Dansk Ornithologisk Forening).

Fig. 143. Regular peak occurrences of Greenshanks in autumn per census area (after Smit, 1977; Busche, 1980; Meltofte, 1980; reports Vadefuglegruppen Dansk Ornithologisk Forening and data from the Institut für Vogelforschung, Wilhelmshaven).

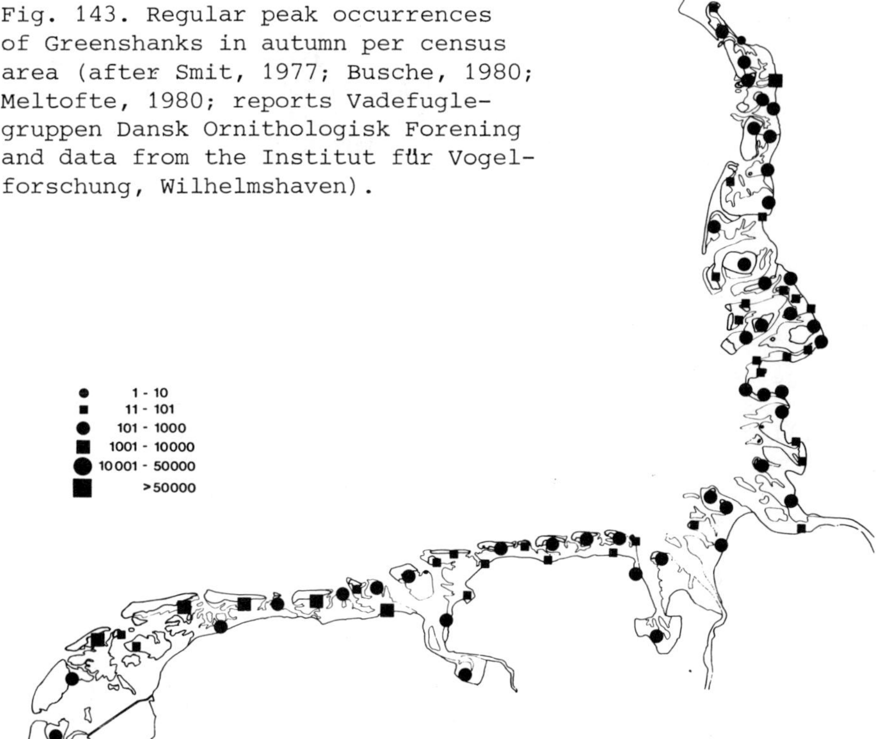

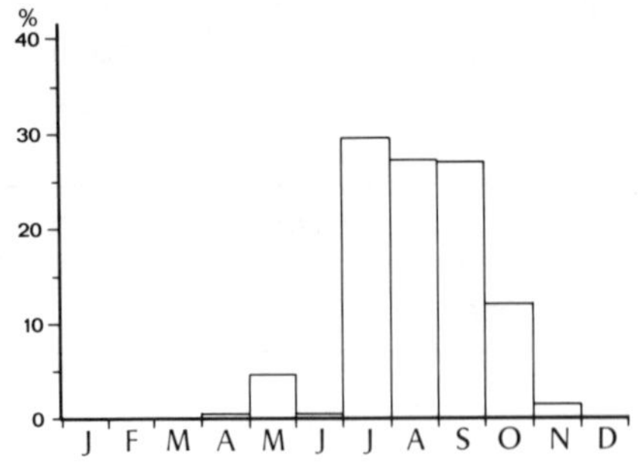

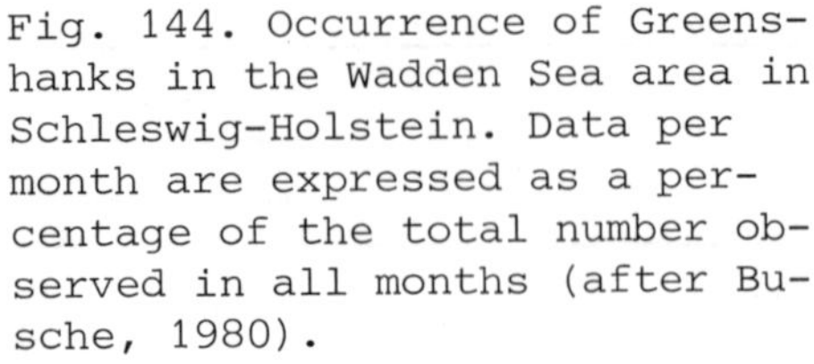

Fig. 144. Occurrence of Greenshanks in the Wadden Sea area in Schleswig-Holstein. Data per month are expressed as a percentage of the total number observed in all months (after Busche, 1980).

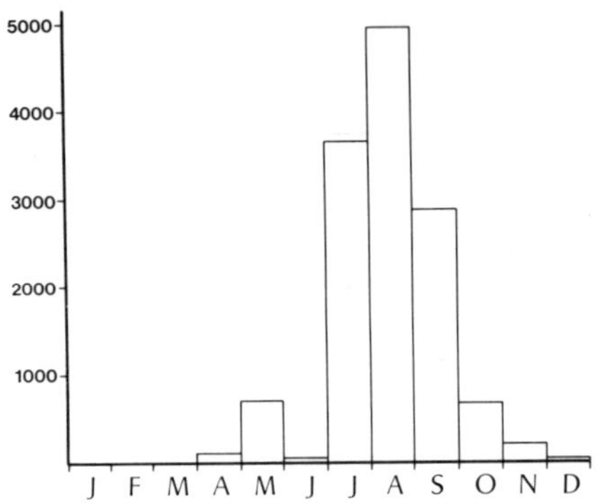

Fig. 145. Mean number of Greenshanks per month in the Dutch part of the Wadden Sea (after data from Smit, 1977).

The first Greenshanks to arrive in Schleswig-Holstein were seen on June 18. Fig. 144 shows that the majority of birds is present from July until October. Numbers in July and August may reach 3000. Generally the birds appear not to stay in the area for a long time. They are not observed from December until March. Spring migration is inconspicuous and peaks in May (Busche, 1980).

Numbers in Niedersachsen are relatively low as well. The phenology pattern resembles that of Schleswig-Holstein. Mean numbers per month for the area Wangerooge, Mellum, Scharhörn, Grosser Knechtsand and Neuwerk from April-September are: April 7; May 74; June 0; July 95; August 277 and September 244 (data: Institut für Vogelforschung, Wilhelmshaven; Smit, 1977). On September 1, 1973 in Schleswig-Holstein and Niedersachsen together 4770 have been counted.

The arrival- and departure pattern in the Dutch part of the Wadden Sea is quite like that in Niedersachsen and Schleswig-Holstein (fig. 145). In autumn on most of the islands 1000-2000 Greenshanks are observed. It is striking that numbers in spring are so low, especially in the Western part of the area. Greenshanks do not show a clear preference for a certain part of the area. Numbers actually counted in the Dutch part of the Wadden Sea are listed in Table 48.

3.26.4 Food

3.26.4.1 Food composition

Food composition was studied in detail by Swennen (1971) in the

Table 48. Numbers of Greenshanks in the Dutch part of the Wadden Sea (after various publications in Watervogels and Limosa and Zegers, in litt.)

Date	Year	Number
January 8	1977	2
January 12	1974	25
January 17	1976	7
January 18	1975	6
April 6	1973	30
April 19	1975	80
May 1	1976	750
May 11	1974	1,080
July 29	1972	5,640
August 22	1963	3,500
August 30	1975	5,600
September 1	1973	4,320
October 19	1974	940
November 13	1976	13
December 29	1966	0

Dutch Wadden Sea. Analysis of pellets shows a preference for *Pomatoschistus microps* and *Crangon crangon* during summer. Later on in the season polychaetes and small specimens of *Carcinus* are taken. The disappearance of the main prey items during the winter, must be considered as an important reason for the annual fluctuation and decrease of numbers present during late autumn and early winter.

3.26.4.2 Feeding activities

Greenshanks feed mainly in shallow water and creeks with a sandy bottom. Also more muddy sites are frequented, especially when areas with shallow water are present there as well. As high tide roosts Greenshanks prefer large open salt marshes, with a short vegetation, sometimes the shores of ponds and creeks or reclaimed polders very close to the tidal flats. The species is very shy. It often roosts together with Redshank and Dunlin.

3.26.4.3 Total food consumption

No information available from the area.

References

Boere, G.C., 1976. The significance of the Dutch Wadden Sea in the annual life cycle of arctic, subarctic and boreal waders. Part 1. The function as a moulting area. Ardea 64: p. 210-291.

Busche, G., 1980. Vogelbestände des Wattenmeeres von Schleswig-Holstein. Kilda, Greven (in press).

Glutz von Blotzheim, U.N., K.M. Bauer & E. Bezzel, 1977. Handbuch der Vögel Mitteleuropas, Vol. 7. Akademische Verlagsgesellschaft, Wiesbaden: 895 pp.

Meltofte, H., 1980. Fugle i Vadehavet. Vadefugletaellinger i Vadehavet 1974-1978. Miljøministeriet, Fredningsstyrelsen, København: 50 pp.

Meltofte, H., S. Pihl & B.M. Sørensen, 1972. Efteraarstraekket af Vadefugle (Charadrii) ved Blaavandshuk 1963-1971. Dansk Orn. Foren. Tidsskr. 66: p. 63-69.

Nethersole-Thompson, D. & M., 1979. Greenshanks. Poyser, Berkhamsted: 275 pp.

Piersma, T., M. Engelmoer, W. Altenburg & R. Mes, 1980. A wader-expedition to Mauritania. Wader Study Group Bull. 29: p. 14.

Prater, A.J., 1976. The distribution of coastal waders in Europe and North Africa. In: M. Smart (ed.). Proceedings Intern. Conference on the Conservation of Wetlands and Waterfowl, Heiligenhafen 1974. IWRB, Slimbridge: p. 255- 271.

Smit, C.J., 1977. On the occurrence of 32 bird species in the Danish, German and Dutch Wadden Sea. Unpubl. report Intern. Wadden Sea Working Group, part 3: 174 pp.

Swennen, C., 1971. Het voedsel van de Groenpootruiter, Tringa nebularia tijdens het verblijf in het Nederlandse Waddenzeegebied. Limosa 44: p. 71-83.

Voous, K.H., 1960. Atlas of European birds. Nelson, London: 284 pp.

3.27 TURNSTONE (*ARENARIA INTERPRES* L.)
G.C. Boere & C.J. Smit

Da: Stenvender; G: Steinwälzer; Du: Steenloper

3.27.1 Distribution

3.27.1.1 Breeding area

The Turnstone is breeding circumpolarly holearctic, mainly in the tundra but also in the boreal and temperate climate zones. In Europe it is a rare breeding bird along the coast in Scandinavia and the Baltic Sea. Two subspecies are described: *Arenaria interpres interpres* breeding in Eurasia, NE Canada, Greenland, Spitsbergen and *Arenaria interpres morinella* breeding in North-America (fig. 146). Ringing results have shown that birds migrating through or wintering in the Wadden Sea originate from the Eurasian breeding population, probably as far as the Taymyr peninsula, and from the breeding area in NW Canada and Greenland (fig. 147). Only birds belonging to *A.i. interpres* occur in the Wadden Sea. *A.i. morinella* winters in Central- and South-America (Nettleship, 1973; Morrison, 1975, 1976, 1977; Glutz et al., 1977; Branson et al., 1978).

3.27.1.2 Migration routes

The Turnstone is a long-distance migratory species. After the breeding season it concentrates along seashores, but migration probably also takes place overland. The exact routes from the breeding area in Central Russia to Western Europe are not known. Observations of individual birds inland in Europe refer to first calendar-year birds.

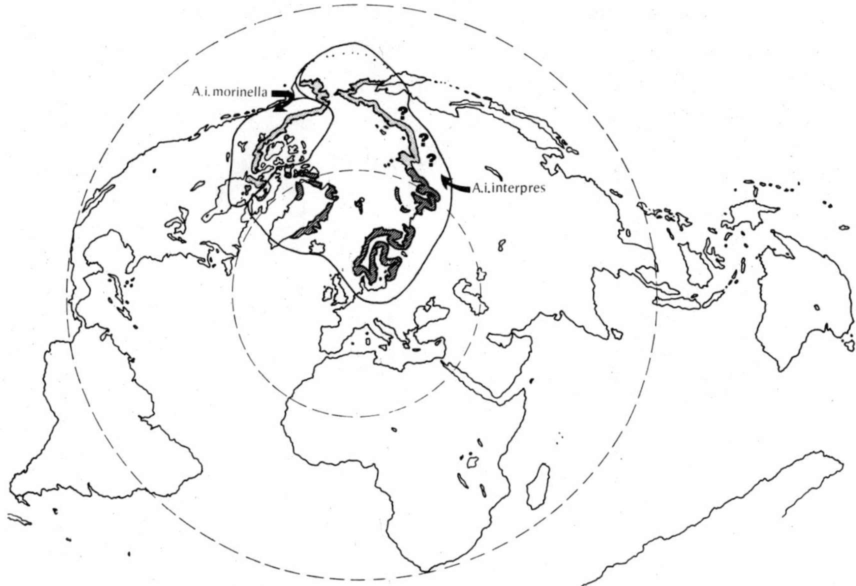

Fig. 146. Breeding area of the Turnstone (after Voous, 1960). The
dark shaded area indicates approximately the region where birds
visiting the Wadden Sea originate from.

3.27.1.3 Wintering areas

In winter the species is almost cosmopolitan. It can be observed
on every seashore, even on the smallest ocean islands. Larger concentrations occur at suitable areas of very different type such as coastal shallows, sandy beaches, but also harbours where fish garbage can
be found, seal hunting stations and even cocopalm plantations (Bailey,
1967; Thompson, 1973; Thompson & De Long, 1967).

The species winters as far south as South-Africa, where large numbers spend the Northern summer as well, and as far north as the frost
line in Scandinavia.

3.27.1.4 Moulting areas

Large parts of the wintering area serve as a moulting place for the
adult post-nuptial moult. The Wadden Sea is an important moulting area.
Other concentrations of moulting birds have been described for areas
in Europe as well as in North Africa, south to the Banc d'Arguin (Mauritania) (Dick, 1975; Boere, 1976; Pienkowski et al., 1976).

3.27.2 Annual cycle

3.27.2.1 Migration

From the first July decade when the first groups arrive in the Wadden Sea, numbers increase to a mid-August peak. At Blaavandshuk autumn
migration of adults peaks by the end of July (Meltofte et al., 1972).

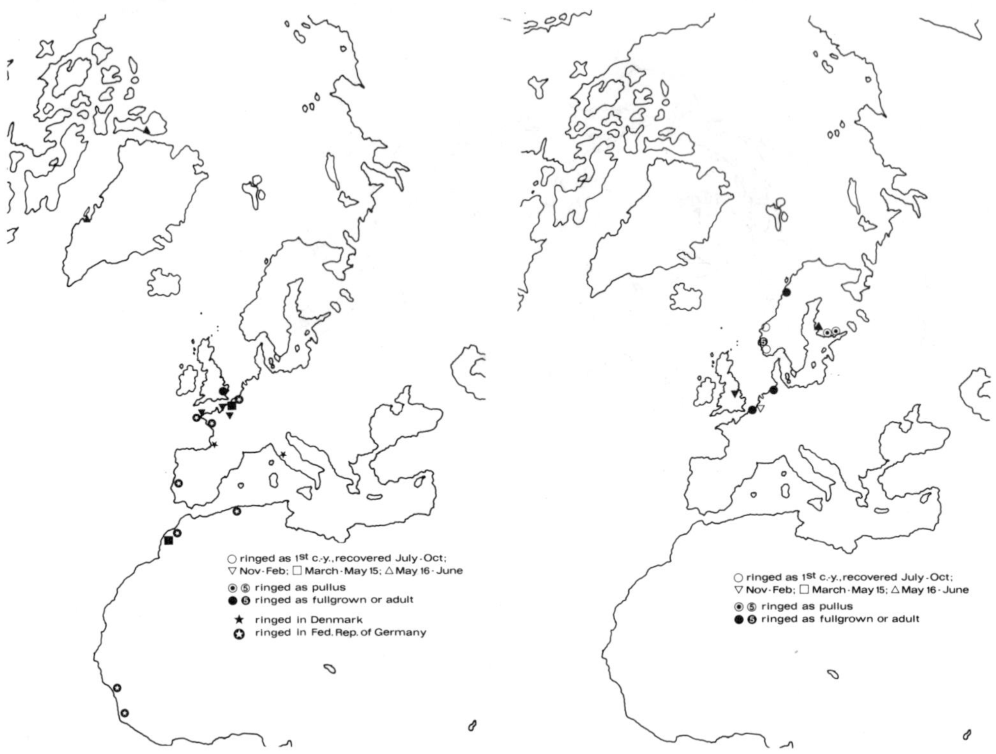

Fig. 147a. Recoveries of Turnstones ringed in the Dutch part of the Wadden Sea, Denmark and the Fed. Rep. of Germany (based on published records from 1911-1976).

Fig. 147b. Ringing places of Turnstones recovered in the Dutch part of the Wadden Sea (based on published records from 1911-1976).

The first birds to arrive probably are non-breeders, birds of the Scandinavian-Baltic-Russian populations and also some second calendar-year birds. The majority of these birds has left the Wadden Sea area by the end of September to winter in Africa. The populations breeding in Greenland and NE Canada are known to leave the breeding area from mid-July on. There are indications that part of this population migrates via the SW Norwegian coast southward towards W Jutland and other parts of W Europe (Salomonsen, 1979; Meltofte & Rabøl, 1977). These populations winter along the W European and NW African coasts (Glutz et al., 1977). Short distance movements because of cold spells occur during winter. Generally these birds do not move further than the edges of the frost areas.

In the Wadden Sea spring migration starts early April but numbers are not as high as in autumn. Peak numbers occur during May and migration is over by the end of May. Small numbers stay in the Wadden Sea during summer.

3.27.2.2 Moult

The adult post-nuptial moult starts in the Wadden Sea by late July and early August and is completed at the end of October. Second calendar-year birds start their moult at the end of May and have completed it by late August or early September. First calendar-year birds moult body feathers spread over a long period from the end of August until December. Pre-nuptial moult in spring, by which the birds receive their breeding plumage, involves a great part of the body feathers. It takes place from the end of February until the end of May. The first birds showing full breeding plumage can be observed by late April (Boere, 1976).

3.27.2.3 Weight changes

Table 49 shows the mean weights per month of birds trapped in the Dutch part of the Wadden Sea. Other information is not available at the moment. The strong increase in weight during May suggests a long non-stop flight to the breeding areas.

3.27.3 Numbers

3.27.3.1 Population size

Prater (1976) estimates the size of the population wintering in NW Europe and N Africa at 23,000, which in his eyes is a highly inaccurate number.

Table 49. Mean monthly weights, sexes combined, of adult and first calendar-year Turnstones, caught on the island of Vlieland from December 1971-December 1975 (Boere, in Glutz et al., 1977). Weights of birds in October, November and December are marked with an ✶. These birds were caught by dazzle netting and are probably not comparable with the others as this method tends to be more successful for catching birds not in optimal condition.

month	adult birds			first calendar-year birds		
	mean	s.d.	n	mean	s.d.	n
January	107.7	8.2	10			
February	113.1	10.5	12			
March	104.7	5.5	24			
April	113.1	8.6	9			
May	144.5	17.2	130			
June	104.5	3.5	2			
July	111.4	11.2	147	100.5	7.8	2
August	114.5	14.0	356	109.0	16.2	51
September	110.2	10.8	19	110.7	9.6	40
October	107.5	6.6	4✶	102.4	9.2	5
November	108	6.2	3✶	111.9	6.1	14
December	110	9.3	4✶	118.0	11.8	9

More than a century ago the Turnstone was a breeding bird on some Wadden Sea islands in Schleswig-Holstein and Denmark (Glutz et al., 1977). In 1949 the species possibly bred on the island of Südfall (Goethe, 1952).

3.27.3.2 Numbers per area

Compared to other wader species Turnstones are not numerous. Therefore they can be missed easily during a wader count, especially because sometimes they roost on dike slopes, groynes and harbour dams.

In the Danish Wadden Sea Turnstones never occur in large numbers. Numbers in the whole Danish part of the Wadden Sea in autumn never exceed about 300, in spring generally less than 200 are present. From December until February only a few are there (Meltofte, 1980; Vadefuglegruppen Dansk Ornithologisk Forening).

Also in Schleswig-Holstein large numbers are never seen. Numbers rise in July and peak in August. By this time up to 2000-2500 may be present in the area. Numbers drop in September and October and again in November. In October in some areas (Hooge, Trischen) an influx of Turnstones has been recorded, resulting in relatively high numbers (Todt & Hillenbrand and Kappes in Busche, 1980). Probably these are newly arriving birds from Greenland. In mild winters some hundreds of Turnstones are present in the area. Numbers rise somewhat in March and again in May. The number of summering birds amounts to about 100-200 (Busche, 1980).

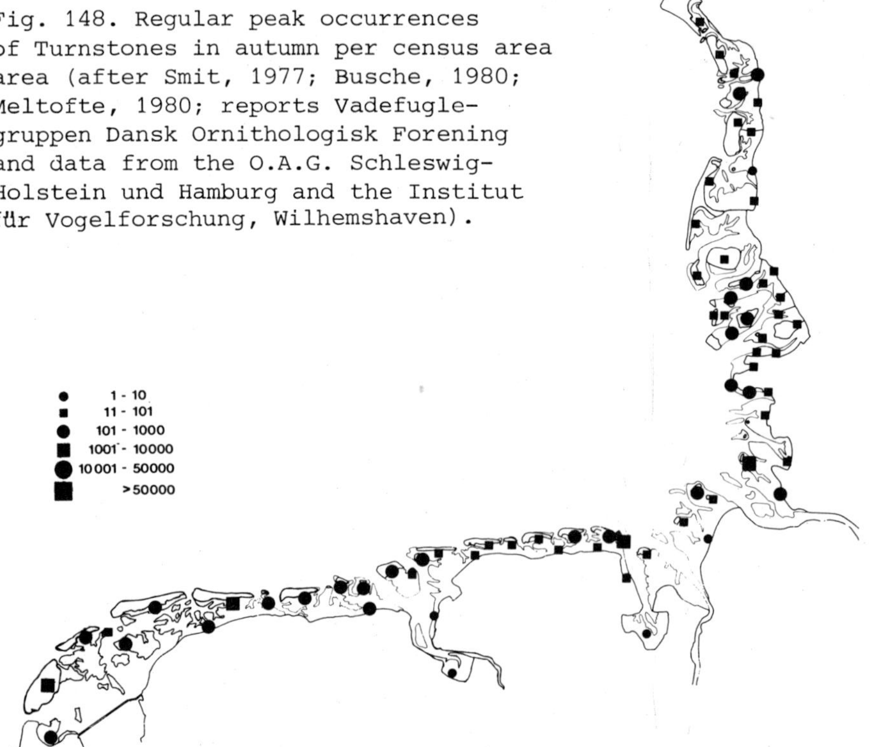

Fig. 148. Regular peak occurrences of Turnstones in autumn per census area area (after Smit, 1977; Busche, 1980; Meltofte, 1980; reports Vadefuglegruppen Dansk Ornithologisk Forening and data from the O.A.G. Schleswig-Holstein und Hamburg and the Institut für Vogelforschung, Wilhemshaven).

Table 50. Numbers of Turnstones in the Dutch part of the Wadden Sea (after various publications in Watervogels and Limosa and Zegers, in litt.).

Date	Year	Number
January 8	1977	2,030
January 12	1974	1,770
January 17	1976	1,720
January 18	1975	910
April 6	1973	1,340
April 19	1975	2,090
May 1	1976	1,645
May 11	1974	3,345
July 29	1972	6,720
August 22	1963	1,800
August 30	1975	3,250
September 1	1973	2,390
October 19	1974	875
November 13	1976	2,560
December 29	1966	1,300

Small numbers occur in the Wadden Sea in Niedersachsen. On April 27, 1975 474 Turnstones were counted in the area, in the beginning of May 1976 226, in September 1973 in Schleswig-Holstein and Niedersachsen together 1350. Turnstones are scarce along the mainland coast, but occur in relatively larger numbers in the Spiekeroog-Wangerooge-Oldeoog area (Grosskopf, 1968; Smit, 1977). Numbers increase from mid-July to peak in August, in some years as late as October. Even in mid-winter flocks of 200-300 may still be present on Wangerooge. In spring numbers increase somewhat, generally to peak in May (Grosskopf, 1968). On Scharhörn numbers in winter generally are very low (Temme, 1974).

In the Dutch part of the Wadden Sea Turnstones never occur in large numbers, though they are more common as compared to other parts of the Wadden Sea area, especially in autumn. In winter they can be rather numerous in some places. Fig. 148 shows mean numbers per month in the area. Numbers counted in the Dutch part of the Wadden Sea are listed in Table 50. Fig. 148 shows that maximum numbers on some islands in autumn sometimes are over 1000.

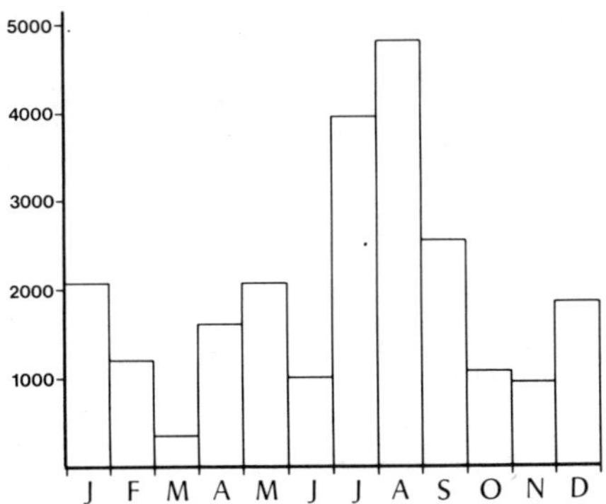

Fig. 149. Mean number of Turnstones per month in the Dutch part of the Wadden Sea (after data from Smit, 1977).

3.27.4 Food

3.27.4.1 Food composition

The number of invertebrate species taken as food varies enormously, depending on place and time of the year. Information is available from the German and Dutch Wadden Sea and includes especially small crustaceans (*Corophium*, *Gammarus*, *Carcinus*, *Balanus* and isopods). Under bad feeding conditions during winter Turnstones eat carrion and human refuse (bread, fish, meat, potatoes etc.) (Glutz et al., 1977).

3.27.4.2 Feeding activities

Main feeding areas are various types of tidal flats (sand as well as mud) and mussel beds, but also artificial rocky shores such as dikes and piers.

High tide roosts are very variable, even under normal tide conditions, and include sandy beaches, shell ridges, dikes slopes, short grassed salt meadows, harbour equipment and potato fields. Generally they are situated close to the water edge.

3.27.4.3 Total food consumption

No information available

References

Bailey, R.S., 1967. Migrant waders in the Indian Ocean. Ibis 109: p. 437-439.

Boere, G.C., 1976. The significance of the Dutch Waddenzee in the annual life cycle of arctic, subarctic and boreal waders. Part 1. The function as a moulting area. Ardea 64: p. 210-291.

Branson, N.B.J.A., E.D. Ponting & C.D.T. Minton, 1978. Turnstone migrations in Britain and Europe. Bird Study 25: p. 181-187.

Busche, G., 1980. Vogelbestände des Wattenmeeres von Schleswig-Holstein. Kilda, Greven (in press).

Dick, W.J.A. (ed.), 1975. Oxford and Cambridge Mauritanian Expedition 1973. Cambridge: 79 pp.

Goethe, F., 1952. Der Bestand an See- und Küstenvögeln an den wichtigsten Brutplätzen in der deutschen Bucht. J. Orn. 93: p. 199-202.

Glutz von Blotzheim, U.N., K.M. Bauer & E. Bezzel, 1977. Handbuch der Vögel Mitteleuropas, Vol. 7. Akademische Verlagsgesellschaft, Wiesbaden: 895 pp.

Grosskopf, G., 1968. Die Vögel der Insel Wangerooge. Abhandl. Vogelkunde 5. Institut für Vogelforschung, Wilhelmshaven: 293 pp.

Meltofte, H., 1980. Fugle i Vadehavet. Vadefugletaellinger i Vadehavet 1974-1978. Miljøministeriet Fredningsstyrelsen, København: 50 pp.

Meltofte, H., S. Pihl & B.M. Sørensen, 1972. Efteraarstraekket af vadefugle (Charadrii) ved Blaavandshuk 1963-1971. Dansk Orn. Foren. Tidsskr. 66: p. 63-69.

Meltofte, H. & J. Rabøl, 1977. Vejrets indflydelse paa efteraarstraekket af vadefugle ved Blaavandshuk, med et forsøg paa en analyse af traekkets geografiske oprindelse. Dansk Orn. Foren. Tidsskr. 71:

p. 43-63.

Morrison, R.J.G., 1975. Migration and morphometrics of European Knot
and Turnstone on Ellesmere Island, Canada. Bird Banding 46: p. 290-
301.

Morrison, R.J.G., 1976. Further records, including the first double-
journey recovery of Europeanbanded Ruddy Turnstones on Ellesmere
Island NWT. Bird Banding 47: p. 274.

Morrison, R.J.G., 1977. Migration of arctic waders wintering in Europe.
Polar Record 18: p. 475-486.

Nettleship, D.N., 1973. Breeding ecology of Turnstones Arenaria interpres at Hazen Camp, Ellesmere Island, N.W.T. Ibis 115: p. 202-217.

Pienkowski, M.W., P.J. Knight, D.J. Stanyard & F.B. Argyle, 1976. The
primary moult of waders on the Atlantic coast of Morocco. Ibis 118:
p. 347-365.

Prater, A.J., 1976. The distribution of coastal waders in Europe and
North Africa. In: M. Smart (ed.). Proceedings Intern. Conf. on Conservation of Wetlands and Waterfowl, Heiligenhafen 1974. IWRB, Slimbridge: p. 255-271.

Salomonsen, F., 1979. Fra Zoologisk Museum XXV. Dansk Orn. Foren. Tidsskr. 73: p. 191-206.

Smit, C.J., 1977. On the occurrence of 32 bird species in the Danish,
German and Dutch Wadden Sea. Unpubl. report Intern. Wadden Sea Working Group, part 3: 174 pp.

Temme, M., 1974. Vogelfreistätte Scharhörn. Jordsand Mitt. 3 (1-4):
p. 5-165.

Thompson, M.C. & R.L. Delong, 1967. The use of cannon- and rocket-
projected nets for trapping shorebirds. Bird Banding 38: p. 214-
218.

Thompson, M.C., 1973. Migratory patterns of Ruddy Turnstones in the
Central Pacific region. The Living Bird 12: p. 5-23.

Voous, K.H., 1960. Atlas of European birds. Nelson, London: 284 pp.

3.28 BLACK-HEADED GULL *(LARUS RIDIBUNDUS L.)*
F. Goethe

Da: Haettemaage; G: Lachmöwe; Du: Kokmeeuw, Kapmeeuw

3.28.1 Distribution

3.28.1.1 Breeding area

The Black-headed gull has a trans-palaearctic distribution, breeding in boreal, temperate, Mediterranean, steppe, and to a small extent
even in the desert climate zone (fig. 150). Since the second half of
the 19th century the Black-headed gull population from NW Europe has
largely expanded its range to the North and Northwest (Voous, 1960;
Vaurie, 1965), perhaps as a consequence of an amelioration of the climate. No subspecies are distinguished.

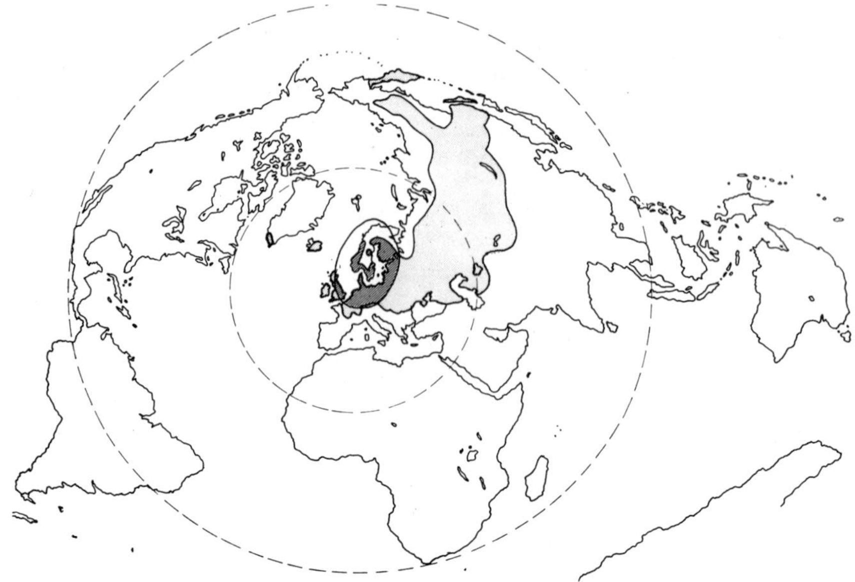

Fig. 150. Breeding area of the Black-headed gull (after Voous, 1960; Vaurie, 1965; Haftorn, 1971; Salomonsen, 1973). The dark shaded area represents the region where birds visiting the Wadden Sea originate from.

3.28.1.2 Migration routes

Breeding birds from W and N Norway partly cross the North Sea to the wintering areas on the British Isles. Part of the breeding birds from this area and the majority of those breeding in S Norway follow the coastline to migrate southward (Haftorn, 1971). Black-headed gulls breeding along the Baltic partly follow the coastline, partly migrate in a broad front in W and S directions, some following rivers and canals (Schüz, 1971). Breeding birds from Germany and the Netherlands may follow the coastline to reach winter quarters. Partly however these birds disperse in S and W directions as well. The same applies for Danish breeding birds. Black-headed gulls breeding in Switzerland, Czechoslovakia and Poland partly follow rivers to the north to reach the Baltic and North Sea (Niethammer, 1942; Milenz, 1961; Schüz, 1971 and unpubl. data from the Institut für Vogelforschung, Wilhelmshaven).

3.28.1.3 Wintering areas

Black-headed gulls breeding in the Western, Central and Northern part of Norway are mostly migratory, wintering especially in countries bordering the North Sea. As compared to those breedings in S Norway relatively many winter on the British Isles. Some Norwegian birds go as far south as Morocco and Senegal. Part of the Norwegian breeding birds winters along the westcoast of the country itself, north to Trondheim (Haftorn, 1971). Danish breeding birds are partly resident, partly winter on the British Isles, NW Germany, the Netherlands, Belgium, France, Spain, Portugal and NW Africa. In all these countries the majority winters along the coasts but as compared to other gull species also rather often along rivers, lakes and large cities inland. Recoveries of ringed Black-headed gulls in Southern Europe come from the Atlantic as well as from the Mediterranean coast (fig. 151). Black-headed gulls breeding in East Denmark on the average migrate further away from their breeding colonies than those breeding in the Western and Northern part of the country (Andersen-Harild, 1971).

German and Dutch birds more or less show the same wintering distribution as the Danish breeding birds (Niethammer, 1942; Eykman et al., 1949). First calendar-year British Black-headed gulls generally disperse southward. After their first winter they become much more sedentary (Radford, 1962). Black-headed gulls from Central Europe and the Baltic winter along coasts, rivers and lakes and canals all over Western, Central and Southern Europe, including the whole Wadden Sea area (Schüz, 1971).

3.28.1.4 Moulting areas

No distinct moulting areas occur. During the presence of large concentrations of birds in the Wadden Sea after the breeding season body moult of first calendar-year birds and a complete post-nuptial moult of adults and immatures occurs.

Fig. 151. Recoveries of Black-headed gulls, ringed in Jutland and Fyn (Denmark) from 1959-1969 (after Andersen-Harild, 1971). Figures refer to the number of recoveries in the specific area.

3.28.2 Annual cycle

3.28.2.1 Migration

Soon after fledging first calendar-year Norwegian Black-headed gulls disperse in southerly directions. Soon afterwards, still in July, they can be found along the British and continental West European coasts (Haftorn, 1971). Danish birds have been recorded on the same area in August (Salomonsen, 1972, 1973). Also first calendar-year birds from Finland, Sweden, Poland and Estonian SSR have been recovered in the Wadden Sea within two months after fledging. Young Dutch birds, but also a part of the adults, have been recovered along the French Atlantic coast in September and in NW Spain in October (Eykman et al., 1949). British first calendar-year birds behave rather identical as those from N and NE Europe. As soon as they are able to fly they disperse (Radford, 1962). Numbers in the Wadden Sea increase already in July. Peak numbers are recorded in the three months following.

Adult birds return to the breeding colonies in March and early April. Second- and third calendar-year birds show migratory movements north- and eastward too, but these are less conspicuous. These birds summer partly in the vicinity of their native breeding colonies, partly near the wintering areas and partly in between. The Wadden Sea acts as an important summering area for immature Black-headed gulls.

3.28.2.2 Moult

A partial moult of first calendar-year birds into the first winter plumage takes place from July until December. Adult birds undergo a complete post-nuptial moult from July, sometimes already June, until October or November. Adult birds receive their breeding plumage through a moult of body feathers between January and March. This prenuptial moult often continues until April or May (Dwight, 1925; Niethammer, 1942).

3.28.2.3 Weight changes

Weight changes of Black-headed gulls from the surroundings of Wilhelmshaven and the Western part of the Dutch Wadden Sea are listed in Table 51.

3.28.3 Numbers

3.28.3.1 Population size

The Black-headed gull expanded its range considerably in the past 100 years. The first breeding record for Norway took place in 1867, nowadays they already breed in the north of the country. In 1975 in the Jaeren colony alone 9500 pairs were breeding (1919 70 pairs, increasing to 1345 in 1960). The first nest on Iceland was found in 1911, and in Italy and Spain in 1960. A recent colonization is S Greenland. Nearly everywhere populations increased considerably. In England and Wales in 1938 16,000 pairs were breeding, in 1973 72,000-78,000. In 1969-1970 the population on the British Isles was estimated at 75,000 pairs. In 1930 25,000 pairs were breeding in

Table 51. Weights (in grams) of Black-headed gulls in the German and Dutch part of the Wadden Sea, sexes combined. German data refer to unpublished data from the Institut für Vogelforschung, Wilhelmshaven; Dutch data are from Swennen (in litt.)

area	age	period	mean	s.d.	n	range
Wilhelmshaven	adult	Jan–Feb	224.0	61.6	10	156–310
Wilhelmshaven	2 cy	Jan–Feb	178.3	46.5	3	125–210
Wilhelmshaven, Wanger-ooge, Jadebusen	adult	Jul–Sep	211.6	37.7	7	162–260
Wilhelmshaven, Hamburg Wangerooge	adult	Nov–Dec	212.3	69.4	3	165–292
Western part of the Dutch Wadden Sea	1st yr	Dec–Feb	280.3	35.8	111	195–374
	2nd "	Dec–Feb	294.7	33.9	72	220–381
	adult	Dec–Feb	289.9	23.3	289	217–360
	2 cy	Mch–Apr	253.7	42.5	6	225–310
	3 cy	Mch–Apr	256.8	31.4	5	220–295
	adult	Mch–Apr	276.9	33.9	49	205–345
	2 cy	May–Jul	252.6	42.7	9	225–352
	3 cy	May–Jul	238.9	25.5	20	205–285
	adult	May–Jul	248.4	24.6	324	195–327
	2 cy	Aug–Sep	254.9	31.1	17	205–310
	3 cy	Aug–Sep	278.9	27.0	9	220–309
	adult	Aug–Sep	264.1	26.5	26	210–305
	2 cy	Oct–Nov	274.2	33.9	6	227–314
	3 cy	Oct–Nov	269.6	22.8	7	233–308
	adult	Oct–Nov	268.7	22.3	31	220–306

the Netherlands, in 1961 70,000-100,000 and in 1975-1976 more than 200,000 pairs (Spaans, 1979), 67,000 of these in the Wadden Sea area (Spaans, in litt.). In Linnulahet (Estonian SSR) in 1914 1000 pairs were breeding, in 1967 6000. In other parts of this area an increase

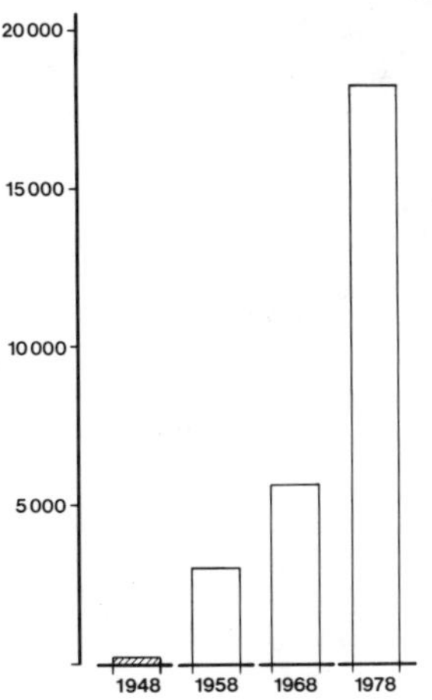

Fig. 152. Increase of the Black-headed gull population in the German Wadden Sea area (after Goethe, 1969).

was noticed as well, for example in Soitsjärvele from 1 pair in 1930
to 1000 in 1954 and in Vaika from 1 pair in 1919 to 400 in 1937. In
the whole Estonian Gulf of Finland area in 1974-1975 16,000 pairs were
breeding, near Leningrad another 9000. The number of breeding pairs in
Sweden was estimated at 300,000 (early nineteenseventies), in Finland
in 1955 at 8000 (all data if not indicated otherwise from Møller, 1978).
In the German Democratic Republic in 1973 about 57,000 pairs were breeding (Klafs & Stübs, 1979), in the Federal Republic of Germany in about
1975-1976 60,000 pairs, 19,000 of these in the Wadden Sea area (after
Goethe, 1969 and unpublished reports Institut für Vogelforschung,
Wilhemshaven). Fig. 152 shows the increase of the Black-headed gull
population in the German Wadden Sea area. An exception in this list is
Denmark where a population decrease has been noticed since about 1964.
Along the coast in 1960 210,000 pairs were breeding, in 1974 110,000
were left. As the inland population in 1974 amounted to 100,000 pairs
the whole Danish population in this year was 210,000 (Møller, 1978).
Somewhat more than 5000 of these were breeding in the Wadden Sea area
(source: Dansk Ornithologisk Forening). Fig. 153 and Table 52 show
the distribution of the population in the Wadden Sea area.

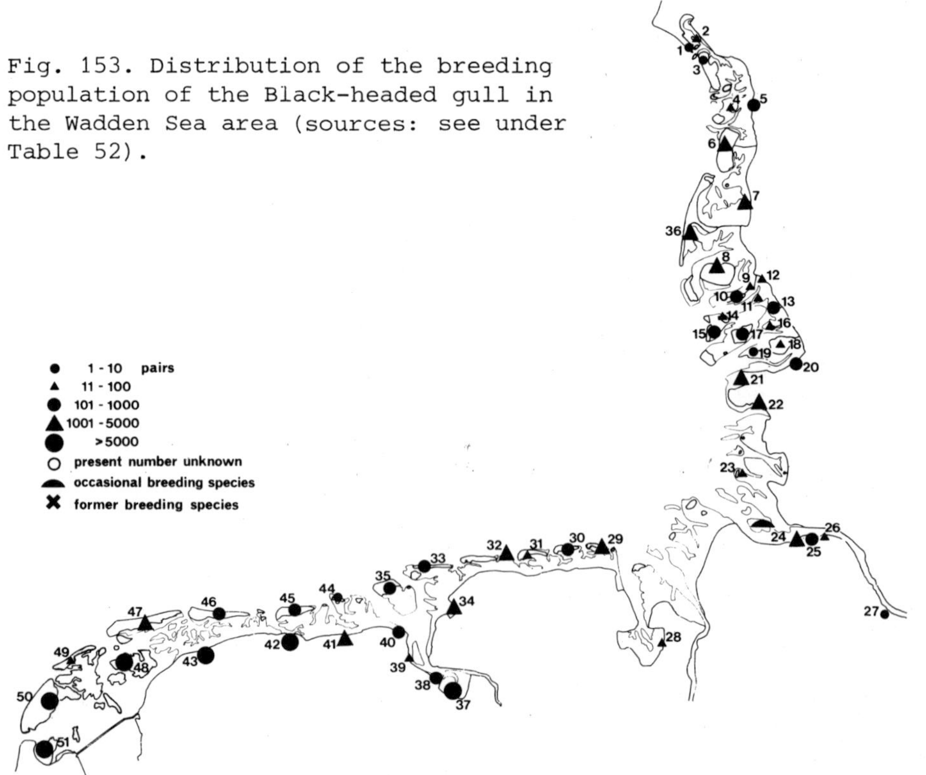

Fig. 153. Distribution of the breeding
population of the Black-headed gull in
the Wadden Sea area (sources: see under
Table 52).

Table 52. Breeding population of the Black-headed gull in the Wadden
Sea area. The number of breeding pairs is given in parenthesis, figures
without parenthesis indicate the number of the colony in fig. 153 (References: Møller, Dansk Ornithologisk Forening; Schutzstation Wattenmeer; Deutscher Bund für Vogelschutz; Verein Jordsand; Mellumrat; Institut für Vogelforschung, Wilhelmshaven; Müller; Raddatz; Bauamt für
Küstenschutz, Norden; Rijksinstituut voor Natuurbeheer, Leersum; Spaans).

Denmark (data from 1974)

1 Skallingen (5), 2 Langli (63), 3 Fanø (9), 4 Mandø (40), 5 Hviding
Holme (600), 6 Romø (1500), 7 Højer (3150).

Germany (data from 1976-1977)

8 Föhr (1052), 9 Oland (59), 10 Langeness (480), 11 Gröde (75), 12
Hauke Haienkoog (40), 13 Hamburger Hallig (115), 14 Hooge (20), 15
Norderoog (1000), 16 Nordstrandischmoor (68), 17 Pellworm (250), 18
Nordstrand (20), 19 Südfall (3), 20 Simonsberg (1000), 21 Tümlauer
Bucht (2500), 22 Grüne Insel (1109), 23 Trischen (40), 24 Hullen
(1650), 25 Nordkehdingen I (620), 26 Belumer Aussendeich (62), 27
Lühesand (6), 28 Jadebusen (100), 29 Wangerooge (2370), 30 Spieker-
oog (300), 31 Langeoog (40), 32 Baltrum (1200), 33 Juist (355), 34
Leybucht (2300), 35 Borkum (800), 36 Sylt (1500).

Netherlands (data from 1976-1977)

37 Dollard (6500), 38 Punt van Reide (350), 39 Eems (50), 40 Eemshaven (145), 41 Coast province of Groningen (1885), 42 Lauwersmeer
(8500-10,000), 43 Ferwerderadeel (6000-7000), 44 Rottumerplaat (some),
45 Schiermonnikoog (600), 46 Ameland (150), 47 Terschelling (2100),
48 Griend (6000), 49 Vlieland (22), 50 Texel (10,000), 51 Balgzand
(5500).

3.28.3.2 Numbers per area

Hardly no data are available from the Danish part of the area. On
April 19, 1980 8960 have been counted in the area, on September 13,
1980 18,700 (Meltofte, in litt.).

The phenology of Black-headed gulls in the Wadden Sea area in
Schleswig-Holstein is shown in fig. 154. Numbers increase in July and
culminate in September when usually about 75,000 are present. In January
numbers have dropped considerably. Maximum numbers in this time of the
year are 7000-8000. In cold winters, probably only some thousands remain. Some thousands summer in the area and undergo moult there. By
this time of the year the number of Black-headed gulls in the area is
about 20,000, including the breeding birds (Busche, 1980).

In the Wadden Sea area in Niedersachsen Black-headed gulls become
more numerous in June. Maximum numbers occur from August to October.
In the area around Wangerooge, Mellum, Knechtsand, Neuwerk and Schar-

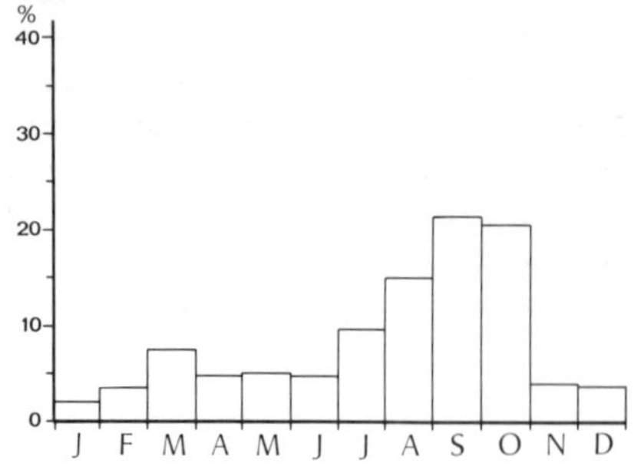

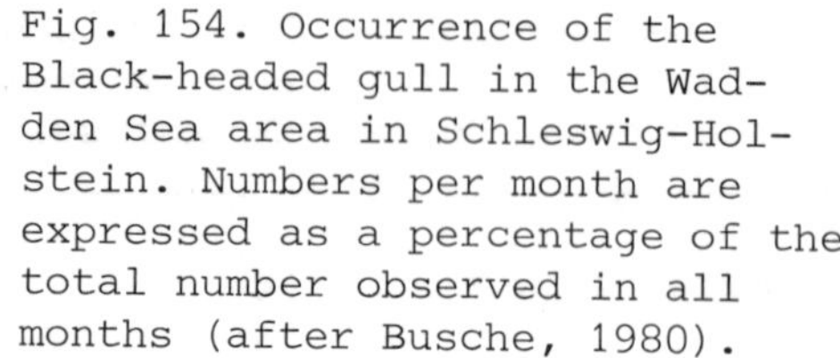

Fig. 154. Occurrence of the Black-headed gull in the Wadden Sea area in Schleswig-Holstein. Numbers per month are expressed as a percentage of the total number observed in all months (after Busche, 1980).

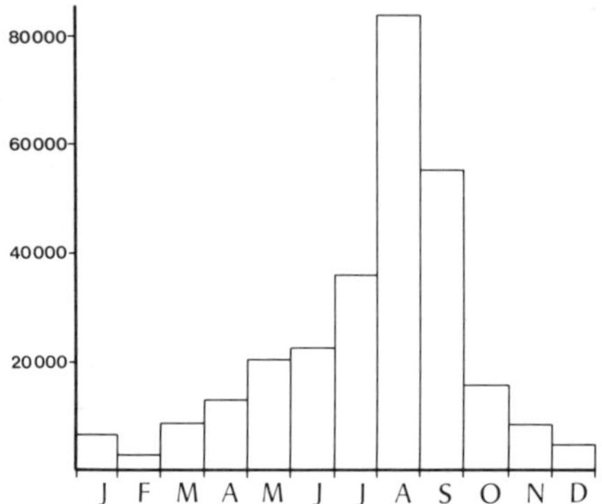

Fig. 155. Mean number of Black-headed gulls per month in the Dutch part of the Wadden Sea based on counts from the shore (after data from Smit, 1977).

hörn counts from the shore yielded mean numbers of 6000-9000 in this period (data: Institut für Vogelforschung, Wilhelmshaven; Smit, 1977). In winter numbers drop considerably. The birds concentrate in harbours. In January 1966 in Cuxhaven 3350 and in Bremerhaven 1275 have been counted; in Wilhelmshaven in January 1973 3100. Mean numbers in Hamburg for the winters 1964-1965 until 1966-1967 were about 50,000 (Lübben, 1973; Rosentreter, 1973; Eggers, 1974; Panzer & Rauhe, 1978).

Fig. 155 shows that, as far as can be registered by counts from the shore, Black-headed gulls in the Dutch part of the Wadden Sea are most numerous from July till September. In winter however they are partly living far from the shores. In the Western part of the Dutch Wadden Sea Swennen (in litt.) counted mean numbers of about 45,000 from October until March. More or less the same is true for the Eastern part of the Wadden Sea where large numbers occur near the sedimentation fields along the Groningen coast in winter (Swennen, in litt.). These birds often are not registered during counts from the shore. Fig. 156 shows the distribution of Black-headed gulls in the Wadden Sea area as far as could be registered by means of counts from the shore.

3.28.4 Food

3.28.4.1 Food composition

Black-headed gulls foraging on the tidal flats of the Wadden Sea mainly take crustaceans (like *Crangon crangon*), polychaetes and to a

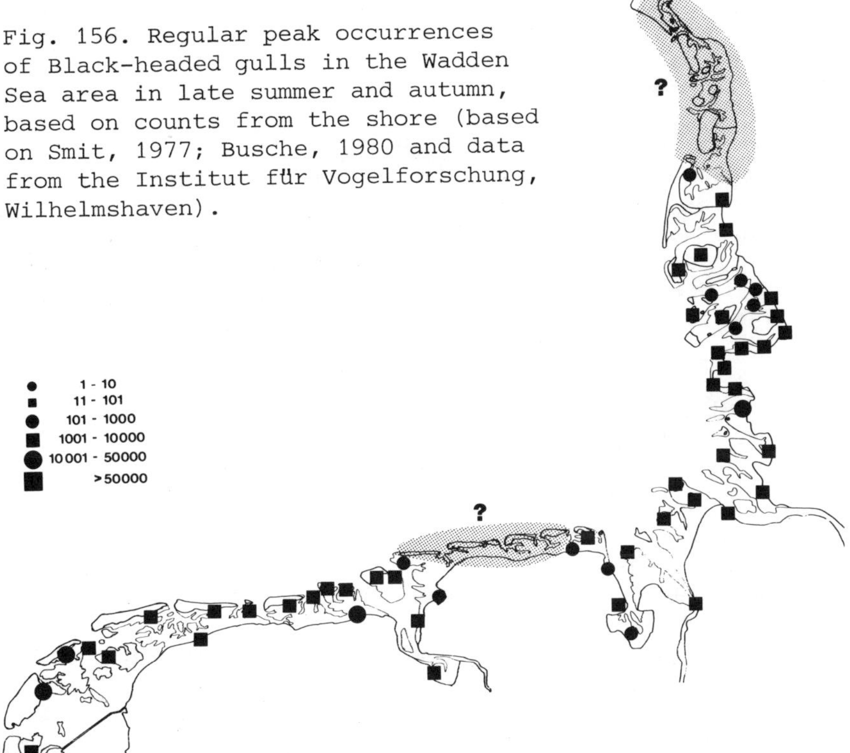

Fig. 156. Regular peak occurrences of Black-headed gulls in the Wadden Sea area in late summer and autumn, based on counts from the shore (based on Smit, 1977; Busche, 1980 and data from the Institut für Vogelforschung, Wilhelmshaven).

lesser extent molluscs. Near the island of Ameland in September some were observed to feed only on *Nereis diversicolor* (Kersten & Piersma, in litt.). During high tide, in large gullies and large areas with open water like for instance the Western part of the Dutch Wadden Sea, they actively forage on fish. Most important prey species are Smelt (*Osmerus eperlanus*), Sprat (*Sprattus sprattus*) and Herring (*Clupea harengus*). Apart from food obtained by foraging in the Wadden Sea Black-headed gulls take a wide variety of prey organisms by foraging inland. They often occur on agricultural land, pastures as well as arable land, to take insects, earthworms and snails. They are also frequently observed on refuse dumps and in harbours, in the German part of the Wadden Sea also often in sewage farms. They also follow ships for refuse (Vernon, 1970, 1972; Eggers, 1974; Isenmann, 1978; Spaans, in litt.; Swennen, in litt.).

3.28.4.2 Feeding activities

Black-headed gulls in the Western part of the Dutch Wadden Sea often plunge dive to catch fish (Swennen, in litt.). When foraging on arable land the birds often follow the plough. When feeding on flying insects they may hover just over the surface or circle at great altitudes. In pastures trampling is a frequently used technique to catch earthworms. Like Common gulls they often pursue Lapwings, Golden plovers and other bird species to rob preys. In Sandwich tern colonies they have been observed to rob fishes brought to the terns' young, in some occasions even considerable amounts (Rooth, 1965). Black-headed gulls have been observed to forage in the Wadden Sea by night (Groenendaal, 1975). In some places this is frequently observed when lamp light is available, for instance in harbours (Goethe, unpubl.). Inland foraging by night has been observed in moonlit nights as well (Ebbinge & Ebbinge-Dalmeyer, 1974). By night Black-headed gulls may roost on the Wadden Sea as well as on isolated places on islands and along the mainland coast. Before going to the sleeping places on the Wadden Sea the birds often come ashore to drink fresh water (Swennen, in litt.). During high tide the gulls may come ashore to form high tide roosts. They may also roost on the Wadden Sea itself.

3.28.4.3 Total food consumption
No data available.

References

Andersen-Harild, P., 1971. Danske Haettemaagers (Larus ridibundus) vinterkvarter. Dansk Orn. Foren. Tidsskr. 65: p. 109-115.

Busche, G., 1980. Vogelbestände des Wattenmeeres von Schleswig-Holstein. Kilda, Greven (in press).

Dwight, J., 1925. The Gulls (Laridae) of the world. Bull. American Mus. Nat. Hist. 52: p. 63-401.

Ebbinge, B. & D. Ebbinge-Dallmeyer, 1974. Nachtelijk spiering-vissen door Kapmeeuwen. De Levende Natuur 77: p. 280.

Eggers, J., 1974. Vorkommen und Herkunft der Lachmöwe (Larus ridibundus) im Hamburger Raum im Vergleich zu Sturm-, Silber- und Mantelmöwe (Larus canus, L. argentatus, L. marinus). Hamburger Avifaun. Beitr. 12: p. 95-144.

Eykman, C., P.A. Hens, F.C. van Heurn, C.G.B. ten Kate, J.G. van Marle, M.J. Tekke & T.G. de Vries, 1949. De Nederlandse Vogels, Vol. 3. Wageningse Boek- en Handelsdrukkerij, Wageningen: 407 pp.

Haftorn, S., 1971. Norges fugler. Universitetsforlaget, Oslo: 862 pp.

Isenmann, P., 1978. La décharge d'ordures ménagères de Marseille comme habitat d'alimentation de la Mouette rieuse Larus ridibundus. Alauda 46: p. 131-146.

Goethe, F., 1969. Die Einwanderung der Lachmöwe Larus ridibundus in das Gebiet der deutschen Nordseeküste und ihrer Inseln. Bonn. Zool. Beitr. 20: p. 164-170.

Groenendaal, M., 1975. Meeuwen op Schiermonnikoog. Schierboek 5: p. 133-161.

Klafs, G. & J. Stübs, 1979. Die Vögel Mecklenburgs. Fischer, Jena: 385 pp.

Lübben, N., 1973. Winterliche Beobachtungen an Silber- und Lachmöwen im südlichen Stadtgebiet von Wilhelmshaven. Unpubl. report University of Oldenburg: 79 pp.

Milenz, K., 1961. Über Zugwege und Winterquartiere Mecklenburgischen Lariden (Larus argentatus Pontoppidan, L. canus L., L. ridibundus L., Sterna hirundo L.). In: H. Schildmacher, Beiträge zur Kenntnis der deutschen Vögel. Jena: p. 189-245.

Møller, A.P., 1978. Maagernes Larinae yngleudbredelse, bestandsstør-
    relse og - aendringer i Danmark, med supplerende oplysninger om for-
    holdene i det øvrige Europa. Dansk Orn. Foren. Tidsskrift 72: p.
    15-39.

Niethammer, G., 1942. Handbuch der Deutschen Vogelkunde, Vol. 3. Aka-
    demische Verlagsgesellschaft, Leipzig: 568 pp.

Panzer, W. & H. Rauhe , 1978. Die Vogelwelt an Elb- und Wesermün-
    dung. Heimatbund Männer vom Morgenstern, Bremerhaven: 336 pp.

Radford, M.C., 1962. British ringing recoveries of the Black-headed
    Gull. Bird Study 9: p. 42-55.

Rooth, J., 1965. Over Sterns en Kaapmeeuwen. De Levende Natuur 68:
    p. 265-275.

Rosentreter, M., 1973. Winterliche Beobachtungen an Silber- und Lach-
    möwen im nördlichen Stadtgebiet von Wilhelmshaven. Unpubl. report
    University of Oldenburg: 90 pp.

Salomonsen, F., 1972. Fugletraekket og dets gaader Munksgaard, Køben-
    havn: 362 pp.

Salomonsen, F., 1973. Abstract of talk on 84th annual meeting Deutsche
    Ornithologen Gesellschaft (1972), Saarbrücken. J. Orn. 114: p. 383-
    384.

Schüz, E., 1971. Grundriss der Vogelzugskunde. Parey, Berlin: 390 pp.

Smit, C.J., 1977. On the occurrence of 32 bird species in the Danish,
    German and Dutch Wadden Sea. Unpubl. report Intern. Wadden Sea
    Working Group, part 3: 174 pp.

Spaans, A.L., 1979. Kokmeeuw. In R.M. Teixeira (ed.). Atlas van de Ne-
    derlandse broedvogels. Natuurmonumenten, 's-Graveland: p. 166-167.

Vaurie, C., 1965. The birds of the Palaearctic Fauna. Non-Passeriformes.
    Witherby, London: 763 pp.

Vernon, J.D.R., 1970. Feeding habitats and food of the Black-headed
    and Common Gulls. Part 1, feeding habitats. Bird Study 17: p. 287-
    296.

Vernon, J.D.R., 1972. Feeding habitats and food of the Black-headed
    and Common Gull. Part 2, food. Bird Study 19: p. 173-186.

Voous, K.H., 1960. Atlas of European birds. Nelson, London: 284 pp.

3.29 COMMON GULL *(LARUS CANUS* L.*)*
   F. Goethe

Da: Stormmaage; G: Sturmmöwe; Du: Stormmeeuw

3.29.1 Distribution

3.29.1.1 Breeding area
   The Common gull has a trans-palaearctic and nearctic distribution,
breeding in the boreal, temperate, steppe and to a limited extent even
in the tundra climate zones (fig. 157). During the past 50 years an
extension of the breeding range has been recorded in the North Atlantic region, probably due to an increase of mean annual temperatures
(Voous, 1960; Vaurie, 1965). There are four subspecies.

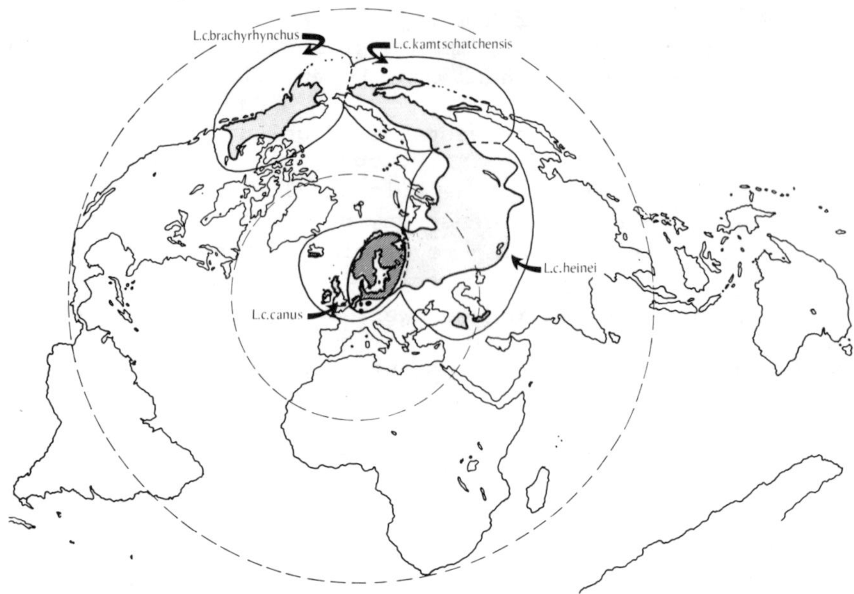

Fig. 157. Breeding area of the Common gull (after Voous, 1960). The dark shaded area represents approximately the region where birds visiting the Wadden Sea originate from.

3.29.1.2 Migration routes

Common gulls breeding in Central and North Norway cross the North Sea to reach the wintering areas around the British Isles and the Fär-öer islands (Haftorn, 1971). Birds breeding in the Southern part of Norway, Denmark, Germany, Sweden, Finland, the Baltic and the USSR migrate south- and westward, probably mainly following the coastline. These birds may thereby pass through the Wadden Sea area.

3.29.1.3 Wintering areas

Norwegian breeding birds are almost strictly migratory. The majority of those breeding in North and Central Norway winters around the British Isles and Färöer, those from the Southern part of the country may be found all along the North Sea coast from Scotland and Denmark in the north till the Channel in the south and Ireland in the west. A smaller part migrates further south to the Iberian peninsula (Haftorn, 1971). Only a small part of the Danish Common gulls winters in Danish waters. The other ones are found between Schleswig-Holstein in the north, the Bay of Biscay in the south and England and Wales in the west. A few recoveries originate from Ireland, Scotland and the Mediterranean (Sørensen, 1977). Breeding birds of Swedish, Finnish, Estonian and German populations winter along the coasts of the North Sea and along the Atlantic coast of France and the Iberian peninsula. Finnish and Estonian Common gulls are numerous winter visitors in Danish waters. Compared to these birds those breeding in the Western part of the Baltic appear to migrate more often to Spain and Portugal.

First calendar-year birds on the average move further south than adults
(Niethammer, 1942; Milenz, 1961; Babbe, 1964; Schüz & Weigold, 1931
and unpubl. ringing results of the Institut für Vogelforschung, Wilhelmshaven). Part of the Dutch Common gulls winters in the Netherlands.
Recoveries of Dutch-ringed birds also come from the Atlantic coasts
of France, Spain and Portugal (Braaksma, 1964). British Common gulls
are probably largely sedentary (Radford, 1960). In the whole wintering
area recoveries far inland are rare.

3.29.1.4 Moulting areas

No specific moulting areas can be distinguished as moult of primaries takes about $4\frac{1}{2}$ months. Post-nuptial moult starts in the breeding
area and continues while the birds migrate towards the wintering areas.
Part of the birds finish it there. Birds present in the Wadden Sea in
summer and early autumn moult primaries in this area.

3.29.2 Annual cycle

3.29.2.1 Migration

In general autumn migration of Common gulls has the form of a slow
south- and westward dispersal starting already in summer and continu-
ing through autumn. Some first calendar-year birds leave the area
where they were hatched already very soon. Norwegian birds have been
found in NW Europe already in August. Adults and most of the juveniles
however leave in September (Haftorn, 1971), Danish birds leave in August, September and October. The majority arrives in the Netherlands
in September, in Belgium and France in November, and on the British
Isles by December, juveniles and immatures on the average somewhat
earlier than the adults (Sørensen, 1977). Peak numbers in the Wadden
Sea occur from July until October.

In periods of cold spells in winter Common gulls move further south
towards the coasts of SW France, Spain and Portugal. In the cold winter of 1962-1963 Danish Common gulls have been recovered in this area
while in normal winters recoveries mainly come from the English and
French coast (Sørensen, 1977). During periods of severe cold numbers
in the Dutch part of the Wadden Sea may increase sharply as well.

Spring migration starts in March, continues through April and proceeds rather rapidly. By this time the first birds arrive in the breeding colonies again. Summering immatures may be found all over the Wadden Sea area. According to Vernon (1969) large numbers summer along
the coast of East Anglia (England) as well. In Denmark immature Common gulls do not arrive until June (Meltofte, in litt.).

3.29.2.2 Moult

First calendar-year Common gulls undergo a partial moult of body
feathers between August and October. Adult birds moult primaries from
about mid-May until mid-October, second calendar-year birds on the average some weeks earlier (Walters, 1978). Body feathers are moulted from
June until October. A partial moult in all birds takes place from March
to May.

Table 53. Weights in grams of Common gulls caught in the Wadden Sea area. German data refer to unpublished data from the Institut für Vogelforschung, Wilhelmshaven; data from Texel refer to cannon net catches by Swennen (in litt.); data from Vlieland are catches by Boere (in litt.). M = males, F = females.

area	sex	age	period	mean	s.d.	n	range
Wilhelmshaven	M,F	adult	Jan-Feb	396.3	90.6	8	280-530
Mellum	M	adult	May	407.5	17.7	2	395-420
Texel	M,F	1st cy	Jul-Sep	410.0	43.3	3	
Texel	M,F	adult	Jul-Sep	392.3	48.9	10	320-452
Texel	M,F	1st cy	Nov-Mch	367.1	48.5	17	285-465
Texel	M,F	2nd cy	Nov-Mch	395.4	57.7	17	305-485
Texel	M,F	3rd cy	Nov-Mch	449.5	7.6	4	443-460
Texel	M,F	adult	Nov-Mch	385.5	46.0	87	305-492
Vlieland	M,F	2nd & 3rd cy	March	413.3	61.1	3	360-480
Vlieland	?	2nd cy	April	390.0		1	
Vlieland	?	2nd cy	May	350.0		1	
Vlieland	M,F	1st & 2nd cy	August	369.0	32.1	11	315-410
Vlieland	M,F	1st & 2nd cy	September	372.0	26.6	7	330-400
Vlieland	?	1st & 2nd cy	October	387.5	60.1	2	345-430

3.29.2.3 Weight changes

   Weights of birds caught in the Wadden Sea area are listed in Table 53. The heavy weight of a Common gull from Cuxhaven (685 g; Heinroth, 1922) is probably due to an error.

3.29.3 Numbers

3.29.3.1 Population size

   A survey of the recent breeding distribution in W and N Europe is given by Møller (1978). These data, added with some more recent figures, show that Common gulls have increased in many areas during the 20th century:
- Ireland: breeding bird since 1955.
- Norway: breeding along the coast as well as inland. Breeding inland, even far up in the highlands, has been confirmed already in the 18th century. The total population size is unknown. In the Troms-district in 1930 about 20,000 pairs were breeding (Haftorn, 1971).
- Sweden: 250,000 pairs.
- Finland: 5,000 pairs (in 1955).
- Letland (USSR): 300 pairs.
- Estland (USSR): 4000 pairs.
- Leningrad surroundings (USSR): 760 pairs.

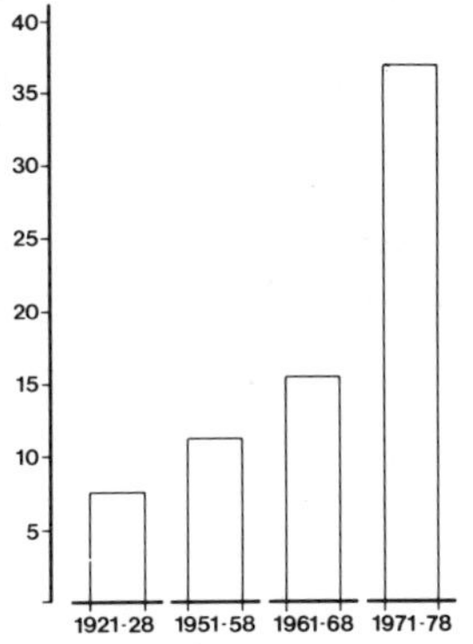

Fig. 158. Number of breeding pairs of the Common gull on the island of Mellum from 1921-1978 (data: Institut für Vogelforschung, Wilhelmshaven).

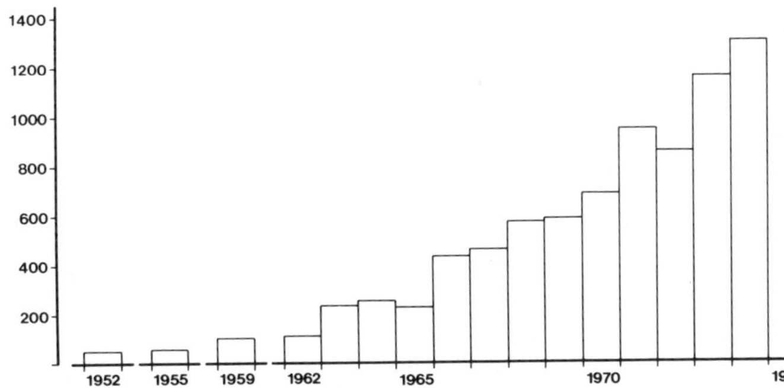

Fig. 159. Number of breeding pairs of the Common gull on the island of Texel from 1952-1974 (after Dijksen & Dijksen, 1977).

pairs. Afterwards a gradual decrease set in to 40,000 pairs in 1974.
- Denmark: From 1960 to 1969 a steady increase, probably to about 65,000. In the Danish Wadden Sea area about 650 pairs were breeding (after recent data from Dansk Ornithologisk Forening)
- German Democratic Republic: The most important colony (Langenwerder) in 1969 had 10,000 pairs. In 1970 in the whole country 13,000 pairs were breeding after which year population regulations took place. In 1975 8000 pairs were breeding in the country.
- German Federal Republic: In 1961 4200 pairs were breeding, in about 1975 5700. Of these somewhat less than 1100 pairs were breeding in the Wadden Sea area (data: Institut für Vogelforschung, Wilhelmshaven). Fig. 158 shows the population increase on the island of Mellum.

- The Netherlands: In 1962 662 pairs were breeding, in about 1976 the number had increased to 7000. Of these 2500 were breeding on the Wadden Sea islands, 4300 in the dunes along the mainland coast, and 140 in the Delta area, (Spaans, 1979). Fig. 159 shows the increase of the population breeding on the island of Texel.
- Britain and Ireland: In 1972 12,425 pairs were breeding, the great majority of these in Scotland (11,620) and Ireland (790). Only 8 pairs were breeding in England, even fewer (6) in Wales (Cramp et al., 1974).
    Breeding far inland has been confirmed in Mecklenburg, Schleswig-Holstein, Niedersachsen, W Central and S Germany (Rheinwald, 1977), Switzerland and E France (if not indicated otherwise all data originate from Møller, 1978). Fig. 160 and Table 54 show size and the dis-

Table 54. Breeding population of the Common gull in the Wadden Sea area. The number of breeding pairs is given in parenthesis, the figures without parenthesis indicate the number of the colony in fig. 160 (references: Møller (Dansk Ornithologisk Forening); Schutzstation Wattenmeer; Deutscher Bund für Vogelschutz; Verein Jordsand, Mellumrat, Institut für Vogelforschung, Wilhelmshaven; Müller; Raddatz; Bauamt für Küstenschutz; Research Institute for Nature Management, Leersum).

Denmark (data from 1974)

1 Varde Å♀ (4), 2 Skallingen (10), 3 Langli (400), 4 Fanø Nord (20), 5 Mandø (30), 43 Rømø (110), 44 Ballum Enge (60-70), 45 Højer Forland (10).

Germany (F.R.) (data from 1976-1977)

6 Uthörn (3), 7 Sylt (30), 8 Föhr (18), 9 Oland (2), 10 Amrum (250), 11 Langeness (29), 12 Gröde (10), 13 Hamburger Hallig (10), 14 Japsand (1), 15 Norderoog (10), 16 Nordstrandischmoor (9), 17 Südfall (4), 18 Trischen (9), 19 Hullen (41), 20 Nordkehdingen I (26), 21 Belumer Aussendeich (14), 22 Pagensand (70), 23 Lühesand (300), 24 Scharhörn (10), 25 Knechtsand (8), 26 Jadebusen (5), 27 Mellum (31), 28 Wangerooge (3), 29 Spiekeroog (4), 30 Langeoog (60), 31 Norderney (30), 32 Leybucht (1), 33 Juist (33), 34 Memmert (6), 35 Borkum (50).

Netherlands (data from 1976-1977)

36 Rottumeroog (53), 37 Rottumerplaat (7), 38 Schiermonnikoog (500), 39 Ameland (45), 40 Terschelling (150), 41 Vlieland (26), 42 Texel (1324).

---

tribution of the population breeding in the Wadden Sea area.

3.29.3.2 Numbers per area

As gulls have not been included in counts in the Danish part of the Wadden Sea up till 1980 hardly no information on numbers is available. On April 19, 1980 5140 have been counted in the area, on September 13, 1980 10,070 (Meltofte, in litt.).

In the Wadden Sea area in Schleswig-Holstein Common gulls in autumn are most numerous in September and October (fig. 161) though in some areas peak numbers occur in August. Numbers are difficult to obtain but in this time of the year 40,000-45,000 may be present in the area. In mild winters up to 30,000 can be present, on January 12, 1975 along the Eiderstedt peninsula 20,000 have been counted. In cold winters numbers may drop to some thousands. Spring passage can be noticed from March until May. Common gulls observed in May and June are mostly summering second- and third calendar-year birds (Busche, 1980).

In the Wadden Sea area in Niedersachsen numbers start to increase

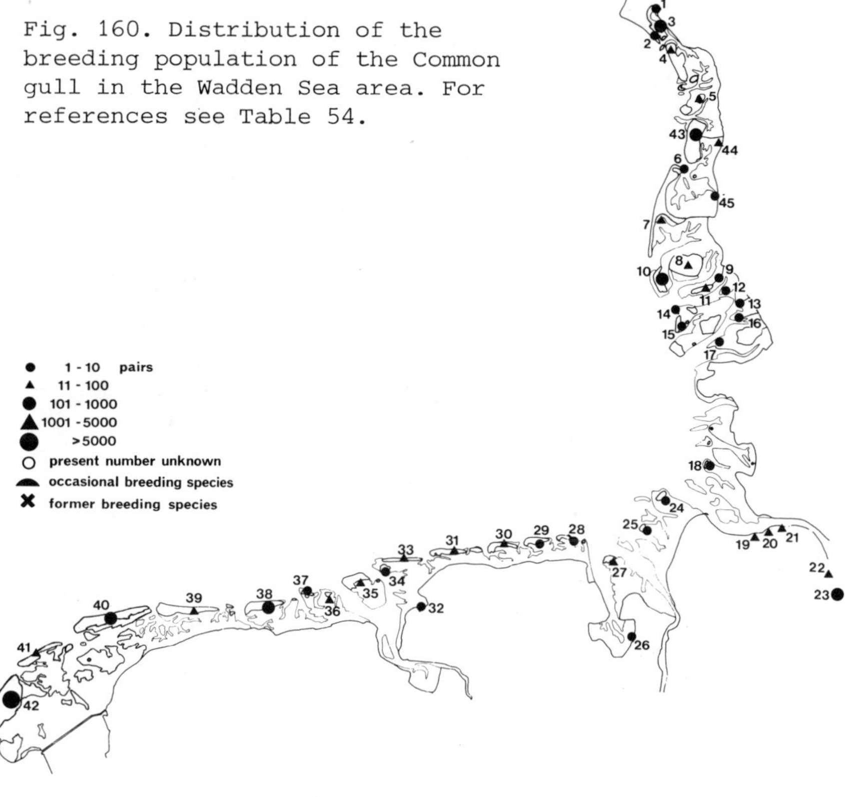

Fig. 160. Distribution of the breeding population of the Common gull in the Wadden Sea area. For references see Table 54.

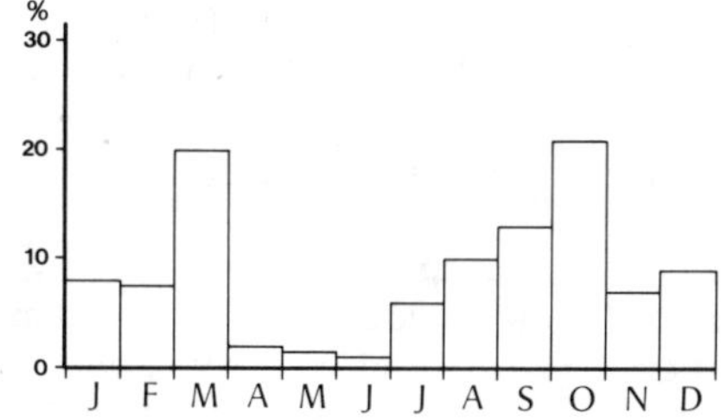

Fig. 161. Occurrence of the Common gull in the Wadden Sea area in Schleswig-Holstein. Numbers per month are expressed as a percentage of the total number observed in all months (after Busche, 1980).

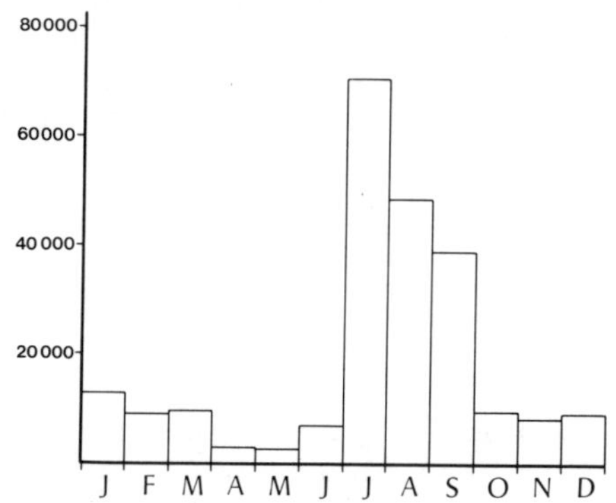

Fig. 162. Mean number of Common gulls per month in the Dutch part of the Wadden Sea based on counts from the shore (after data from Smit, 1977).

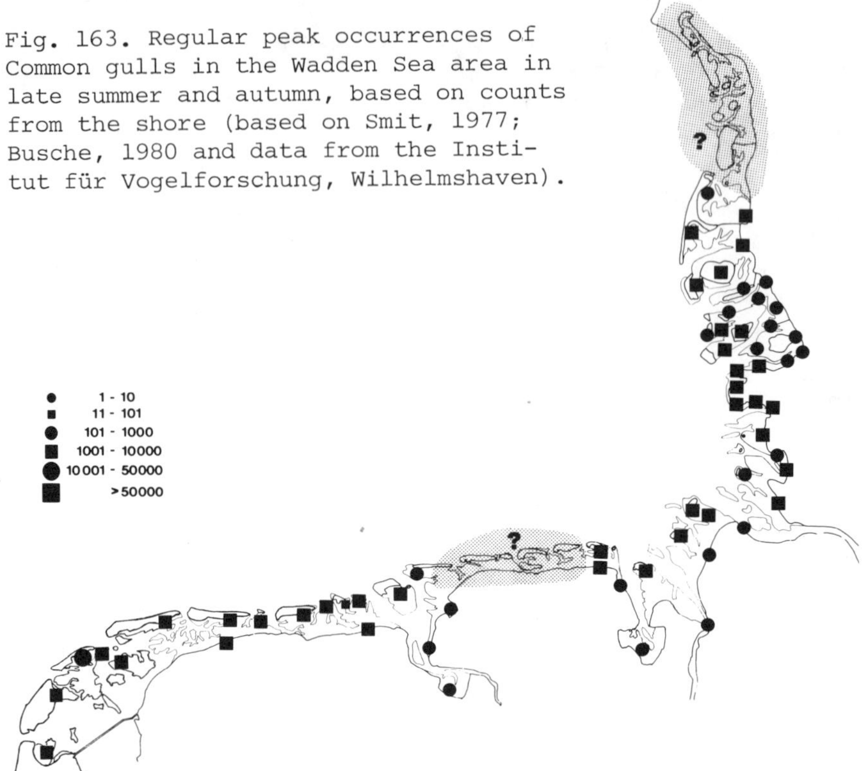

Fig. 163. Regular peak occurrences of Common gulls in the Wadden Sea area in late summer and autumn, based on counts from the shore (based on Smit, 1977; Busche, 1980 and data from the Institut für Vogelforschung, Wilhelmshaven).

in July. In the area around Wangerooge, Mellum, Knechtsand, Scharhörn and Neuwerk from August-September 5000-10,000 are present (data: Institut für Vogelforschung, Wilhelmshaven; Smit, 1977). In winter probably some thousands are present but simultaneous counts do not exist. In harbours numbers in winter amount to approximately 400 (Cuxhaven) and 500 (Bremerhaven) (Panzer & Rauhe, 1978).

The approximate amount of Common gulls per month, as registered by counts from the shore, in the Dutch part of the Wadden Sea is shown in fig. 162. Counts from ships between Den Helder and the Terschelling tidal watershed, representing the whole Western part of the Dutch Wadden Sea, yielded the following numbers per month: October 6600; November 4200; December 6400; January 15,600; February 104,000; March 7200 (Swennen in litt.). In January 1979, during a cold spell, on the island of Terschelling 17,000 Common gulls were counted (Spaans, in litt.).

Fig. 163 shows the distribution in the Wadden Sea area in autumn.

3.29.4 Food

3.29.4.1 Food composition
Common gulls rather often are found foraging inland in pastures or in arable land, especially when it has just been ploughed. Compared

to other gull species it is less often seen in sewage farms, refuse dumps and harbours. Their diet will therefore be less variable. In the Wadden Sea Common gulls take a certain variety of molluscs, crustaceans, worms and fish. When feeding inland they take earthworms, insects, insect larvae, snails and sometimes small mammals (Von Törne, 1940). No specific studies on the feeding ecology of Common gulls have been carried out in the Wadden Sea area.

3.29.4.2 Feeding activities

By following plough and cattle earthworms and a large variety of insects and insect larvae are taken. On some occasions they chase flying insects (Odonata, Coleoptera, Diptera). Like other gulls they use foot paddling when trying to catch earthworms. Like Herring gulls and Crows they drop cockles and mussels on hard surfaces to break shells. They rather often pursue Lapwings, Black-headed gulls, Golden plovers and other bird species to steal food. They may follow ships for refuse as well.

Flocks of Common gulls may roost on the Wadden Sea and on isolated places along the edges of islands and salt marshes by night. There are however several observations of Common gulls foraging by night (Groenendaal, 1974). During high tide part of the birds roosts on salt marshes, beaches and islands. They may also stay on the Wadden Sea to roost.

3.29.4.3 Total food consumption

No data available.

References

Babbe, R., 1964. Funde beringter Sturmmöwen (Larus canus) vom Graswarder bei Heiligenhafen, Schleswig-Holstein. Auspicium 2: p. 61-86.

Braaksma, S., 1964. Het voorkomen van de Stormmeeuw (Larus canus L.). Limosa 37: p. 58-95.

Busche, G., 1980. Vogelbestände des Wattenmeeres von Schleswig-Holstein. Kilda, Greven (in press).

Cramp, S., W.R.P. Bourne & D. Saunders, 1974. The Seabirds of Britain and Ireland. Collins, London: 287 pp.

Dijksen, A.J. & L.J. Dijksen, 1977. Texel vogeleiland. Thieme, Zutphen, 246 pp.

Haftorn, S., 1971. Norges fugler. Universitetsforlaget, Oslo: 862 pp.

Heinroth, O., 1922. Beziehungen zwischen Vogelgewicht, Eigewicht, Gelegegrösse und Brutdauer. J. Orn. 70: p. 172-285.

Groenendaal, M., 1974. Meeuwen op Schiermonnikoog. Schierboek 5: p. 133-161.

Milenz, K., 1961. Über Zugwege und Winterquartiere mecklenburgischer Lariden (Larus argentatus Pontoppidan, L. canus L., L. ridibundus L., Sterna hirundo L.). In: H. Schildmacher, Beiträge zur Kenntnis der deutschen Vögel. Jena: p. 189-245.

Møller, A.P., 1978. Maagernes Larinae yngleudbredelse, bestandsstørrelse og -aendringer i Danmark, med supplerende oplysninger om forholdene i det øvrige Europa. Dansk Orn. Foren. Tidsskr. 72: p. 15-39.

Niethammer, G., 1942. Handbuch der Deutschen Vogelkunde, Vol. 3. Akademische Verlagsgesellschaft, Leipzig: 568 pp.

Oelke, H., 1968. Vögel auf dem Grossen Knechtsand. Falke 15: p. 342-351 & 372-377.

Panzer, W. & H. Rauhe, 1978. Die Vogelwelt an Elb- und Wesermündung. Männer vom Morgenstern, Bremerhaven: 336 pp.

Radford, M.C., 1960. Common gull movements shown by ringing returns. Bird Study 7: p. 81-93.

Rheinwald, G., 1977. Atlas der Brutverbreitung westdeutscher Vogelarten. Dachverband Deutscher Avifaunisten: 54 pp.

Schüz, E. & H. Weigold, 1931. Atlas des Vogelzuges nach den Beringungsergebnissen bei palaearktischen Vögeln. Abh. Vogelzugforschung 3, Berlin.

Smit, C.J., 1977. On the occurrence of 32 bird species in the Danish, German and Dutch Wadden Sea. Unpubl. report. Intern. Wadden Sea Working Group, part 3: 174 pp.

Sørensen, L.H., 1977. An analysis of Common gull (Larus canus) recoveries recorded from 1931 to 1972 by the Zoological Museum in Copenhagen. Gerfaut 67: p. 133-160.

Spaans, A.L., 1979. Stormmeeuw. In R.M. Teixeira (ed.). Atlas van de Nederlandse broedvogels. Natuurmonumenten, 's-Graveland: p. 168-169.

Törne, H. von, 1940. Einiges über die Ernährung der Sturmmöwe auf Schleimünde. Deutsche Vogelwelt 65: p. 155-159.

Vaurie, C., 1965. The birds of the Palaearctic Fauna. Non-Passeriformes. Witherby, London: 763 pp.

Vernon, J.D.R., 1969. Spring migration of the Common Gull in Britain and Ireland. Bird Study 16: p. 101-107.

Voous, K.H., 1960. Atlas of European birds. Nelson, London: 284 pp.

Walters, J., 1978. The primary moult in four Gull species near Amsterdam. Ardea 66: p. 32-47.

3.30 HERRING GULL (*LARUS ARGENTATUS* PONTOPPIDAN)
   F. Goethe

Da: Sølvmaage; G: Silbermöwe; Du: Zilvermeeuw

3.30.1 Distribution

3.30.1.1 Breeding area

The Herring gull has a distribution in the NW Palaearctic and Nearctic, breeding in temperate, boreal, and tundra climate zones (Voous, 1960). The breeding area is shown in fig. 164. According to Vaurie (1965) two groups of Herring gulls occur in Europe, Asia and North-America. Birds belonging to the Northern group breed in NW Europe eastward through Siberia to the Bering Sea and in North-America. Birds of the Southern group nest on the Azores, Madeira and the Canaries and through the Mediterranean area across Asia to Mongolia and Western Manchuria. These two groups are separated except in the basin of the Ob

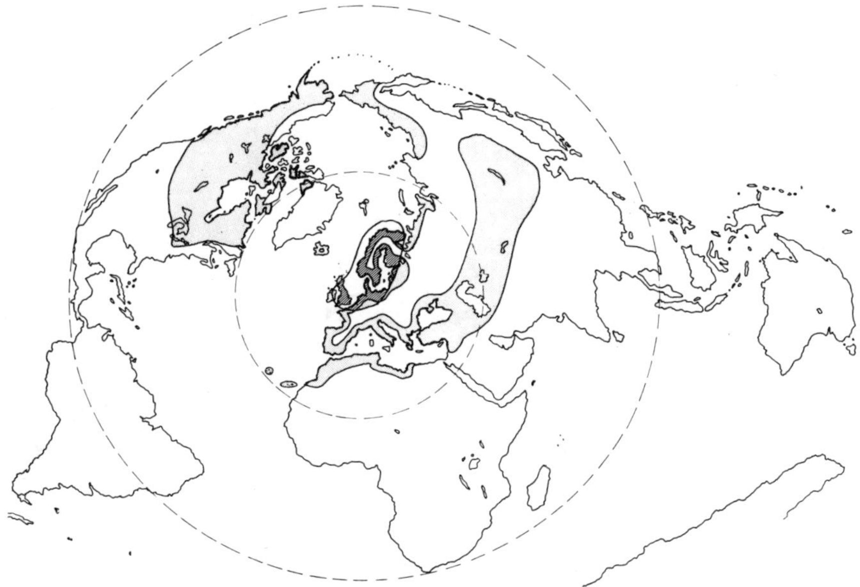

Fig. 164. Breeding area of the Herring gull (after Voous, 1960; Vaurie, 1965; Haftorn, 1971). The dark shaded area represents approximately the region where birds visiting the Wadden Sea originate from. No subspecies are indicated because the taxonomy of the Herring gull is still in discussion.

where the occasionally intermingle.

In NW Europe Herring gulls are breeding predominantly along coasts and on islands from Northern Scandinavia and the Eastern Baltic south to the Bay of Biscaya. They settled in Northern Ireland recently. According to Vaurie (1965) all NW European Herring gulls can be lumped to *L.a. argentatus*. Other authors consider birds of the Danish population to belong to *L.a. argentatus* and those of the Dutch and the British population to *L.a. argenteus* and birds of the German population and the Danish part of the Wadden Sea to a hybrid population of *argentatus* and *argenteus* (Stegmann, 1934; Goethe, 1961; Barth, 1975; Kuschert, 1979).

30.3.1.2 Migration routes

Herring gulls from colonies in the Wadden Sea area are partly resident, partly they show some dispersion. Generally they more or less follow the coastline though inland observations not too far from the coast are common. Herring gulls from colonies on the Dutch Wadden Sea islands chiefly remain within a 100-150 km distance from the place where they hatched. These birds stay partly along the coast but disperse as well to inland sites where they visit refuse dumps, cities and villages (Spaans, 1971). Some birds follow estuaries and rivers far inland. A German Herring gull has been recorded in Southern France, probably by following Rhine and Rhône (Goethe, 1973). Recoveries of

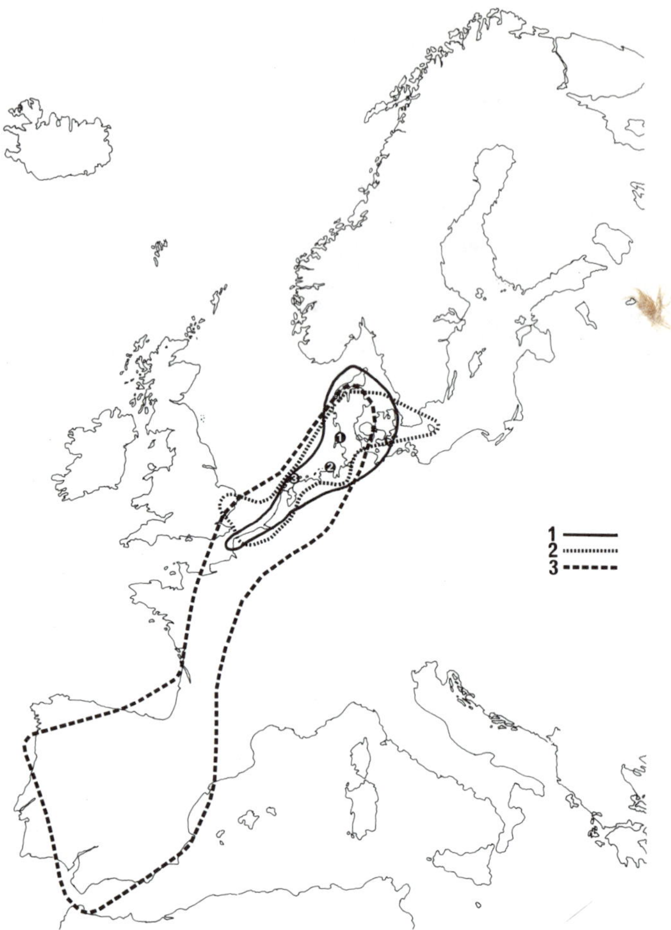

Fig. 165. Areas over which Herring gulls from three colonies on Wadden
Sea islands disperse. 1 indicates Jordsand (Denmark), 2 Langeoog
(Niedersachsen), 3 Terschelling (Netherlands) (after Paludan, 1953;
Reichmann, 1977 and several Dutch ringing reports in Ardea 2-15 and
Limosa 10-40).

Herring gulls ringed on the Dutch Wadden Sea islands show a slight pre-
dominance from Southwestern directions (Spaans, 1971). Recoveries of
the German Wadden Sea population come slightly more from Eastern directions (Goethe, 1956; Drost & Schilling, 1940). These differences may
be attributed to the location of the colonies with regard to favourable feeding areas like cities and harbours. Birds from Northern Scandinavia, Finland and the USSR show true migratory movements. These
birds probably mainly follow the coastline (Olsson, 1958; Haftorn,
1971; Schüz, 1971).

3.30.1.3 Wintering areas

Herring gull populations from Denmark (Jørgensen, 1973), the German Federal Republic (Goertz, 1971; Goethe, 1973; Goertz & Goethe, 1969), the Netherlands (Spaans, 1971) and the British Isles (Harris, 1964) are mostly resident. Wintering takes place in the breeding areas. A part of these populations disperses and may be recovered at rather large distances from the places where they were hatched (fig. 165). Part of the birds breeding in Scandinavia is resident as well, a part is more or less migratory. Birds from the coastal area of Central and Southern Norway have been recovered in winter along the Norwegian, Danish, South Swedish and German coasts, birds from Northern Norway migrate further south and may be found all over the North Sea area from Central Norway and Scotland to France. Birds from the USSR partly winter along the Norwegian coast (Haftorn, 1971). Most important wintering areas for Swedish and Finnish birds are the Danish isles (Olsson, 1958). The Wadden Sea is another rather important wintering area for Finnish birds (Cebulla, pers. comm.). Only a small part of the Baltic population breeding along the coast of the German Democratic Republic enters the Wadden Sea (Nehls, 1971). There are some recoveries of Swedish birds and some of birds from the Murmansk coast in the Wadden Sea. Under bad feeding conditions in winter a relatively large part of the Dutch Wadden Sea population migrates inland to forage on refuse dumps. Spaans (1971) estimated that under favourable foraging conditions 32% of the Dutch Wadden Sea winter population used refuse dumps inland. Under unfavourable conditions this percentage increased to at least 77%.

3.30.1.4 Moulting areas

Large flocks of immature Herring gulls use the Wadden Sea as a moulting area. Breeding birds have no distinct moulting areas. Herring gulls breeding in the Wadden Sea area start moulting flight feathers already during the breeding period. By the end of July the "club-places" (Tinbergen, 1953) at the edges of the gulleries begin to look like moulting places because of the wing and body feathers, which may be found there in large quantities (Goethe, 1956). After the breeding season moult continues. Like other coastal areas the Wadden Sea therefore acts as an important moulting area.

3.30.2 Annual cycle

3.30.2.1 Migration

Dispersal of birds from the breeding colonies in the Wadden Sea area starts in July and August. According to several authors (for example Goethe, 1956; Jørgensen, 1973; Drost & Schilling, 1940; Eykman et al., 1949) Herring gulls show a tendency for greater dispersal in first autumn and winter than in subsequent years. Other authors (for example Spaans (1971)), indicate that this does not hold everywhere. Apparently in the Dutch part of the Wadden Sea a change in behaviour of first calendar-year birds took place showing a smaller migratory tendency in recent years. According to Spaans (1971) this phenomenon is due to more favourable food conditions in winter.

Migration of Norwegian breeding birds starts in August and continues till November. Many first calendar-year birds are already in Southern Sweden, Denmark or the British Isles by September-October (Haftorn, 1971). Numbers increase in the Wadden Sea in July, August and September. Along the Eastcoast of the British Isles peak numbers are recorded in December and January (Prater, 1976).

In Scandinavia Herring gulls already return to the breeding area by the end of February (Haftorn, 1971), in the German part of the Wadden Sea sometimes already in early February (Goethe, 1973). In the Dutch part of the Wadden Sea, as well as in many places in Germany, Britain, Denmark, South- and Central Norway, egg-laying starts by the end of April or early May (Spaans & Spaans, 1975).

3.30.2.2 Moult

Juvenile Herring gulls moult body feathers between August and November to receive their first winter plumage. Immature and adult birds undergo a partial moult from January to April and a complete moult from May to October. During their fourth calendar-year (that means when nearly 3½ years of age) the birds attain their mature plumage. Occasionally three years old birds are breeding already in nearly mature plumage. On the other hand 5 years old birds with remains of the immature plumage are recorded (Dwight, 1925; Goethe, 1956; Drost & Schilling, 1940).

3.30.2.3 Weight changes

Weights of Herring gulls ringed on the island of Texel in their first year are listed in Table 55. Numbers of other age classes ringed on Texel are too small to give representative figures.

Weights of breeding birds from Schleswig-Holstein were for males (n=63) 955-1365 g with a mean of 1169.6 g and for females (n=76) 805-1055 g with a mean of 936.5 g (Kuschert, 1979).

Breeding birds of the island of Mellum, yielded following figures: males (n=80) 890-1240 with a mean of 1051 g and females (n=80) 720-1080 g with a mean of 958 g (Goethe, 1961).

Table 55. Weights in grams of Herring gulls in their first year caught on the island of Texel (data: Swennen, in litt.).

	mean	s.d.	n	range
August	817.4	99.9	49	607-1015
September	819.1	112.5	72	565-1110
Oktober	863.3	110.0	28	720-1165
November	980.4	119.0	95	735-1234
December	956.2	120.5	62	635-1274
January	1012.3	125.9	15	820-1275
February	913.6	101.2	20	705-1115
March	935.0	110.6	14	760-1115

3.30.3 Numbers

3.30.3.1 Population size

Population sizes in some European countries are summarized in Møller (1978). In England in 1969-1970 about 334,000 were breeding, in
Norway the numbers amounts to at least 10,000. The number of breed-
ing pairs in Sweden (around 1975) was estimated at 180,000, that in
Finland in 1958 at about 1400; but increased considerably since then.
The number in Denmark in 1974 was about 62,300, in 1976 the population
breeding in the Danish Wadden Sea area was about 2400 pairs (data:
Dansk Ornithologisk Forening). In the German Wadden Sea area in 1977
26,500 pairs were breeding. The total number in the German Federal
Republic (inland populations and Baltic coast included) is about 26,700
(Kuhn, in litt.; Kuschert, in litt.; Institut für Vogelforschung,
Wilhelmshaven). The number breeding along the coast of the German
Democratic Republic in 1977 amounted to about 1000 pairs, after drastic
population regulations (Nehls, in litt.). The total number of breeding
pairs in the Netherlands in 1977 was about 53,000, 37,600 of these
breeding in the Dutch Wadden Sea (Table 56; Spaans, 1979). The population breeding in the whole Wadden Sea area therefore amounts to about
66,400 pairs. The distribution of this population is shown in fig. 166
and Table 57.

In many areas the population increased considerably. In the Eastern
part of the Dutch Wadden Sea there was an annual increase of 12.1%
between 1968 and 1977, in the Western part the increase was even 14.2%
(Spaans, 1979). An increase has also been recorded on Iceland, in Denmark (from 3000 pairs in 1920 to 62,300 pairs in 1974), Norway, Sweden,
Finland (in some places an increase has been recorded of 2.3-19.4 times
between 1949 and 1963) (Møller, 1978). The German population tenfold-
ed from 1906 until 1930. A further strong increase in the German Wadden Sea area has perhaps been weakened by limitation of adults (nar-
cotizing and hunting) and by putting down artificial eggs. In the
Netherlands such measures have been banned since 1967. Management poli-
cy is restricted there at avoiding locally that Herring gulls take over
breeding sites in tern colonies.

Table 56. Number of breeding pairs of the Herring gull in the Netherlands (after: Spaans, 1979).

	1968	1969	1970	1971	1972	1973	1974	1975	1976	1977
Eastern part Dutch Wadden Sea area	3200	3600	5200	5200	5400	5700	5800	8500	7600	8000
Western part Dutch Wadden Sea area	9200	9800	11,000	12,800	14,300	17,900	20,700	21,000	27,500	29,600
Dunes mainland coast	1500	1500	1500	1800	1900	2000	2400	3000	4000	5100
Delta area	2700	2500	2500	2700	2600	3200	4700	6400	8700	9700
Total (approximately)	17,000	17,000	20,000	22,000	24,000	29,000	34,000	39,000	48,000	53,000

Table 57. Breeding population of the Herring gull in the Wadden Sea
area. Number of breeding pairs in parentheses, figures without parentheses indicate the number of the colony in fig. 166. After data from
A.P. Møller, H.U.S. Møller and Rattenburg (Dansk Ornithologisk Forening); Schutzstation Wattenmeer; Deutscher Bund für Vogelschutz; Verein
Jordsand; Mellumrat; Institut für Vogelforschung, Wilhelmshaven; Prokosch; Müller; Raddatz; Bauamt für Küstenschutz, Norden; Reseach Institute for Nature Management, Leersum; Spaans).

Denmark (situation 1974)

1 Langli (915); 2 Skallingen (3); 3 Fanø (5); 4 Mandø (180); 5 Jordsand
(1300); 45 Rømø (2-4); 46 Højer Forland (6).

Germany (F.R.) (situation 1977)

6 Uthörn (3), 7 Sylt (1), 8 Föhr (4), 9 Amrum (900), 10 Langeness (178),
11 Gröde (300), 12 Hamburger Hallig (90), 13 Hooge (60), 14 Japsand
(7), 15 Norderoog (100), 16 Nordstrandischmoor (250), 17 Nordstrand
(200), 18 Südfall (60), 19 Süderoog (300), 20 Tümlauer Bucht (3), 21
Grüne Insel (100), 22 Trischen (1226), 23 Pagensand (40), 24 Lühesand
(36), 25 Knechtsand (15), 26 Bremerhaven (15), 27 Augustengroden
(117), 28 Wilhelmshaven (15), 29 Mellum (4000), 30 Spiekeroog (180),
31 Langeoog (4400), 32 Baltrum (1300), 33 Norderney (200), 34 Memmert
(10,000), 35 Juist (150), 36 Lütje Hörn (2000), 37 Borkum (40), 47
Oland (21), 48 Pellworm (5).

Netherlands (situation 1976/1977)

38 Rottumeroog (1900), 39 Rottumerplaat (1000), 40 Schiermonnikoog
(3450), 41 Ameland (528), 42 Terschelling (15050), 43 Vlieland (9940),
44 Texel (4560), 49 Lauwersmeer (31), 50 Griend (1).

3.30.3.2 Numbers per area

From the Danish part of the Wadden Sea area hardly any figures are
available. A count on April 19, 1980 yielded 9230 birds, another one
on September 13, 1980 18,700 (Meltofte, in litt.).

In the Wadden Sea area in Schleswig-Holstein numbers increase from
July-October when birds from breeding populations in the North join
those of the local breeding population. Numbers in the whole area in
this time of the year may rise to 30,000-50,000. Numbers in mild- and
normal winters decrease to 8000-10000. In March more than 20,000 may
be present again. In the months following numbers decrease to about
6000 probably all belonging to the breeding population of the area
(Busche, 1980).

In Niedersachsen numbers in July may be very high in some areas.
On July 26, 1964 more than 40,000 Herring gulls were counted on
Knechtsand (Oelke, 1968). In 1971 from July-September 10,000-25,000
were present here (Kollek & Nikolaus in Goethe, 1973). Like in other
parts of the Wadden Sea outside the breeding colonies flocks of imma-

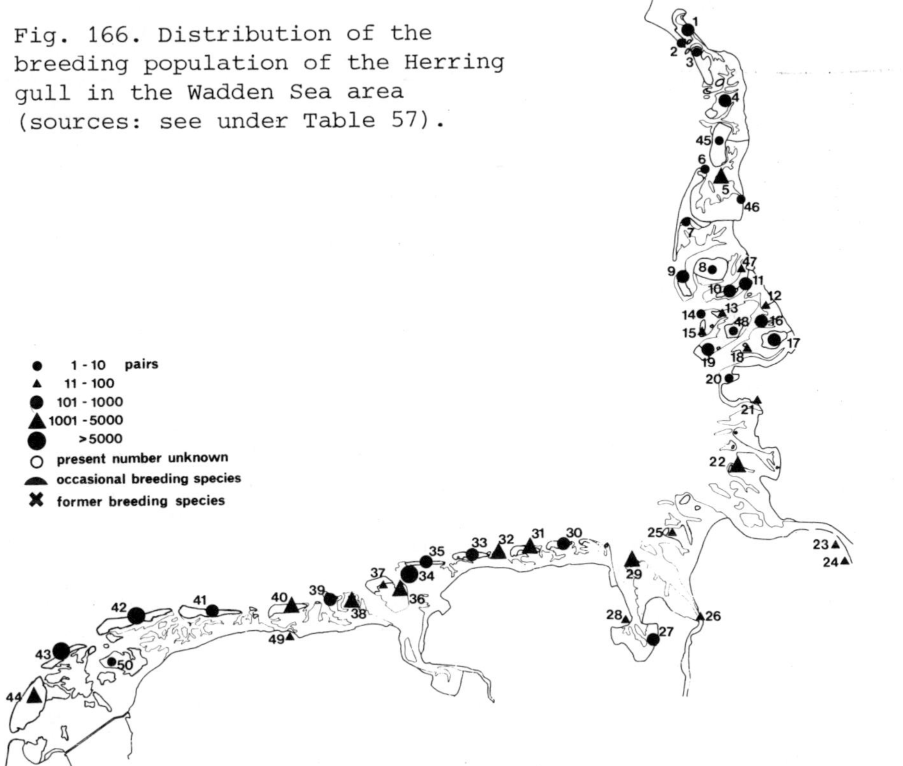

Fig. 166. Distribution of the breeding population of the Herring gull in the Wadden Sea area (sources: see under Table 57).

tures are present from April till June. In July numbers increase because adults appear more frequently outside the colonies. In August and September the same happens with first calendar-year birds (Grosskopf, 1968; Goethe, 1973).

Results of aerial surveys in the whole German part of the Wadden Sea are: January 11, 1964: 13,360; January 19, 1965: 38,740 and February 4, 1969: 25,200 (Goethe, unpubl.). These data are minimum figures as not all areas, particularly not in Schleswig-Holstein, could be visited. Two counts on islands and along the mainland coast of Niedersachsen, main fishery harbours not included, gave: January 19, 1969: 26,230 and February 4, 1962: 31,100 (Goethe, 1973).

In winter large concentrations may occur in harbours. On January 19, 1969 6400 Herring gulls were present in Cuxhaven and 2950 in Bremerhaven. Generally in both cities the winter population amounts to about 8000 (Panzer & Rauhe, 1978).

Numbers present in the Dutch part of the Wadden Sea are shown in fig. 167 and Table 58. Spaans (1971) estimated minimum and maximum numbers in the winter of 1966-1967 at 21,000 and 37,000, respectively, and in the winter of 1967-1968 at 14,000 and at least 28,000. Numbers on refuse dumps on the mainland in the three Northern provinces Friesland, Groningen and Drente are low in the breeding season but increase already considerably shortly afterwards. From October 1967 until April 1968 wintering numbers fluctuated from 24,000 to at least 47,000. Dur-

Table 58. Numbers of Herring gulls in the Dutch part of the Wadden Sea as determined by counts from the shore (after various publications in Watervogels and Limosa and Zegers, in litt.).

Date	Year	Number
January 8	1977	17,100
January 12	1974	38,330
January 17	1976	26,080
January 18	1975	22,290
April 6	1973	21,900
July 29	1972	53,690
August 30	1975	76,300
September 1	1973	33,080
October 19	1974	25,035
November 13	1976	33,240
December 29	1966	38,000

ing periods with tides lower than average in the Wadden Sea, when large areas of cockle beds are exposed, numbers foraging on the refuse dumps, which were situated on the mainland, not too far from the Wadden Sea, dropped. Probably those gulls foraging on refuse dumps in the provinces Friesland, Groningen and the Northern part of the province of Drente roost on the Wadden Sea by night. Those Herring gulls foraging in the Southern part of the province of Drente probably roost on the IJsselmeer.

Fig. 168 shows the distribution of the birds in autumn in the Wadden Sea.

3.30.4 Food

3.30.4.1 Food composition

Food of immature and adult Herring gulls mainly consists of marine invertebrates like *Mytilus*, *Carcinus*, *Asterias*, *Cerastoderma* and *Mya* and fish. A large variety of other prey species has been found less

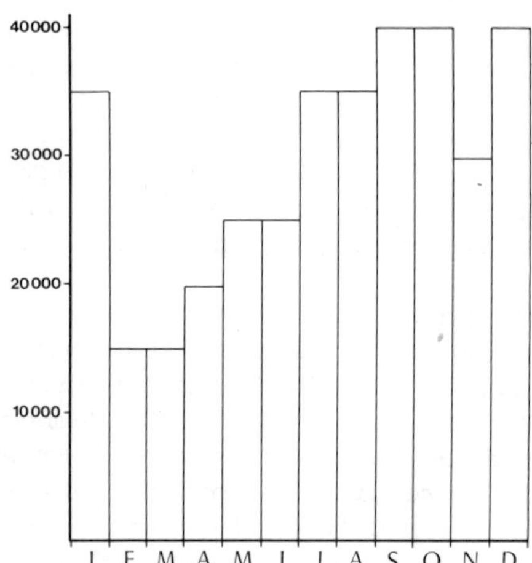

Fig. 167. Mean number of Herring gulls in the Dutch part of the Wadden Sea based on counts from the shore (after data from Smit, 1977).

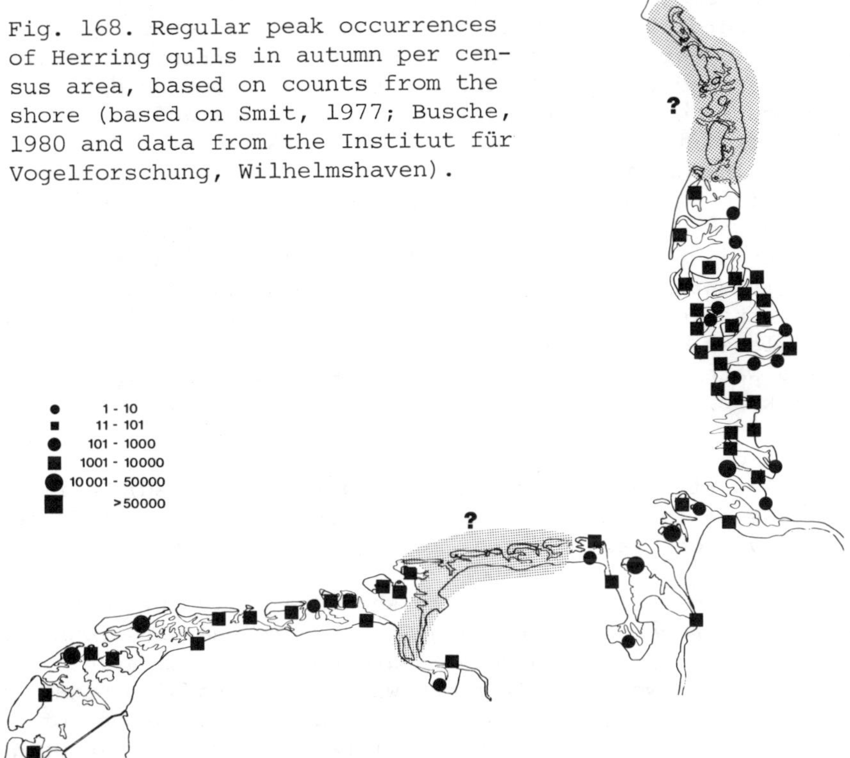

Fig. 168. Regular peak occurrences of Herring gulls in autumn per census area, based on counts from the shore (based on Smit, 1977; Busche, 1980 and data from the Institut für Vogelforschung, Wilhelmshaven).

frequently. Of these *Crangon*, *Macropipus*, *Nereis*, *Littorina* and small mammals were most common. Garbage and vegetable matter and carrion were taken as well (Spaans, 1971).

Food composition varies from place to place depending on whatever is available for the birds. Meyering (1954), Goethe (1956) and Ehlert (1961) found a somewhat different diet in the German part of the Wadden Sea, *Asterias* often being absent.

Considerable differences appeared to exist in food composition in the course of the year as well. These are also attributed to changes in abundance, accessibility and attractiveness of prey species (compare Spaans, 1971).

Food of chicks consists mainly of fish (clupeids, gadids, flatfish), to a lesser extent of marine invertebrates and garbage. Most of the Gadidae and flatfish came from discards of commercial fishery, comprising about one-fifth to one-third of the total food supply. As garbage also provided about one-fourth of it, food obtained through human activities comprised about 50% of the total food supply to the young (Spaans, 1971). On the island of Mellum young Herring gulls were fed with fish (42%), molluscs (24%), crustaceans (14%) and meat garbage (12%) (Hinrichs in Goethe, 1973).

3.30.4.2 Feeding activities

Without human activities Herring gulls should be feeding in coastal areas in the North Sea and in the Wadden Sea, especially on the tidal flats, along the borders of gullies and creeks, and in wrack-beds on the beaches. Due to agricultural activities of man Herring gulls may forage on pastures and farmland. Further they feed along shipping routes, in harbours, cities, villages, on refuse dumps and sewage plants. Like Common gulls they sometimes drop cockles and mussels on hard surfaces to break shells.

During high tide Herring gulls may roost on salt marshes, sandflats and inland. They may roost on the Wadden Sea too. During night large numbers may roost on the Wadden Sea, on bare sandflats and on isolated places on salt marshes. Gulls foraging on refuse dumps on the mainland or along estuaries and harbours far inland carry out flights towards roosting places on lakes as well.

For drinking water Herring gulls prefer fresh water though they are adapted to drink sea water.

3.30.4.3 Total food consumption

Spaans (1971) estimated food intake of a wild (not caged) Herring gull per low tide period at 167 ml cockle flesh. This corresponds to 294 cockles of over one year old and 2165 cockles of only a few months old.

Full-grown Herring gulls in captivity showed a daily intake of 365 g Horse-mackerels *(Trachurus)*, or 212 g Sprat *(Sprattus)* and young Herring *(Clupea)*, or 76 g bread. These data are not quite comparable because birds with fish put on weight, but those which were fed with bread lost weight. In captivity the total food (fish) intake of two groups of captive chicks during the first six weeks of life was 9.2 kg fresh weight (2.5 kg dry weight) and 8.4 kg (2.4 kg) respectively per chick (Spaans, 1971).

References

Barth, E.K., 1975. Taxonomy of Larus argentatus and Larus fuscus in North-Western Europe. Ornis Scand. 6: p. 49-63.

Busche, G., 1980. Vogelbestände des Wattenmeeres von Schleswig-Holstein. Kilda, Greven (in press).

Drost, R., E. Focke & G. Freytag, 1961. Entwicklung und Aufbau einer Population der Silbermöwe Larus a. argentatus. J. Orn. 102: p. 404-429.

Drost, R. & L. Schilling, 1940. Über den Lebensraum deutscher Silbermöwen, Larus a. argentatus Pontopp., auf Grund von Beringungsergebnissen. Vogelzug 11: p. 1-22.

Dwight, J., 1925. The Gulls (Laridae) of the World. Bull. Americ. Mus. Nat. Hist. 52, III: p. 63-401.

Ehlert, W., 1961. Weitere Untersuchnungen über die Nahrungswelt der Silbermöwe (Larus argentatus) auf Mellum. Vogelwarte 21: p. 48-50.

Eykman, C., P.A. Hens, F.C. van Heurn, C.G.B. ten Kate, J.G. van Marle, M.J. Tekke & T.G. de Vries, 1949. De Nederlandsche Vogels, Vol. 3. Wageningse Boek- en Handelsdrukkerij, Wageningen: 407 pp.

Goertz, M., 1971. Ringfunde deutscher Nordsee-Silbermöwen (Larus argentatus) Teil 3: Beringungen auf der Insel Spiekeroog. Auspicium 4: p. 303-310.

Goertz, M. & F. Goethe, 1969. Ringfunde deutscher Nordsee-Silbermöwen (Larus argentatus) Teil 1: Beringungen auf der Insel Memmert. Auspicium 3: p. 305-317.

Goethe, F., 1956. Die Silbermöwe. Neue Brehm Bücherei, Vol. 182. Ziemsen, Wittenberg-Lutherstadt: 95 pp.

Goethe, F., 1961. Zur Taxionomie der Silbermöwe (Larus argentatus) im südlichen deutschen Nordseegebiet. Vogelwarte 21: p. 1-24.

Goethe, F., 1973. Die Silbermöwe - Larus argentatus in Niedersachsen. In: H. Ringleben & H. Schumann (ed.). Aus der Avifauna von Niedersachsen. Wilhelmshaven: p. 25-46.

Grosskopf, G., 1968. Die Vögel der Insel Wangerooge. Abh. Vogelk. 5. Institut für Vogelforschung, Wilhelmshaven: 293 pp.

Harris, M.P., 1964. Aspects of the breeding biology of the gulls Larus argentatus, L. fuscus and L. marinus. Ibis 106: p. 432-456.

Haftorn, S., 1971. Norges fugler. Universitetsforlaget, Oslo: 862 pp.

Jørgensen, O.H., 1973. Some results of Herring Gull ringing in Denmark 1958 to 1969. Dansk Orn. Foren. Tidsskr. 67: p. 53-63.

Kuschert, H., 1979. Die Silbermöwe (Larus argentatus) in Schleswig-Holstein. Ein Beitrag zur Diskussion über ihre taxonomische Stellung. Festschrift Inselstation Helgoland. Abh. Geb. Vogelkund. 6: p. 87-112.

Meyering, M.P.D., 1954. Zur Frage der Variationen in der Ernährung der Silbermöwe, Larus argentatus Pontopp. Ardea 42: p. 163-175.

Møller, A.P., 1978. Maagernes Larinae yngleudbredelse, bestandsstørrelse og -aendringer i Danmark, med supplerende oplysninger om forholdene i det øvrige Europa. Dansk Orn. Foren. Tidsskr. 72: p. 15-39.

Nehls, H., 1971. Funde an der deutschen Ostseeküste beringter Silbermöwen (Larus argentatus). Auspicium 4: p. 193-226.

Oelke, H., 1968. Vögel auf dem Knechtsand. Falke 15: p. 342-351 & 372-377.

Olsson, V., 1958. Dispersal, migration, longevity and death causes of Strix aluco, Buteo buteo, Ardea cinerea and Larus argentatus. Acta Vertebr. 1: p. 91-189.

Paludan, K., 1953. Nogle resultater af Københavns Zoologiske Museums Ringmaerkning af Larus argentatus. Vidensk. Medd. Dansk naturh. For. 115: p. 181-204.

Panzer, W. & H. Rauhe, 1978. Die Vogelwelt an Elb- und Wesermündung. Heimatbund Männer vom Morgenstern, Bremerhaven: 336 pp.

Prater, A.J., 1976. Birds of estuaries enquiry 1973-1974. British Trust for Ornithology, Royal Soc. Prot. Birds, Wildfowl Trust: 48 pp.

Reichmann, K.H., 1971. Ringfunde deutscher Nordsee-Silbermöwen (Larus argentatus) Teil 2: Beringungen auf Langeoog. Auspicium 4: p. 273-302.

Schüz, E., 1971. Grundriss der Vogelzugskunde. Parey, Berlin: 391 pp.

Spaans, A.L., 1971. On the feeding ecology of the Herring Gull Larus argentatus Pont. in the northern part of The Netherlands. Ardea 59: p. 73-188.

Spaans, A.L., 1979. Zilvermeeuw. In: R.M. Teixeira (ed.). Atlas van
    de Nederlandse broedvogels. Natuurmonumenten, 's-Graveland: p. 172-
    173.
Spaans, M.J. & A.L. Spaans, 1975. Enkele gegevens over de broedbiolo-
    gie van de Zilvermeeuw Larus argentatus op Terschelling. Limosa 48:
    p. 1-39.
Stegman, B., 1934. Über die Formen der grossen Möwen ("subgenus Larus")
    und ihre gegenseitigen Beziehungen. J. Orn. 82: p. 340-380.
Tinbergen, N., 1953. The Herring Gull's World. Collins, London: 355 pp.
Vaurie, C., 1965. The Birds of the Palaearctic Fauna. Non-Passeriformes.
    London: 763 pp.
Voous, K.H., 1960. Atlas of European birds. Nelson, London: 284 pp.

3.31 SANDWICH TERN (*STERNA SANDVICENSIS* LATHAM)
J. Rooth

Da: Splitterne; G: Brandseeschwalbe; Du: Grote Stern

3.31.1 Distribution

3.31.1.1 Breeding area
Sandwich terns have a very scattered, holarctic and South-American
distribution, breeding in temperate, Mediterranean, steppe, desert
and probably savana climatic zones. It breeds along sea coasts with
clear water and a predominantly sandy bottom, in extensive delta re-

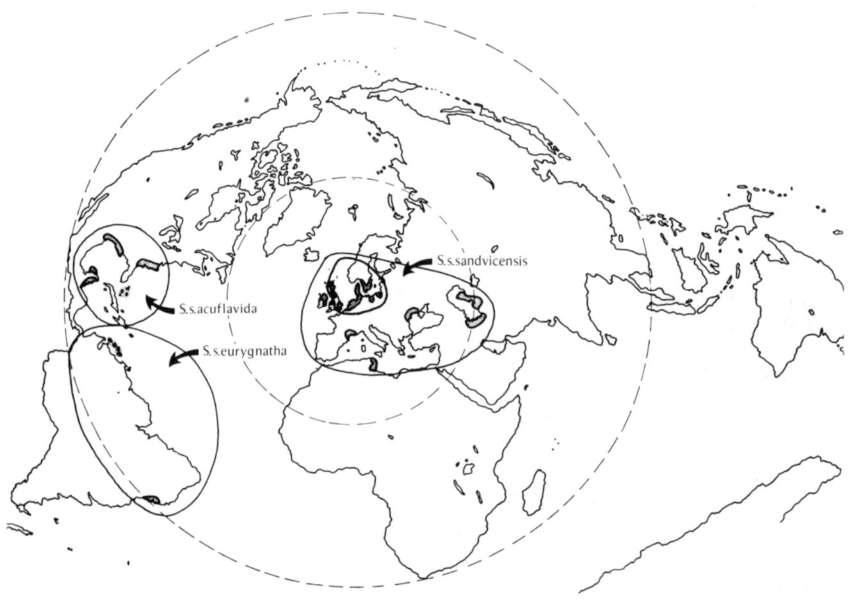

Fig. 169. Breeding area of the Sandwich tern (after Voous, 1960).
The dark shaded area represents the area where birds visiting the
Wadden Sea originate from.

gions and along rocky coasts with flat and mostly calcareous coastal
islets (Voous, 1960). It nests on sandbanks, coastal plains, flat
coastal islands, sand dunes with sparse or without vegetation, generally in colonies.

In NW Europe the species is breeding in France, Great-Britain, Ireland, the Netherlands, the German Federal- and Democratic Republic,
Denmark and Sweden. There are three subspecies (fig. 169).

3.31.1.2 Migration routes

The NW European breeding birds migrate along the North Sea and the
Atlantic coasts of Europe and Africa (fig. 170). As far as known Sandwich terns from the Danish East coast and Sweden follow the Danish
coast line and do not cross Jutland or Schleswig-Holstein (Müller,
1959).

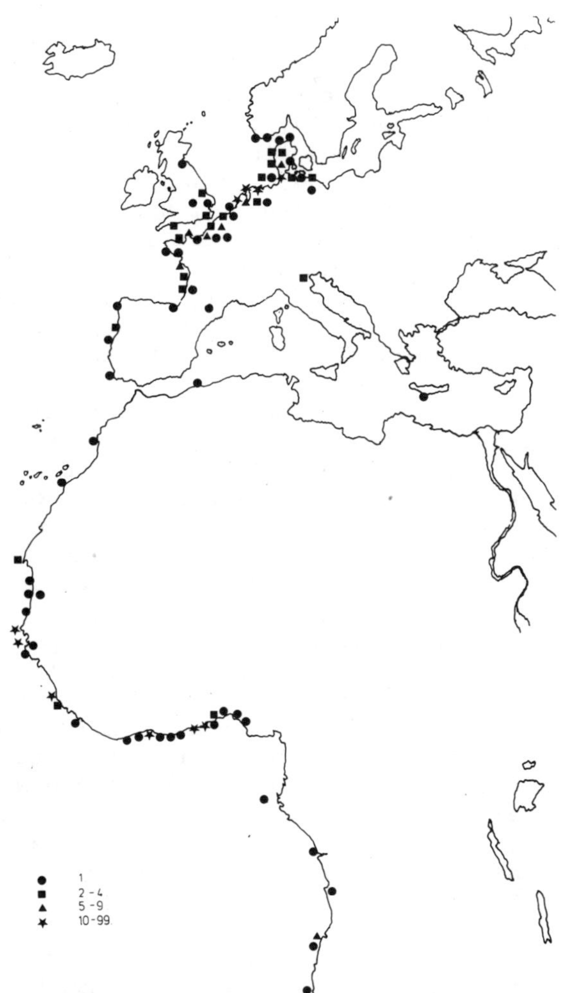

Fig. 170. Recoveries of
Sandwich terns ringed as
pullus in the Dutch
Wadden Sea area (data:
Vogeltrekstation Arnhem,
unpubl.).

3.31.1.3 Wintering areas

The main wintering area for the birds breeding in the Wadden Sea is the coastal area with adjacent ocean in West Africa between 5° and 16° N. A smaller part winters further south, mostly along the Angolan coast. Recently there are also observations of wintering birds in the SW part of the Netherlands (Ouweneel, 1975).

3.31.1.4 Moulting areas

Post-nuptial moult takes place in the breeding period and afterwards when the birds are migrating towards the wintering area or even in the wintering area itself (end of May-December). Pre-nuptial moult takes place when birds carry out spring migration (February-April) (Witherby et al., 1952).

3.31.2 Annual cycle

3.31.2.1 Migration

Adult Sandwich terns arrive in the North Sea by the end of March. Generally however they do not arrive in the Western part of the Wadden Sea until the second April decade (Veen, 1977), on Wangerooge not until the end of April (Grosskopf, 1968). Egg laying on Griend generally starts in the first days of May (Veen, 1977), on Wangerooge between May 6 and 12 (Grosskopf, 1968), on Scharhörn from May 6 onwards (Temme, 1967). Egg laying may continue until the end of June, probably by birds which have lost their eggs in earlier attempts in the same or other colonies. From the extensive work by Müller (1959) it appears that birds breeding in Great Britain, France, the Netherlands, Germany, Denmark and Sweden have the same migratorial behaviour. After fledging juvenile Sandwich terns show a pre-migratory dispersal and may be found in coastal waters around North Sea and Baltic generally occurs in August, especially Dutch birds however have been recovered along the French Atlantic coast already in July. The moment of departure however may vary strongly between birds. In September the majority is found along the French, Spanish and Portugese coast, in October the first arrive in African waters, some enter the Mediterranean. By this time however juvenile Sandwich terns may still be present in the Wadden Sea area or even in Danish waters. The following summer is spent around the African west coast. Third calendar-year birds spend summer along the West coasts of Europe and Africa, in their fourth year probably all are present in West European waters. At the end of their fourth year Sandwich terns start breeding. Also adult birds show some pre-migratory dispersal, most of them have left Europe by October.

Sandwich terns breeding in Denmark and Sweden pass through the Wadden Sea from July-September (data: Zoologisk Museum, København). After mid-August numbers decrease throughout the whole area, the last Sandwich terns leave the Wadden Sea by mid-October. The first birds arrive in West Africa by October (Speek, 1969).

3.31.2.2 Moult

A complete moult in adults takes place from May-December. Moult be-

Table 59: Weights in grams of Sandwich terns caught near the island of Vlieland from December 1971-December 1975 (after Boere, in litt.).

	age	mean	s.d.	n	range
April	older than 1 cy	282.5	3.5	2	280-285
August	1st cy	219.4	13.6	17	200-245
August	older than 1 cy	237.0	14.4	30	215-275
September	1st cy	241.3	8.1	3	234-250

gins on the head. Some birds begin to show white on fore-head or crown by the end of May, others do so early June often while still breeding. Another almost complete moult takes place from February-April, though probably not all primaries are moulted in spring. Juvenile body-plumage, a varying number of wing-coverts, innermost secondaries and sometimes the central pair of tail-feathers are moulted from August-December. In first summer a complete moult takes place from March-June though outer tail-feathers and sometimes part of the secondaries are not renewed (Witherby et al., 1952).

3.30.2.3 Weight changes
The only information on Sandwich tern weights is listed in Table 59.

3.31.3 Numbers

3.31.3.1 Population size
Sandwich terns prefer isolated areas as a breeding habitat, like islands, sandbanks and peninsulas with a scarce vegetation and the presence of a Black-headed gull colony. The gulls are laying earlier and give the terns protection because of their aggresive behaviour against common predators (Rooth, 1958; Fuchs, 1977; Veen, 1977). Islands and sandbanks in estuaries can change their suitability as breeding place very quickly. This means that some sites are only occupied for a couple of years and that the numbers fluctuate strongly from place to place. Moreover there is an interchange between nesting sites (Nehls, 1969; Cramp et al., 1974). Factors like food supply and pesticides may cause fluctuations in the whole Wadden Sea- or NW European population too.
Since about 1920 no breeding cases of the Sandwich tern have been reported from the Danish part of the Wadden Sea area. In 1977 however small numbers did breed on Rømø (Iversen, in litt.). In the German part Scharhörn is a breeding place of the Sandwich tern since the beginning of this century (Temme, 1967) and still is (Goethe, in litt.). Important colonies have been present on Lütje Hörn from 1950-1962 and on Wangerooge till 1964. The colony of Oldeoog started in 1952 and increased very much recently. In 1968 an important settlement started

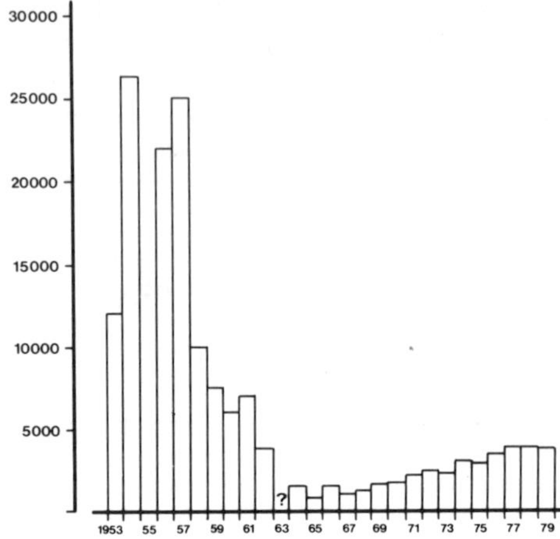

Fig. 171. Number of breeding pairs of the Sandwich tern on the Dutch Wadden Sea islands (Rooth, unpubl.).

on Knechtsand. Since 1955 Trischen has been very important as breeding ground, especially since 1964. On Norderoog nowadays smaller numbers occur. This colony is very old, Naumann (1819) already reported on it. The total breeding numbers in the German part of the Wadden Sea fluctuated in the period 1954-1978 between 2000 and 6000 pairs, with low numbers in the fifties. There is a tendency to increase in the last years.

In the Dutch part of the Wadden Sea the main colony is settled on Griend with Texel, Schiermonnikoog, Rottumerplaat and Rottumeroog sometimes as smaller additional settlements. The numbers fluctuated from 1953-1962 between 6000 and 26,000 pairs. A decrease followed to 3800 pairs in 1962, 1500 in 1964 and only 800 in 1965. This tremendous decrease was mainly due to discharge of the pesticides dieldrin and telodrin by a factory near Rotterdam (Koeman et al., 1967; Koeman & Van Genderen, 1972). The closing of the telodrin factory in 1965 and the improvement of the purification plant by the same factory in 1967 helped to improve the situation (Rooth & Jonkers, 1972). Fig. 171 shows the increase of the Dutch Wadden Sea breeding population afterwards.

The total numbers for the whole Wadden Sea area from 1954-1978 fluctuate between 15,000 and 29,000 pairs, the whole NW European population in the same period between 30,000 and 50,000 (Rooth, unpubl.). Therefore more than half of the NW European population was settled in the Wadden Sea. In 1964, 1965 and 1966 there were 6600, 3500 and 4600 pairs breeding in the Wadden Sea area. The NW European population amounted 20,000-25,000 breeding pairs then. This means that 15-30% was housed in the Wadden Sea. From 1974-1978 the population in the Wadden Sea had increased to 9000 pairs (fig. 172). In the same period the British population increased to more than 13,000 pairs in 1978 (Thomas, in litt.) and in Denmark too there was a recent increase (outside the Wadden Sea) to 4000 pairs (Mardal, 1974). The total population for NW Europe in this period therefore amounted to 30,000-35,000 breeding pairs. This means that the Wadden Sea recently contributes approximately 30% to the NW European population.

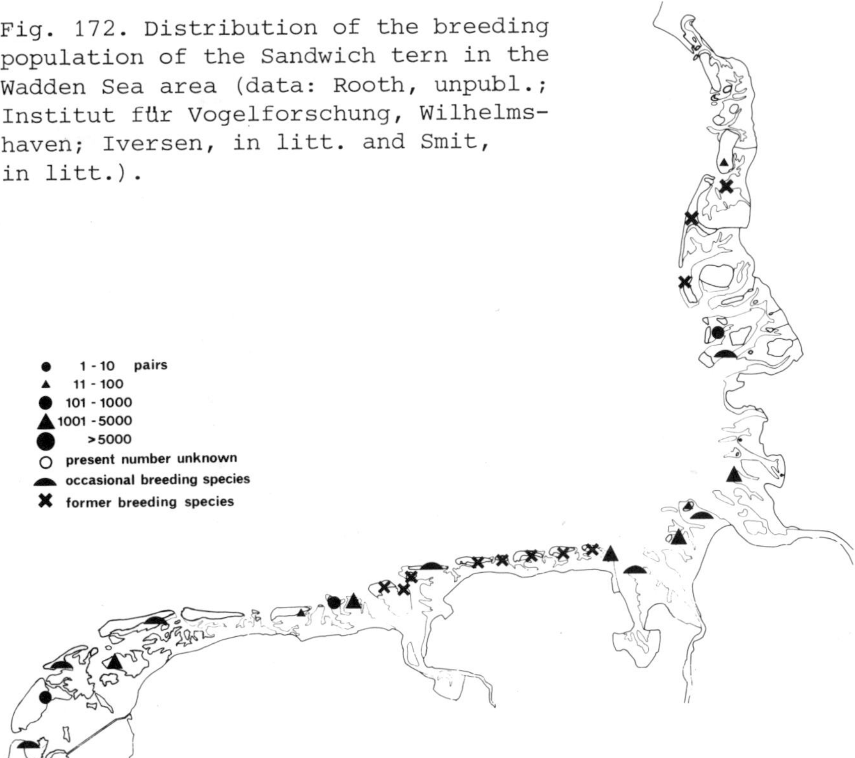

Fig. 172. Distribution of the breeding population of the Sandwich tern in the Wadden Sea area (data: Rooth, unpubl.; Institut für Vogelforschung, Wilhelms- haven; Iversen, in litt. and Smit, in litt.).

3.31.3.2 Numbers per area

Generally with the aid of ground counts insufficient insight can be gained of the total numbers present in a certain area because dur- ing high tide Sandwich terns may forage at sea out of sight for the observers. Fig. 173 gives the results of specific counts of the num- bers present at Scharhörn where at the time of observations 435 pairs were breeding. Fig. 174 shows regular peak occurrences per census area.

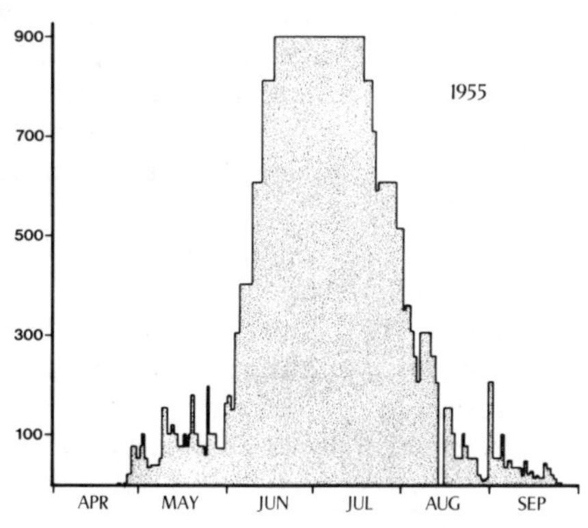

Fig. 173. Numbers of Sandwich terns present on the island of Scharhörn in 1955 (after Temme, 1967).

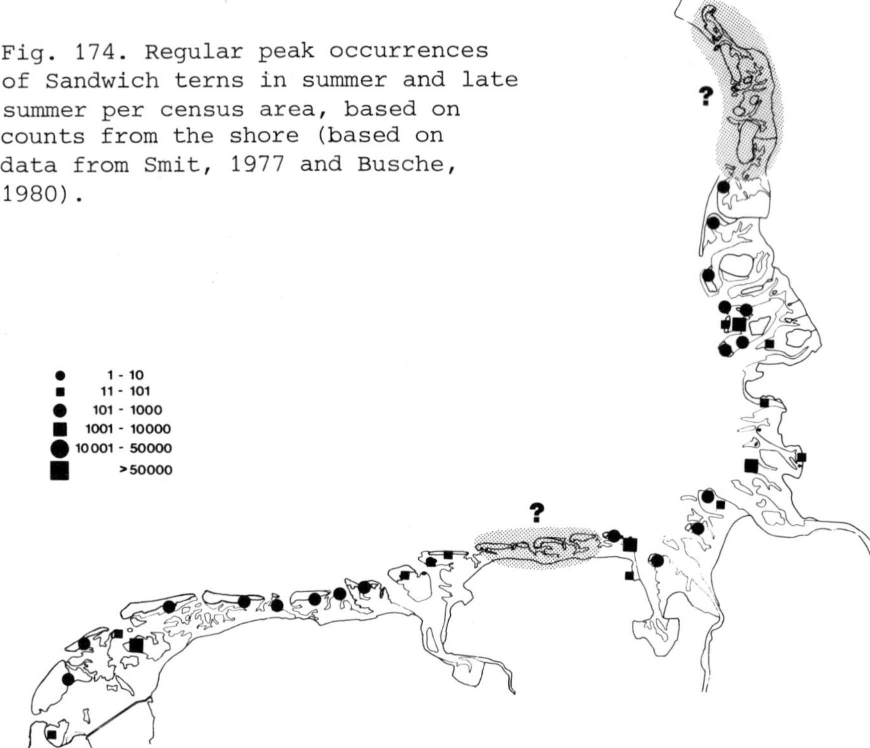

Fig. 174. Regular peak occurrences
of Sandwich terns in summer and late
summer per census area, based on
counts from the shore (based on
data from Smit, 1977 and Busche,
1980).

3.31.4 Food

3.31.4.1 Food composition

Veen (1977) described for the Griend colony the fishes that were
fed to partners and chicks: 96-98% of the fish belonged to four species:
Sand Eels *(Ammodytes tobianus* and *Hyperoplus lanceolatus)*, Herring
(Clupea harengus) and Sprat *(Sprattus sprattus)*. As in many other sea-
bird species Sand Eels play an important role in the Sandwich tern
life-cycle. They spawn in December and January and after six months
in June and July they are about 5 cm long, the previous age class measures 10-20 cm by this time. Both age classes are very important as a
food resource for the chicks. After cold winters and/or a cold spring
Sand Eels do not enter coastal waters so early. In the former Sandwich
tern colony of "De Beer" this resulted in a big or even total mortality of the chicks due to lack of food (Rooth, 1965).

Veen (1977) mentioned that in the Griend colony the newly hatched
chicks are almost exclusively fed with very small clupeids. Young Her-
rings of the 0-group measure 4-5 cm when they enter the Dutch Wadden
Sea between February and April. They have reached a length of about
6 cm in June (Fonds, 1978). Thus, in the hatching period of the chicks
in the Wadden Sea an important condition, viz. the availability of
small fish, can be fulfilled well.

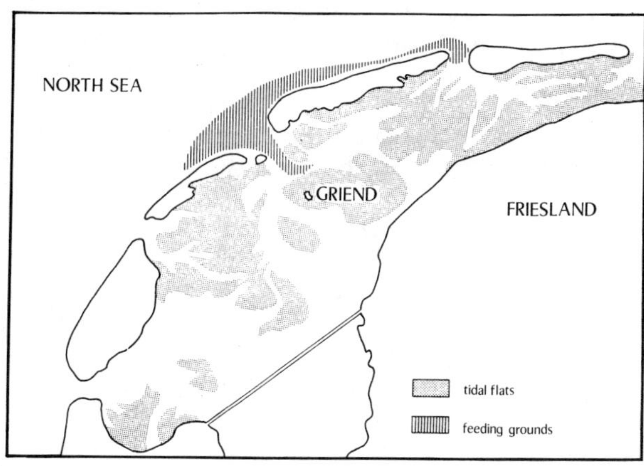

Fig. 175. Feeding grouds of Sandwich terns breeding on the island of Griend (after Veen, 1977).

3.31.4.2 Feeding activities

Sandwich terns mainly catch their food by plunge diving off the North Sea coast. This means that in some colonies they have to cover a distance of 15-25 km to reach their feeding grounds. Those are often places with turbulent water where bottom-living species like Sand Eels come to the surface and are easy to catch (fig. 175).

References

Busche, G., 1980. Vogelbestände des Wattenmeeres von Schleswig-Holstein. Kilda, Greven (in press).

Cramp, S., W.R.P. Bourne & D. Saunders, 1974. The seabirds of Britain and Ireland. Collins, London: p. 142-150.

Fonds, M., 1978. In: N. Dankers, W.J. Wolff & J.J. Zijlstra (ed.). Fishes and fisheries of the Wadden Sea. Report 5 Wadden Sea Working Group: p. 42-77.

Fuchs, E., 1977. Predation and anti-predator behaviour in a mixed colony of terns Sterna sp. and Blackheaded Gulls Larus ridibundus with special reference to the Sandwich Tern Sterna sandvicensis. Ornis Scand. 8: p. 17-32.

Grosskopf, G., 1968. Die Vögel der Insel Wangerooge. Abhandl. Vogelk. 5. Inst. für Vogelforschung, Wilhelmshaven: p. 128-130.

Koeman, J.H., A.A.G. Oskamp, J. Veen, E. Brouwer, J. Rooth, P. Zwart, E. v.d. Broek & H. van Genderen, 1967. Insecticides as a factor in the mortality of the Sandwich tern (Sterna sandvicensis). Mededelingen Rijksfaculteit Landbouwwet. Gent 32: p. 841-854.

Koeman, J.H. & H. van Genderen, 1972. Tissue levels in animals and effects caused by chlorinated hydrocarbon insecticides, chlorinated biphenyl and mercury in the marine environment along the Netherland coast. In: Marine pollution and sea life. Fishing News, Surrey: p. 428-434.

Mardal, W., 1974. Ternegruppen. Feltornithologen 16: p. 4-7.

Müller, H., 1959. Die Zugverhältnisse der europäischen Brandseeschwalben (Sterna sandvicensis) nach Beringungsergebnissen. Vogelwarte 20: p. 91-115.

Naumann, J.F., 1819. Ornithologische Bemerkungen und Beobachtungen als
    Resultate einer Reise durch einen Teil der Herzogtümer Holstein,
    Schleswig und die Inseln der dänischen Westsee. Okens Isens, Vol. 2,
    1845-1861.
Nehls, H.W., 1969. Zur Umsiedlung und Brutreife der Brandseeschwalben,
    Sterna sandvicensis, nach Ringfunde auf Langenwerder. Vogelwarte 25:
    p. 52-57.
Ouweneel, G.L., 1975. Overwinterende Grote Sterns Sterna sandvicensis
    in Nederland. Limosa 48: p. 197-201.
Rooth, J., 1958. Relations between Black-headed Gulls (Larus ridibundus)
    and Terns (Sterna spec.) in the Netherlands. VII Bulletin of the In-
    ternational Committee for Bird Preservation: p. 117-119.
Rooth, J., 1965. Over sterns en kaapmeeuwen. De Levende Natuur 68: p.
    265-275.
Rooth, J. & M.F. Mörzer Bruyns, 1959. De Grote Stern (Sterna s. sand-
    vicensis Lath.) als broedvogel in Nederland. Limosa 32: p. 13-25.
Rooth, J. & D.A. Jonkers, 1972. The status of some piscivorous birds
    in the Netherlands. TNO-nieuws 27: p. 551-555.
Smit, C.J., 1977. On the occurrence of 32 bird species in the Danish
    German and Dutch Wadden Sea. Unpubl. report Intern. Wadden Sea Work-
    ing Group, part 3: 174 pp.
Speek, B.J., 1969. Ringverslag van het vogeltrekstation Nr. 53 (1968).
    Limosa 42: p. 82-109.
Temme, M., 1967. Vogelfreistätte Scharhörn. Jordsand Mitteilungen 3
    (1-4): p. 104-111.
Veen, J., 1977. The Sandwich Tern. Functional and causal aspects of
    nest distribution. Behaviour, Suppl. 20: 193 pp.
Voous, K.H., 1960. Atlas der Europese vogels. Elsevier, Amsterdam: p.
    128.
Witherby, H.F., F.C.R. Jourdain, N.F. Ticehurst & B.W. Tucker, 1952.
    The handbook of British Birds, Vol. 5. Witherby, London: p. 18-24.

3.32 COMMON TERN (*STERNA HIRUNDO* L.)
   J. Rooth

Da: Fjordterne; G: Flussseeschwalbe; Du: Visdief

3.32.1 Distribution

3.32.1.1 Breeding area

The Common tern is a breeding bird in the temperate, Mediterranean,
steppe, desert and locally perhaps in the tundra and tropical rainfo-
rest climatic zones. The distribution is holarctic but not circumpolar,
with small colonisations in the Ethiopian and South-American regions
(fig. 176; Voous, 1960). The species, unlike most of the other tern
species, is not restricted to coastal regions but is also breeding in
inland habitats. It breeds in colonies along rivers, inland lakes and

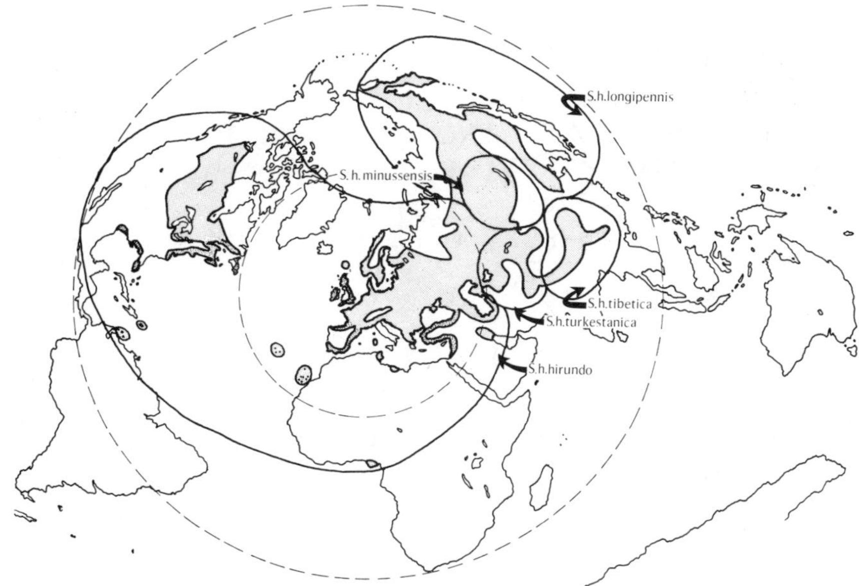

Fig. 176. Breeding area of the Common tern (after Voous, 1960).

pools with shallow fresh water and grassy shores, on salt marshes, on sand-, shell- and pebblestone banks along estuaries and rocky sea coasts and in sand dunes along the coasts. Generally four or five subspecies are distinguished, *S.h. turkestanica* being a doubtful subspecies. Some overlap in distribution occurs with the closely related species *Sterna paradisea*. Birds visiting the Wadden Sea probably originate from Fenno-Scandinavia and countries bordering the Baltic and Wadden Sea.

3.32.1.2 Migration routes

Birds from the population breeding in the Wadden Sea area migrate mostly along the North Sea and Atlantic coasts of Europe and Africa. There are however some recoveries inland.

3.32.1.3 Wintering areas

The birds winter mainly on the African West coast. The largest concentrations of ring recoveries of birds from the Dutch Wadden Sea area originate from between 3° and 17° (fig. 177). There are some winter records from the Wadden Sea but these are very exceptional.

3.32.1.4 Moulting areas

A complete moult begins in July-August when birds may still be present in the Wadden Sea. This moult is completed in the winter-quarters as late as January-February. A moult of body tail feathers and wing-coverts takes place from February-March (Witherby et al., 1952) when birds are on spring migration towards their breeding places.

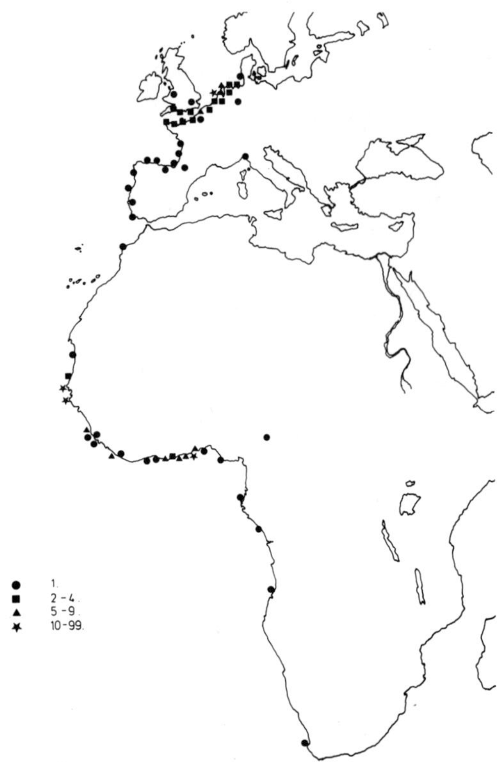

Fig. 177. Recoveries of Common terns ringed as pullus in the Dutch Wadden Sea area (data: Vogeltrekstation Arnhem, unpubl.).

3.32.2 Annual cycle

3.32.2.1 Migration

In the Western part of the Dutch Wadden Sea Common terns arrive in the beginning of April, (Dijksen & Dijksen, 1977) on Wangerooge and Scharhörn generally around April 20 (Temme, 1967; Grosskopf, 1968). The birds start breeding in May and sometimes in June.

Birds from the Wadden Sea population fledge generally in early July but some breeding birds may still be feeding their young on the nests until mid-August. Data from the colony on Wangerooge (Grosskopf, 1968) show that from July on the birds disperse and may be recovered all over the Southern part of the German Bight. Birds from Wangerooge generally depart by mid-August. Scandinavian breeding birds depart southward in July (Haftorn, 1971). In August birds from Northern breeding places (till 63° N) have been recovered in the Wadden Sea. Common terns leave the Dutch Wadden Sea in September and October. The number passing through in France culminates in August and the first part of September, in Spain and Portugal in the second and third September decade. In October and November most birds reach their winter quarters (Grosskopf, 1968). Subadult birds generally spend summer south of their breeding colonies.

3.32.2.2 Moult

A complete moult is beginning in July-August with body, tail and inner primaries and is completed in the winter-quarters even as late as January and February. A second moult of body and tail-feathers and wing-coverts takes place in February and March, but primaries and secondaries do not appear to be moulted twice (Witherby et al., 1952).

3.32.2.3 Weight changes

Information available is listed in Table 60.

3.32.3 Numbers

3.32.3.1 Population size

As ground breeders Common terns prefer isolated places where no ground predators can disturb them. They often form colonies, sometimes together with other terns or Black-headed gulls, but solitary nests and very small colonies do occur. They have bred or still breed on nearly all Wadden Sea islands and on many places along the mainland coast. This means too that they are more spread than the other terns and that less quantitative information is available.

There was a serious decline in the Western part of the Wadden Sea in the end of the fifties and the early sixties. In 1954 the number of Common terns in the Dutch part of the Wadden Sea was estimated at 30,000-35,000, in 1957 at 12,000-15,000 and in 1971 at 4650-5150 pairs (Osieck, 1972). In the seventies numbers increased somewhat again (fig. 178). Cause for this decrease mainly were chlorinated hydrocarbons (Rooth & Jonkers, 1972). In the seventies there were about 3000-5000 breeding pairs in the Dutch part of the Wadden Sea, about 2400-5500 in the Niedersachsen, about 850-1250 in the Schleswig-Holstein and 200-250 in the Danish part of the Wadden Sea. The whole population breeding in the Wadden Sea area therefore amounts to about 6400-11,500 pairs, the Wadden Sea area therefore amounts to about 6400-11,500 pairs, which is about 25% of the population breeding in NW Europe. In the first half of the century this number must have been 5-10 times higher, in the Wadden Sea as well as elsewhere in NW Europe (Rooth, unpubl.). There is a tendency that in the Northern part of the Wadden Sea area the proportion of Arctic terns increases and consequently that of Com-

Table 60: Weights in grams of Common terns caught near the island of Vlieland from December 1971-December 1975 (after Boere, in litt.).

	age	mean	s.d.	n	range
May		128	9.9	2	121-135
July	1st, 2nd cy	118	24.1	3	91-138
July older than 1st, 2nd cy		123	6.7	7	114-135
August	1st, 2nd cy	129	11.0	11	110-140
August older than 1st, 2nd cy		133	10.6	48	115-160
September	1st, 2nd cy	131	10.0	3	121-141

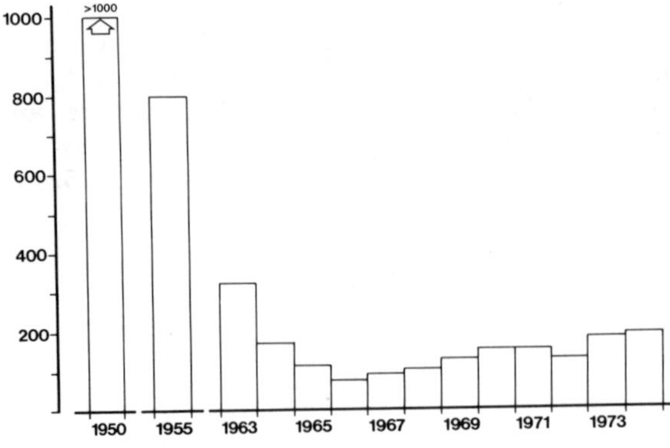

Fig. 178. Number of breeding pairs of the Common tern in the "Schorren" nature reserve on the island of Texel (after Dijksen & Dijksen, 1977)

mon tern decreases. Fig. 179 shows the distribution of the population breeding around the Wadden Sea. The most important colonies occur on Scharhörn, recently with 1300-2800 pairs (Temme, 1967; unpubl. data Institut für Vogelforschung, Wilhelmshaven), and Griend with 1000-1500 pairs (Rooth, 1979).

3.32.3.2 Numbers per area

Generally insufficient knowledge of the total number present in a certain area can be gained with the aid of ground counts, because dur-

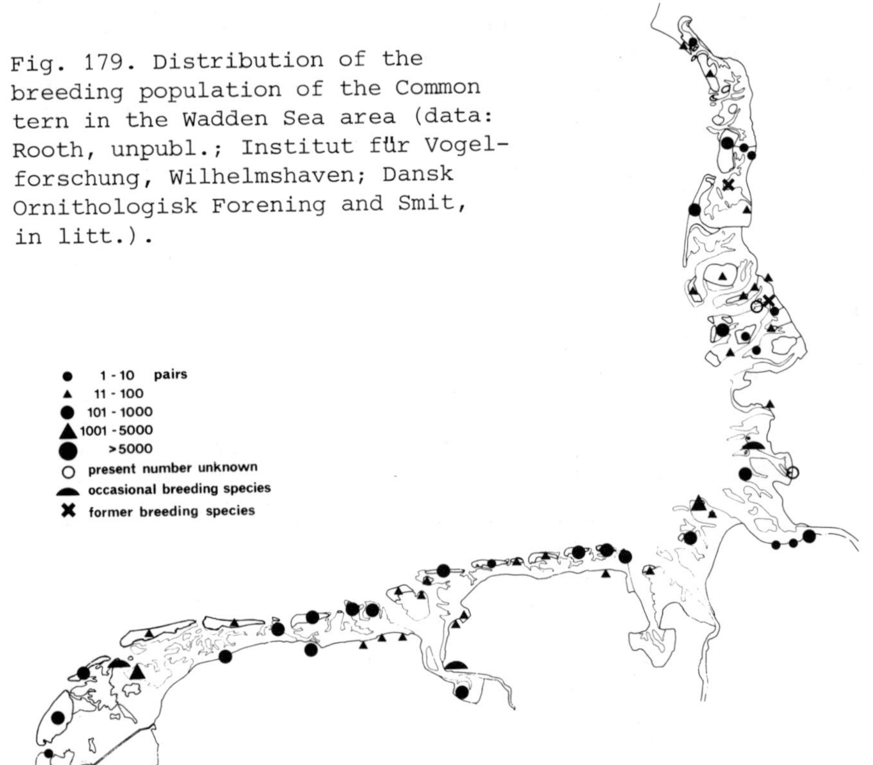

Fig. 179. Distribution of the breeding population of the Common tern in the Wadden Sea area (data: Rooth, unpubl.; Institut für Vogelforschung, Wilhelmshaven; Dansk Ornithologisk Forening and Smit, in litt.).

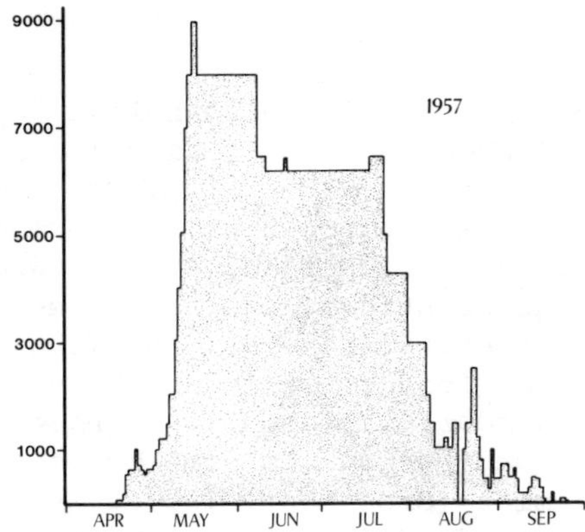

Fig. 180. Numbers of Common terns present on the island of Scharhörn in 1957 (after Temme, 1967).

Fig. 181. Regular peak occurrences of Common terns in summer and late summer per census area, based on counts from the shore (based on data from Smit, 1977 and Busche, 1980).

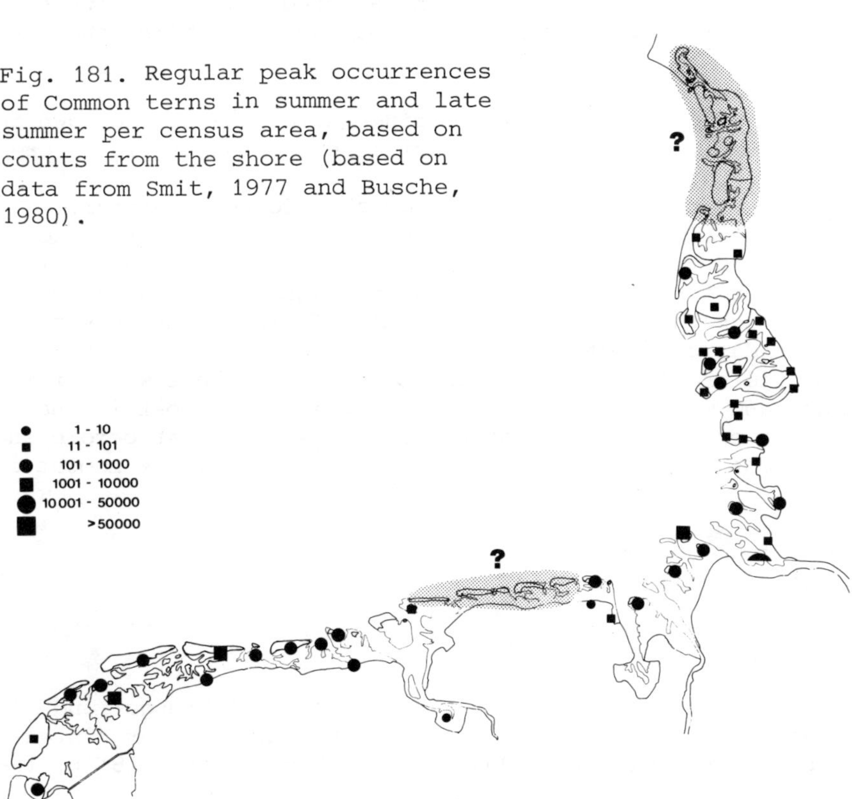

ing high tide Common terns may forage or rest at sea not visible for observers from the shore. Fig. 180 gives the results of counts of the numbers present on Scharhörn where in 1957 3000 pairs were breeding. Fig. 181 shows regular peak occurrences per census area.

3.32.4 Food

3.32.4.1 Food composition

Observations of 251 birds feeding their young at Engelsmanplaat showed that food consisted of 16% *Sprattus sprattus*, 8% *Ciliata mustela*, 6% *Pleuronectus* spec., 26% *Crangon crangon* and 23% *Carcinus maenas* (Mes & Schuckard, 1979). Boecker (1967) found on Wangerooge that food brought to the young consisted of: 75% fish, viz. 52.6% Clupeidae, 4.5% *Pleuronectus* spec., 7.1% Ammodytidae and 5.2% Gasterostidae. The rest consisted of 17.9% crustaceans and 6.8% polychaetes (*Arenicola* and *Nereis* spec.). Temme (1967) found on Scharhörn that food consisted of *Clupea harengus*, *Pleuronectus platessa*, *Ammodytes tobianus*, *Syngnathus rostellatus* and *Agonus cataphractus* but no *Crangon crangon*. In some occasions Common terns have been observed catching mice (Grosskopf, 1968). In the Balgzand area Boer & Monsees (1967) found as food: small freshwater fishes, small flatfishes, *Hyperoplus lanceolatus* and *Osmerus eperlanus*.

3.32.4.2 Feeding activities

Common terns catch fish and other marine animals by diving down into the water, sometimes touching the surface but often partially or completely submerging.

Boecker (1967) determined on Wangerooge that foraging activities were most intensive from two hours before until two hours after low tide. Foraging took mainly place in the broad gullies around the island but also on the tidal flats. During the night no foraging took place. Mes & Schuckard (1976) found that more factors than only the course of the tidal cycle determined foraging activities and that also the course of the breeding season, time of the day, place where foraging took place and water level were important factors. Most intensive foraging activities took place just after sunrise and just before sunset and in periods from two hours after high tide until the moment of low tide.

3.32.4.3 Total food consumption

No information available.

References

Boecker, M., 1967. Vergleichende Untersuchungen zur Nahrungs- und Nistökologie der Flussseeschwalbe (Sterna hirundo L.) und der Küstenseeschwalbe (Sterna paradisea Pont). Bonn. Zool. Beitr. 18: p. 15-126.

Boer, P. & G.R. Monsees, 1967. Avifaunistisch overzicht van het Balgzand en enkele omliggende terreinen. Limosa 40: p. 188-205.

Busche, G., 1980. Vogelbestände des Wattenmeeres von Schleswig-Holstein. Kilda, Greven (in press).

Dijksen, A.J. & L.J. Dijksen, 1977. Texel vogeleiland. Thieme, Zurphen: p. 140-141.

Grosskopf, G., 1968. Die Vögel der Insel Wangerooge. Abhandl. Vogelk. 5. Inst. für Vogelforschung, Wilhelmshaven: p. 109-116.

Haftorn, S., 1971. Norges fugler. Universitetsforlaget, Oslo: 862 pp.

Mardal, W., 1974. Ternegruppen. Feltornithologen 16: p. 4-7.

Mes, R. & R. Schuckard, 1976. Een onderzoek naar verschillen in fourageeraktiviteit tussen Visdief Sterna hirundo en Noordse Stern Sterna paradisaea op de Engelsmanplaat (NL). Verslagen en Technische gegevens Inst. voor Taxon. Zool. (Zoologisch Museum). Report Nr. 11 University of Amsterdam: 36 pp.

Mes, R. & R. Schuckard, 1979. Visdieven, Sterna hirundo en Noordse Sterns, Sterna paradisaea op de Engelsmanplaat (Waddenzee, NL). In: Ned. Orn. Unie: Samenvattingen van de voordrachten, gehouden Zaterdag 3 februari 1979 tijdens de themadag Kolonievogels: p. 4-5.

Osieck, E.R., 1972. De Visdief (Sterna hirundo L.) als broedvogel in Nederland. Vogeljaar 20: p. 130-136.

Rooth, J., 1979. Over de stand van onze Sterns. De Lepelaar 60: p. 12-15.

Rooth, J. & D.A. Jonkers, 1972. The status of some piscivorous birds in the Netherlands. TNO-nieuws 27: p. 551-555.

Smit, C.J., 1977. On the occurrence of 32 bird species in the Danish, German and Dutch Wadden Sea. Unpubl. report Intern. Wadden Sea Working Group, part 3: 174 pp.

Temme, M., 1967. Vogelfreistätte Scharhörn. Jordsand Mitteilungen 3 (1-4): p. 92-100.

Voous, K.H., 1960. Atlas van de Europese vogels. Elsevier, Amsterdam: p. 128.

Witherby, H.F., F.C.R. Jourdain, N.F. Ticehurst & B.W. Tucker, 1952. The Handbook of British Birds, Vol. 5. Witherby, London: p. 28-35.

3.33 LITTLE TERN *(STERNA ALBIFRONS* PALLAS)
J. Rooth

Da: Dvaergterne; G: Zwergseeschwalbe; Du: Dwergstern

3.33.1 Distribution

3.33.1.1 Breeding area

The Little tern is a highly scattered semicosmopolitan bird, scarcely occurring in the South American region, breeding in all climatic zones except in the tundra (Voous, 1960). Generally 7 subspecies are distinguished (fig. 182). In Europe it breeds along all seashores, with exception of large parts of Fenno-Scandinavia, and inland along large rivers as the Rhône, Loire and Danube and from the Baltic states eastward into Asia. The origin of birds visiting the Wadden Sea is somewhat obscure because of the very limited number of recoveries of ringed Little terns. There are indications that birds breeding along the Baltic Sea also pass the Wadden Sea area. Some exchange between populations breeding in Denmark, Sweden and the German part of the Wadden Sea has been ascertained (Temme, 1967 and data Zoologisk Museum, København).

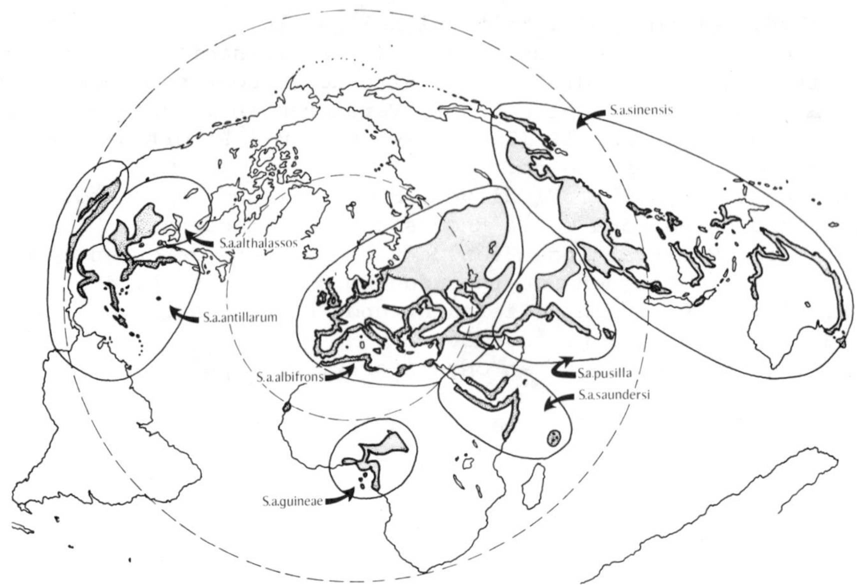

Fig. 182. Breeding area of the Little tern (after Voous, 1960 and Nadler, 1976).

3.33.1.2 Migration routes

The population breeding in the Wadden Sea migrates along the North Sea and Atlantic coasts of Europe (fig. 183 and Cramp et al., 1974).

3.33.1.3 Wintering area

Because of the presence of wintering Little terns along the West-coast of Africa this area is believed to be the wintering area. However, this is not yet confirmed by ring recoveries. The most southern recoveries originate from the Atlantic coast of Spain and Portugal (fig. 183). Presumably these are birds still migrating towards the wintering areas as they were recovered between July and mid-September (data: Vogeltrekstation, Arnhem, unpubl.).

3.33.1.4 Moulting areas

A complete moult takes place twice a year. The birds show a post-nuptial moult from August-December when they may still be present in the Wadden Sea, or already be on their way to the wintering quarters. This moult is completed in the wintering quarters. A second moult takes place from January-March when birds are in the wintering quarters or already on their way towards the breeding areas.

3.33.2 Annual cycle

3.33.2.1 Migration

Little terns arrive in the Dutch Wadden Sea from mid-April on (Dijksen & Dijksen, 1977) around Scharhörn by the end of April (Temme,

Fig. 183. Recoveries of Little
terns ringed as pullus on the
Dutch Wadden Sea islands (data:
Vogeltrekstation Arnhem, unpubl.).

1967). Generally they start breeding about one month later. They often
loose their eggs due to high tides and they can easily resettle and
start a second clutch till the end of June. They leave the Wadden Sea
by the end of the summer until in October. Because only few recoveries
are known of ringed Little terns (fig. 183) unfortunately there is no
detailed information available on timing of migration towards and from
the wintering areas.

3.33.2.2 Moult

In adult Little terns a complete moult takes place from August-December and from January-March. The juvenile body-plumage, lesser and
median wing-coverts, innermost greater, and some innermost secondaries
and central tail feathers are moulted from August-October. In spring
a complete moult takes place as in adults after which birds apparently become very much like adults (Witherby et al., 1952). Until now very few Little terns have been examined in moult.

3.33.2.3 Weight changes

No information available

3.33.3 Numbers

3.33.3.1 Population size

In the Wadden Sea area Little terns prefer shell ridges or beaches with many shells. On sandy beaches the risk is too high that the nests are covered by sand during stormy weather. Little terns are not able to refind this overblown nests as for example Oystercatchers can. The habitat preference provides them with a high number of good breeding sites in the Wadden Sea area. Nearly every island housed Little terns recently or in the past.

In the Dutch part of the Wadden Sea a serious decline in the number of Little terns took place in the end of the fifties which continued in the sixties. This was followed by an increase in the seventies. The cause of this decrease is not quite clear but pollution and pesticides may have played a role (Rooth & Jonkers, 1972).

It is not likely that, like in Great Britain (Cramp et al., 1974), disturbance by recreation was an important factor for the decrease in the Netherlands because the serious decline in the Netherlands to 5% of the population in 15 years was also observed in sanctuaries without human disturbance. Another indication is that fluctuations in the Netherlands do not correspond with those in the German part of the Wadden Sea where since 1950 numbers fluctuated between 150 and 500 pairs without showing a significant decrease. Disturbances caused by tourism are more or less comparable in the two countries. Another argument that

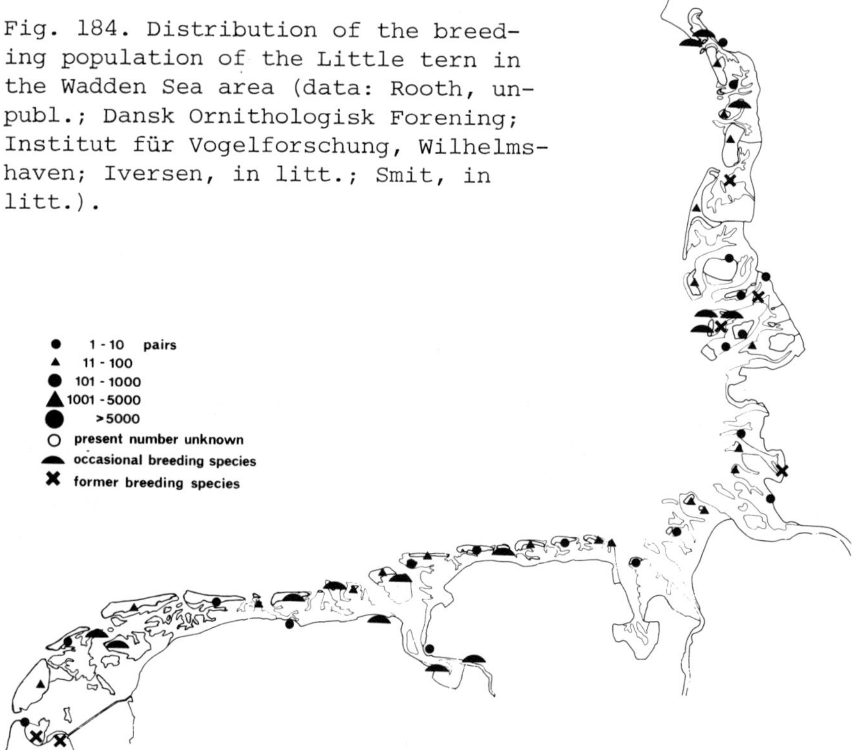

Fig. 184. Distribution of the breeding population of the Little tern in the Wadden Sea area (data: Rooth, unpubl.; Dansk Ornithologisk Forening; Institut für Vogelforschung, Wilhelmshaven; Iversen, in litt.; Smit, in litt.).

the decline in the Netherlands was caused by pollution is given by the
Sandwich tern, whose decline has well been analysed and been proven to
be caused by chlorinated hydrocarbons. The population breeding in the
Dutch part of the Wadden Sea recently increased to about 100-150 pairs
again. The population in the Wadden Sea area in Niedersachsen can be estimated at 100-200 pairs, in Schleswig-Holstein at 150 to more than 200
pairs (data: Institut für Vogelforschung, Wilhelmshaven; Smit, in litt.).
For the Danish part we have only information of the last years, when
on all islands breeding occurred with a maximum number of 100 pairs
(Mardal, pers. comm.). This means that the number of breeding birds
during the last years amounted to about 400-700 pairs in the whole Wadden Sea. The whole NW European population counts about 3500 pairs (Rooth,
unpubl.), so that the Wadden Sea contributes about 15% to this population. Fig. 184 shows the distribution of the breeding population of the
Little tern in the Wadden Sea area.

3.33.3.2 Numbers per area

As in the other tern species numbers determined by ground counts,
generally give insufficient information on the actual numbers present
in a certain area. From some areas in Denmark and Niedersachsen no information is available at all. Fig. 185 summarizes the information by
giving regular peak occurrences. From some guarded reserves in Germa-

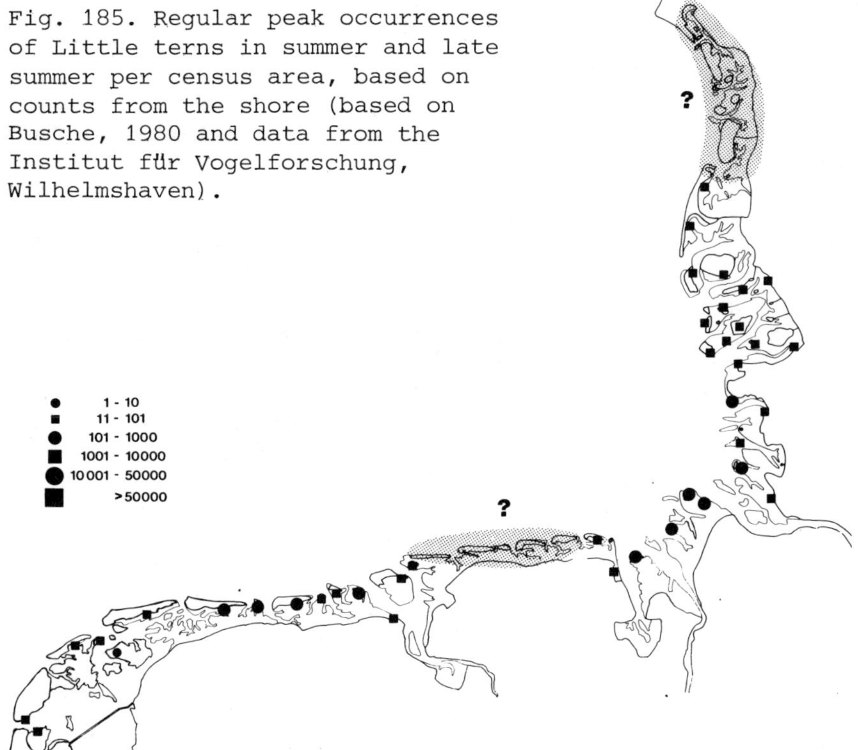

Fig. 185. Regular peak occurrences
of Little terns in summer and late
summer per census area, based on
counts from the shore (based on
Busche, 1980 and data from the
Institut für Vogelforschung,
Wilhelmshaven).

ny information on numbers throughout the summer season exists. It appears that larger concentrations of migrating Little terns may occur sometimes. Near Scharhörn up to 120 birds have been seen in the last April decade, early August numbers were even up to 600. Generally however numbers in August were considerably lower: 50-200. In the period in between only the breeding population was present. In September numbers around Scharhörn never exceeded 100 birds (Temme, 1967).

3.33.4 Food

3.33.4.1 Food composition

There is only very incomplete information from the Wadden Sea area but experiences in other areas in NW Europe make it probable that food mostly consists of fish and crustaceans and sometimes insects as well. Among the fish species Sand Eels *(Ammodytes tobianus* and *Hyperoplus lanceolatus)* and small Plaice *(Pleuronectus platessa)* occur.

3.33.4.2 Feeding activities

Little terns fish by plunge-diving in creeks and shallow pools left by the receding tide. On the Wadden Sea islands they sometimes forage in small fresh- or brackish water ditches.

3.33.4.3 Total food consumption

No data available.

References

Busche, G., 1980. Vogelbestände des Wattenmeeres von Schleswig-Holstein. Kilda, Greven (in press).

Cramp, S., W.R.P. Bourne & D. Saunders, 1974. The seabirds of Britain and Ireland. Collins, London: p. 166-169.

Dijksen, A.J. & L.J. Dijksen, 1977. Texel Vogeleiland Thieme, Zutphen: p. 142-143.

Mardal, W., 1974. Ternegruppen. Feltornithologen 16: p. 4-7.

Nadler, T., 1976. Die Zwergseeschwalbe. Neue Brehm Bücherei, Vol. 495. Ziemsen, Wittenberg-Lutherstadt: 136 pp.

Rooth, J., 1956. Vreemd broedgedrag. De Levende Natuur 59: p. 225-230.

Rooth, J. & D.A. Jonkers, 1972. The status of some piscivorous birds in The Netherlands. TNO-Nieuws 27: p. 551-555.

Temme, M., 1967. Vogelfreistätte Scharhörn. Jordsand Mitt. 3(1-4): p. 102-104.

Voous, K.H., 1960. Atlas van de Europese vogels. Elsevier, Amsterdam: p. 130.

Witherby, H.F., F.C.R. Jourdain, N.F. Ticehurst & B.W. Tucker, 1952. The Handbook of British Birds, Vol. 5: p. 40-43.

4 HABITAT SELECTION AND COMPETITION IN WADING BIRDS
Leo Zwarts

4.1 HABITAT SELECTION AND COMPETITION WITHIN THE SPECIES

The dispersion of birds feeding on the intertidal flats is determined by several factors. Probably the most important one is the density of the available prey species. It was found in several studies that the feeding density depended on the density of the main prey (e.g. Goss-Custard, 1970; Bryant, 1979; Goss-Custard et al., 1977). Fig. 186 shows this relationship for Oystercatchers which feed mainly on *Cerastoderma edule*.

It would be a profitable strategy for birds to select feeding areas where prey density is high, for these are probably nearly always the areas where food intake is maximal (Hulscher, 1976; Goss-Custard, 1977a, 1977b). However, if all birds present would feed on those optimal parts of the feeding area the expected advantages might disappear. Goss-Custard (1977a) suggested two disadvantages as a result of a high feeding density. First, the feeding profitability might decline as a result of social interactions, such as fighting but also more subtle effects like disturbance of each other's search path. Secondly the presence of many birds might depress the density of the preys which are attainable or detectable, because of anti-predator behaviour of the benthic fauna. Both mechanisms have not yet been studied in detail, but a decrease of the average food intake at increasing bird density was found in several studies for Redshank (Goss-Custard, 1976),

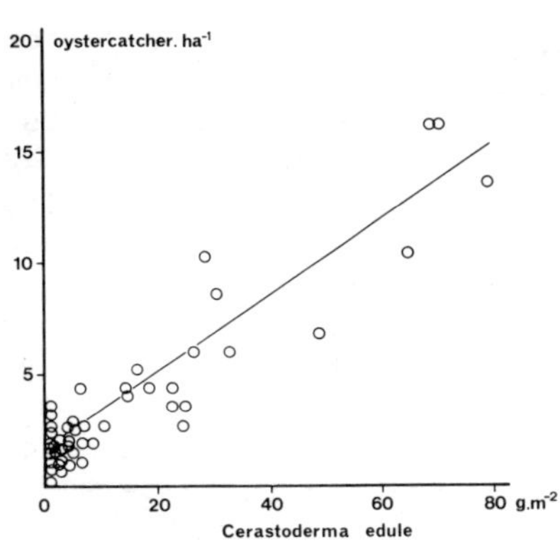

Fig. 186. Feeding density of Oystercatcher as related to the density of the main prey Cerastoderma edule (gram ash-free dry weight .m^{-2}). Macoma balthica and, especially on places where Cerastoderma was rare, Scrobicularia plana were taken too. Bird density is calculated for 232 counts carried out in July and August 1976 during 9 tidal cycles from ebb till flood. The 51 study sites (0.1 ha) were situated along the coast of the province of Groningen (Dutch Wadden Sea).

Curlew (Zwarts, 1978) and Oystercatcher (Koene & Drent, in prep.;
Zwarts & Drent, in prep.).

It seems that the deterioration of the feeding circumstances as a
consequence of a high bird density can be very pronounced, at least
for a part of the birds present, since it was found that the proportion of the birds on the feeding area which actually feed, decreased
at high bird densities. This effect was described for the Teal (Zwarts,
1976) and is shown here for the Oystercatcher on a mussel bed (fig.
187).

It is to be expected that birds tend to avoid such a high feeding
density either by feeding elsewhere, or if they stay by chasing away
other birds. Consequently, there would be a spreading out over less
optimal feeding areas as many birds are present. Such a change in the
dispersion pattern was found indeed. The data collected so far, however suggest that this effect on the dispersion is rather small within a certain feeding area of approximately 1-10 ha but is more pronounced on a larger scale (different feeding areas of 100-1000 ha).

An example of the dispersion on a small scale is given in Table
61. Whenever the density of Oystercatchers was low, most birds fed
where *Cerastoderma edule* was most common. At a high feeding density
the numbers of birds on marginal feeding areas increased relatively
much more than in preferred parts of the feeding areas. The data giv-
en in Table 61 concern a situation where optimal and marginal parts
of the feeding area are situated close to each other. In such a situation the colour-banded Oystercatchers appeared to search for food in
sites with high as well as low prey densities, even if the bird density was very low (Blomert, Engelmoer, Hulsman and Logemann, unpubl.).
The dispersion data suggest that high bird densities stimulate at
least a part of the birds to frequent less optimal parts of the feeding area more often and/or during longer periods.

The density-related effect on dispersion is obvious during receding tide when the birds leave the tidal roost and fly to the different

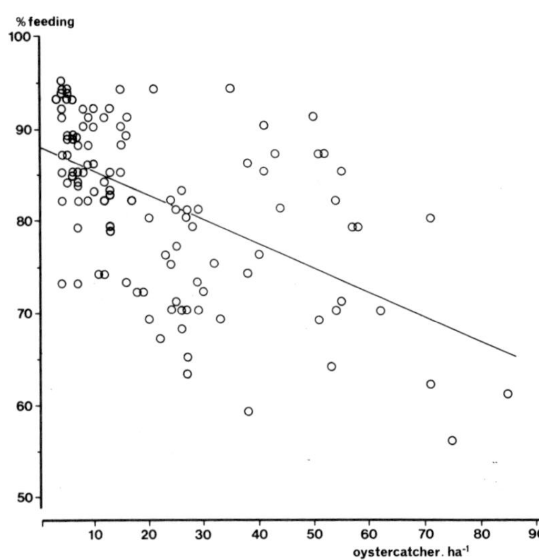

Fig. 187. The effect of Oystercatcher densities (average number during a series of counts
from ebb till flood) on the average feeding percentage during
the low water period. The data
originate from four counting
sites (1 ha) where mussel cover
amounted to 11-26%. The study
was carried out on intertidal
mussel beds near the island of
Schiermonnikoog in 1971-1973.

Table 61. Dispersion of Oystercatchers during low tide over 73 study sites of 0.1 ha each where *Cerastoderma edule* and *Scrobicularia plana* were the main prey. The average number of Oystercatchers (number per 0.1 ha) as related to the density of the two prey species (gram ash-free dry weight per m^2) is shown for three different bird densities. Bird counts were carried out by Bruno Ens from a tower along the Frisian coast in late summer 1978.

| | 100 birds | | | 100-200 birds | | | 200-300 birds | | |
	$C.$ edule g.m^{-2}			$C.$ edule g.m^{-2}			$C.$ edule g.m^{-2}		
S. plana	50	50-100	100-150	50	50-100	100-150	50	50-100	100-150
25 g.m^{-2}	0.11	0.67	1.00	1.68	1.86	2.26	2.67	3.97	3.06
25-50 g.m^{-2}	0.26	0.57	0.77	1.60	2.22	2.24	3.12	4.50	3.16
50-120 g.m^{-2}	0.37	0.40	–	1.94	2.35	–	5.06	5.57	–

low water feeding areas. In general, densities reach a ceiling level on the feeding area close to the roost (fig. 188; see also fig. 192b). Such a feeding area may be preferred because, when other environmental factors are equal, it is profitable to minimize the flight distance between roost and feeding area. Flying two km in stead of one km every incoming and receding tide means an extra cost of about one percent of the daily energy expenditure. The flight distance is an important factor determing the distribution of the wading birds over the flats. Most wader species fly usually no more than some kilometers during each flight between high tide roost and the tidal flats but for Oystercatchers, Curlews and gull species larger distances (up to 10 km) are more common.

Beside minimizing of the flight distance between roost and feeding area, two other factors can be mentioned which could explain the higher bird density near the coast. The first is that in many areas intertidal flats close to the coast are situated at an on average higher level, which guarantees a longer feeding period. Secondly the bottom often contains a higher prey density as related to similar habitat far from the coast (Beukema, 1976).

The conclusion is that the distribution of the wading birds over the intertidal flats depends on the food density and the distance from the feeding area to the roost, but that the dispersion is also dependent on the total number of birds being present.

It seems likely that bird densities are limited by competition for space. It is however not yet possible to answer the key question whether social interactions between birds have such a negative effect on the feeding circumstances of a part of the birds that they limit the wading bird populations on the estuarine feeding areas.

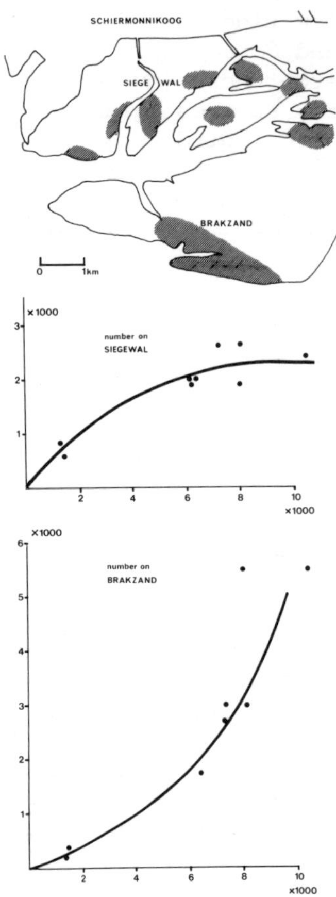

Fig. 188. Dispersion of Common- and
Black-headed gulls over their feeding
areas (hatched) on the tidal flats
south of the Western part of the island
of Schiermonnikoog in August 1967. There
appears to be a levelling off in the
number present on the feeding area
near the roost ("Siegewal") as the total
number increased, but there is no
ceiling level on the "Brakzand" area
far from the roost.

4.2 HABITAT SEGREGATION AND COMPETITION BETWEEN THE SPECIES

Studies on habitat use of estuarine bird species (e.g. Recher, 1966;
Baker & Baker, 1973; Burger et al., 1977) show that the habitat segregation is rather complete. This is achieved by selection of different
prey, by feeding in different habitats and/or by using the habitat
at different times.

Sanderlings feed on beaches, Purple sandpipers along (artificial)
rocky shores and Avocets on a muddy substrate (Wolff, 1969). These
species show no overlap in their distribution. For wader species using the same habitat there appears to be some segregation on micro-
habitat scale. This was studied on a mussel bed near the island of
Schiermonnikoog where 10-15 bird species fed. Oystercatchers, Herring
gulls and Common gulls reached high densities on the mussel beds only.
Knots, Bar-tailed godwits and Dunlins avoided the mussel beds and fed
on the muddy flats between the mussel ridges. Redshanks and Curlew
were common on the bare tidal flats as well as on the mussel beds
(fig. 189).

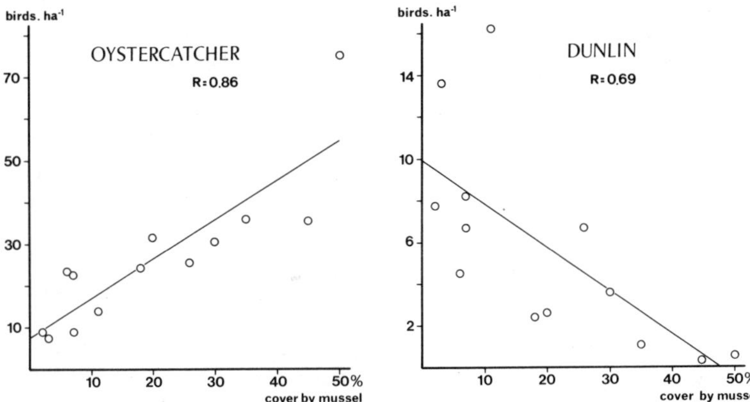

Fig. 189. Average bird density in August as a function of the proportion of the counting site (1 ha) being covered by mussels. The calculated bird density is based upon 4071 counts in 13 study sites which were situated on mussel beds near the island of Schiermonnikoog.

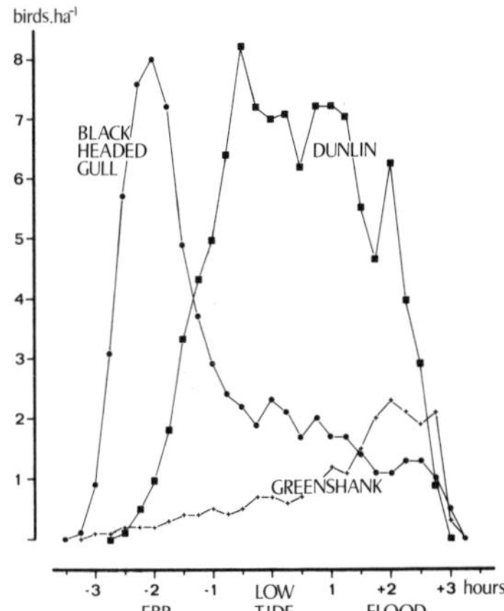

Fig. 190. Average bird density on mussel beds and surrounding tidal flats as a function of the tidal cycle. The graphs are bases upon 7315 counts made between May and Novenber in 13 study sites situated near the island of Schiermonnikoog.

Since most wader species in the Wadden Sea broadly overlap in their migratory periods (see chapter 3), the seasonal segregation can be considered to be rather low in general. Maybe more important is a temporal segregation which is achieved within the tidal cycle. Black-headed gulls feed in the tidal rim during receding tide only, whereas Greenshanks and Spotted redshanks do the same, but only when water comes up. Other species like Redshank and Dunlin feed in the intervening time on the emerged flats. This is shown for some species present on a mussel bed near Schiermonnikoog (fig. 190).

The counts made in this study area were used to quantify for all estuarine bird species the overlap in space (distribution over mussel bed, mud- and sandflat), and in time (seasonal- and tidal occurrence). From this analysis it appeared that the overall overlap was rather low. For instance, Greenshank and Spotted Redshank, though present in the same time of the year and frequenting the musselbank during the same stage of the tidal cycle, exhibited a habitat partitioning in space. Greenshanks were feeding in small creeks whilst Spotted redshanks mainly visited tidal pools between the mussel ridges.

Spatial and temporal segregation between the species can be connected with a different prey selection which explains for instance the distribution of Dunlin and Oystercatcher on a mussel bed (fig. 189): Oystercatchers fed on mussel ridges only, because *Mytilus edulis* was their main prey, whereas Dunlins were present on the flats where their prey species, small worms, were abundant.

The segregation between the estuarine bird species regarding the food is different per habitat. On a mussel bed where many prey species can be found, the overlap in diet during late summer seems to be rather low. Oystercatchers prey on large *Mytilus*, Herring Gulls swallow small *Mytilus*, Curlews feed on *Carcinus maenas*, Greenshanks take *Pomatoschistus microps*, and so on. On the other hand, a nearly complete overlap in food was found on a brackish intertidal mudflat (Ventjager, Dutch Delta area) where *Nereis diversicolor* was the main prey for fifteen estuarine bird species. However, even if bird species take the same prey, there is still some segregation, because larger birds tend to feed on larger prey items. In the Wadden Sea this could be shown

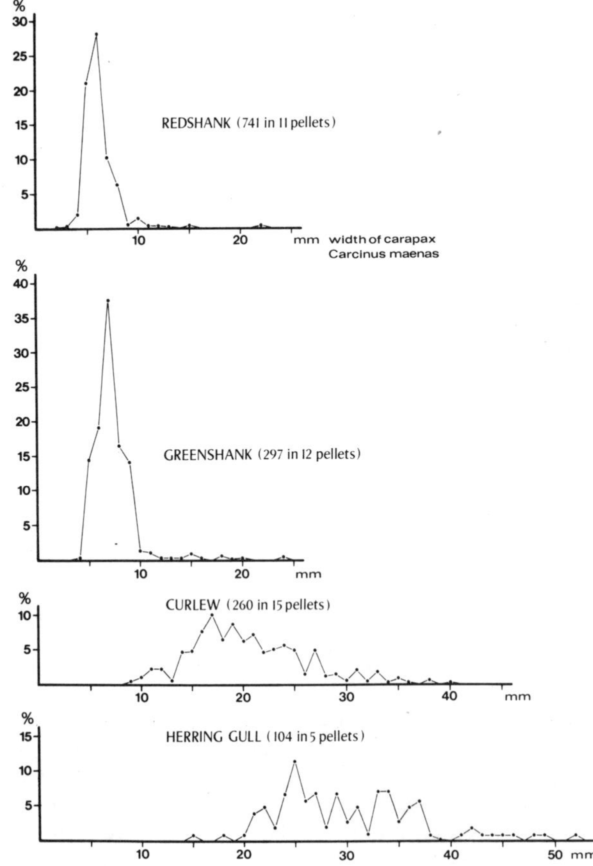

Fig. 191. Size selection by some estuarine bird species feeding on Carcinus maenas, based upon analysis of pellets collected on the same tidal roost along the coast of Friesland in the same period (end of July, 1978).

for four bird species which feed on *Carcinus maenas* (fig. 191).
   The data suggest that there is resource partitioning between the
coexisting species regarding space, time or food. It is assumed generally that this has been brought about by interspecific competition
in the course of evolution. However, it is unclear to what degree the
habitat use as we see it to day is still dependent on the competition
between the estuarine bird species. It is difficult to solve this problem. A description of the resource overlap cannot help us in predict-
ing the significance of competition between species. If bird species
for instant select different prey sizes, it would be wrong to conclude
that there is no competition, for the bird species which selects the

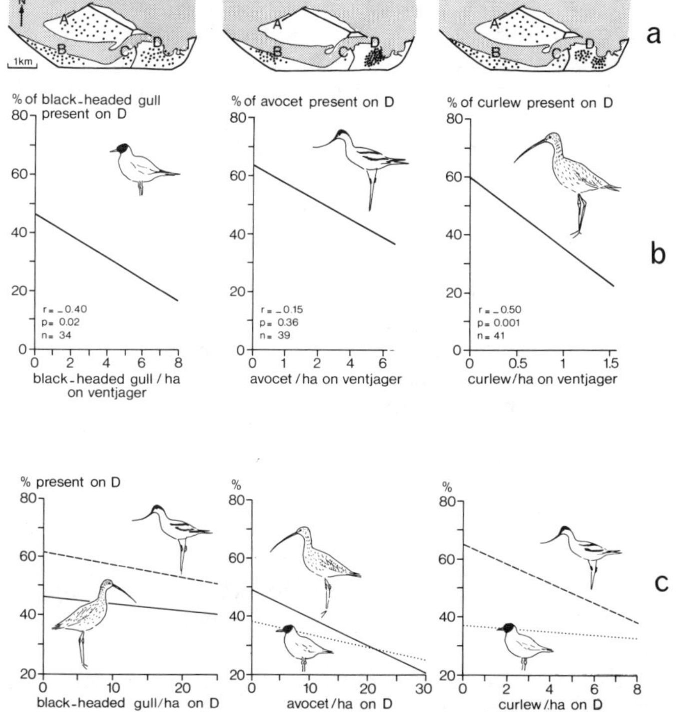

Fig. 192a. Dispersion pattern of Avocet, Curlew and Black-headed gull
on the Ventjager (Dutch Delta area). The map shows the average distribution during 61 low water counts (1965-1969). Each dot is 1% of the
mean total number.
Fig. 192b. The percentage of the birds feeding on area D (see fig.
192a) as related to the total number in the Ventjager area. The prop-
ortional decrease in D suggests that D was a preferred feeding area
for the three species.
Fig. 192c. The proportion of birds present on the entire Ventjager
feeding area which fed on D, plotted as a function of the number of
another species on D. The effect of the tide level and total number
of another species were kept constant in a multiple regression
analysis. The effect of Curlew on Avocet and the other way round were
significant (p=0.01) (after Zwarts, 1978).

smaller prey, takes away future prey for species taking the larger size class. It is also conceivable that bird species are depleting the same food supply, thus being in competition even if there is a complete segregation in time.

Regarding the competition for space, it is possible to investigate the direct effect of a bird species on the food intake and dispersion of another bird species. A lower food intake of a bird as related to a high density of another bird species may be expected because of resource depressing, disturbance of the search path, and/or prey-robbing and fighting. This has not been studied yet, but if this kind of interference occurs it is likely that the distribution of a bird species over the feeding area effects the dispersion of another species. This effect on distribution was found indeed in Curlew, Avocet and Black-headed Gull, which fed on *Nereis diversicolor* and spread out over about the same feeding area (fig. 192a). The same part of the feeding area was preferred by the three species because at increasing numbers the density reached there a ceiling level (fig. 192b). However, if many Curlews were present on this part of the feeding area, relatively few Avocets stayed there. The same effect was found the other way round (fig. 192c). These observations suggest an avoidance of feeding areas otherwise used, when these are occupied by another species.

The ecological significance of competition for space between estuarine bird species is yet unclear. Maybe it occurs on a small scale, by which it has some effect on the dispersion pattern only. But if interspecific competition for space is more common, then habitat shifts or even a decrease of the numbers of certain species may be expected if insufficient space is available.

References

Baker, M.C. & A.E.M. Baker, 1973. Niche relationships among six species of shorebirds on their wintering and breeding ranges. Ecol. Monog. 43: p. 191-212.

Beukema, J.J., 1976. Biomass and species richness of the macro-benthic animals living on the tidal flats of the Dutch Wadden Sea. Neth. J. Sea Res. 10: p. 236-261.

Bryant, D.M., 1979. Effects of prey density and site character on estuary usage by overwintering waders (Charadrii). Est. Coastal Mar. Sc. 9: p. 369-384.

Burger, J., M.A. Howe, D.C. Hahn & J. Chase, 1977. Effects of tide cycles on habitat selection and habitat partitioning by migrating shorebirds. Auk 94: p. 743-758.

Goss-Custard, J.D., 1970. The responses of Redshank (Tringa totanus (L.)) to spatial variations in the density of the prey. J. Anim. Ecol. 39: p. 91-113.

Goss-Custard, J.D., 1976. Variation in the dispersion of Redshank, Tringa totanus, on their winter feeding grounds. Ibis 118: p. 257-263.

Goss-Custard, J.D., 1977a. Predator responses and prey mortality in
    Redshank, Tringa totanus (L.), and a preferred prey, Corophium
    volutator (Pallas). J. Anim. Ecol. 46: p. 21-35.
Goss-Custard, J.D., 1977b. The ecology of the Wash III. Density-relat-
    ed behaviour and the possible effects of a loss of feeding grounds
    on wading birds (Charadrii). J. Appl. Ecol. 14: p. 721-739.
Goss-Custard, J.D., D.G. Kay & R.M. Blindell, 1977. The density of
    migratory and overwintering Redshanks, Tringa totanus (L.) and Cur-
    lew, Numenius arquata (L.) in relation to the density of their prey
    in South-east England. Est. Coastal Mar. Sc. 5: p. 497-510.
Hulscher, J.B., 1976. Localisation of cockles (Cardium edule L.) by
    the Oystercatcher (Haemantopus ostralegus L.) in darkness and day-
    light. Ardea 64: p. 292-310.
Recher, H.F., 1966. Some aspects of the ecology of migrant shorebirds.
    Ecology 47: p. 393-407.
Wolff, W.J., 1969. Distribution of non-breeding waders in an estuarine
    area in relation to the distribution of their food organisms. Ardea
    57: p. 1-28.
Zwarts, L., 1976. Density-related processes in feeding dispersion and
    feeding activity of Teal (Anas crecca). Ardea 64: p. 192-209.
Zwarts, L., 1978. Intra- and interspecific competition for space in
    estuarine bird species in a one-prey situation. Proc. XVII Intern.
    Ornith. Congr. (Berlin) (in press.).

5 THE IMPORTANCE OF THE WADDEN SEA FOR ESTUARINE BIRDS
   C.J. Smit

5.1 THE TOTAL NUMBER OF BIRDS IN THE WADDEN SEA

Birds visiting the Wadden Sea, - ducks, geese, waders, gulls and terns
-, breed in a huge area in Arctic and Subarctic Europe, Asia and North
America. Outside the breeding season however., they are dependent on a
relatively small number of areas (fig. 193). Along the European and
NW African coast the Wadden Sea is the largest of these areas having
a size of about 4000 km^2 of tidal flats. Other comparable areas are
considerably smaller. The Banc d'Arguin in Mauritania with about 500
km^2 of tidal flats, the Delta area in the SW Netherlands with about
300 km^2 of flats and Morecambe Bay and the Wash in England, both having 250-300 km^2, are the next largest. The total surface of tidal
flats in estuaries and along coasts in Europe and North Africa amounts
to about 7000 km^2.

All over the Wadden Sea area maximum numbers of estuarine birds
occur from August to October. Though not yet confirmed sufficiently
by simultaneous counts it is very probable that their number in the
whole Wadden Sea area amounts to 2.5-3.4 million specimens by this time
(Table 62), 1.7-2.3 million of these being waders. This number drops
to less than 1.5 million birds in January and February, including a-
bout 0.6-0.8 million waders, and increases to 1.3-1.9 million in April
and May, including 0.8-1.3 million waders (data computed from Smit,
1977; Busche, 1980; Meltofte, 1980; several reports from Vadefuglegruppen Dansk Ornithologisk Forening and unpublished data of the Institut
für Vogelforschung, Wilhelmshaven). The results of more or less simul-
tanuous counts of the number of waders in the area are listed in Table
63.

Because there is a continuous flow of birds through an area like
the Wadden Sea the actual number using it during a certain time is
much higher than the numbers mentioned here. It could very well be
true that autumn numbers will have to be doubled or tripled to yield
the number of birds actually using the Wadden Sea.

In the Danish part of the Wadden Sea Dunlin is by far the most numerous species. Generally from August-October 235,000-350,000 are present at the same time, during spring migration 75,000-195,000. Other
numerous species are Oystercatcher, Eider and Bar-tailed godwit. In
winter the Eider is the most common species (Joensen, 1974; Meltofte,
1977, 1978, 1979, 1980; Rønnest, 1974; Thelle & Netterstrøm, 1971;
Meltofte et al., 1974; Braae & Meltofte, 1975; Meltofte & Rønnest,
1975.

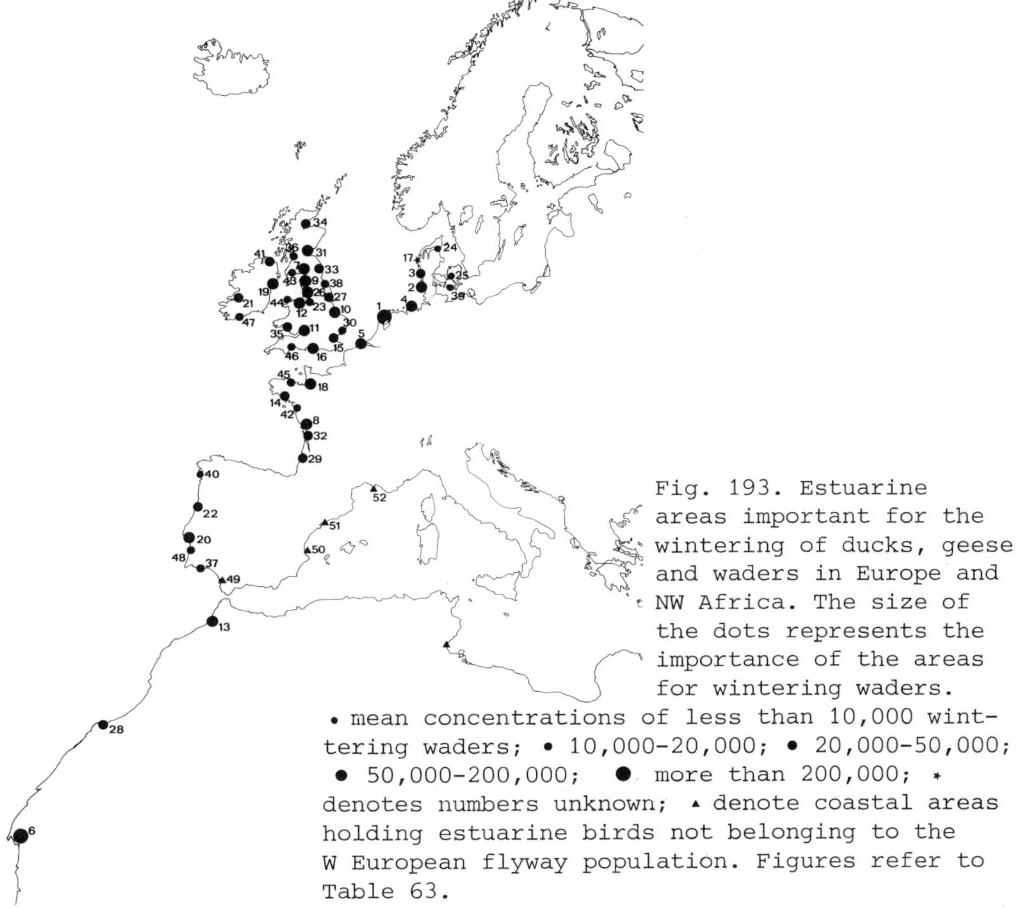

Fig. 193. Estuarine areas important for the wintering of ducks, geese and waders in Europe and NW Africa. The size of the dots represents the importance of the areas for wintering waders.
• mean concentrations of less than 10,000 wintering waders; • 10,000-20,000; • 20,000-50,000; • 50,000-200,000; ● more than 200,000; * denotes numbers unknown; ▲ denote coastal areas holding estuarine birds not belonging to the W European flyway population. Figures refer to Table 63.

In Schleswig-Holstein Dunlins as well as Knots are very numerous at times. Of both species 300,000 are present in early autumn. Oystercatcher (80,000), Black-headed gull (75,000) and Shelduck (60,000) follow. Numbers in winter are rather low, Eider, Oystercatcher, Shelduck and Common gull being the most numerous birds. In spring 200,000-300,000, both of Dunlin and Knot, may be present again (Heldt, 1968; Schlenker, 1968; Busche, 1980; Drenckhahn et al., 1971).

Data concerning Niedersachsen are too fragmentary to give exact information on the total numbers of birds present. Numbers given in Table 62 are estimates based on Temme, 1967; Grosskopf, 1968; Smit, 1977; Panzer & Rauhe, 1978 and unpublished data of the Institut für Vogelforschung, Wilhelmshaven.

Numbers of estuarine birds per month in the Dutch part of the Wadden Sea are shown in fig. 194. Fig. 195 shows the mean number day, averaged over the year. It appears that Oystercatcher (17.8%), Dunlin (17.5%) and Eider (9.3%) are the commonest species. All other species represent less than 10% of the total number. Per year 240 million bird-

Table 62. Approximate number (in millions) of estuarine birds in three seasons in the Wadden Sea. Numbers for Niedersachsen have been estimated. References are cited in the text.

	January–February	April–May	August–October
Denmark	0.1–0.2	0.2–0.4	0.3–0.6
Schleswig-Holstein	0.1–0.3	0.5–0.7	1.1–1.3
Niedersachsen	0.2–0.3	0.15–0.3	0.3–0.5
The Netherlands	0.6–0.8	0.4–0.5	0.8–1.0
Totals	1.1–1.5	1.3–1.9	2.5–3.4

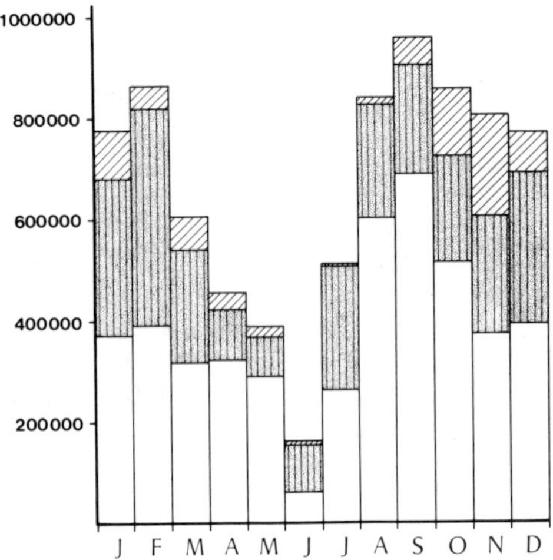

Fig. 194. Mean number of estuarine birds per month in the Dutch part of the Wadden Sea. White columns denote waders, light shaded columns grebes, diving ducks, Shelduck, gulls and terns and hatched columns dabbling ducks and geese (computed after data from Smit, 1977 added with unpublished data on the occurrence of diving ducks and gulls from Swennen).

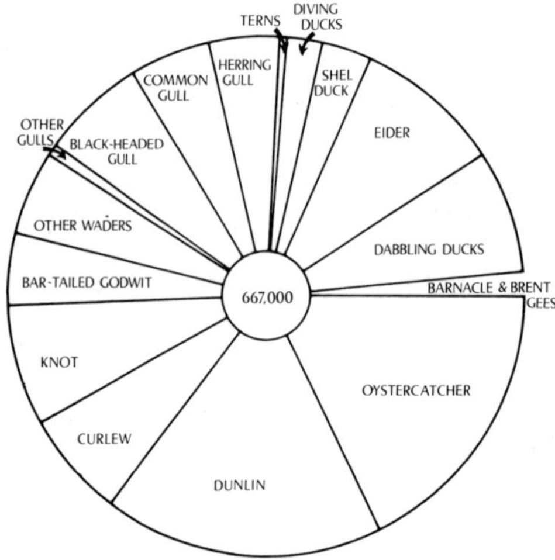

Fig. 195. The number of birds per day, averaged over the year, in the Dutch part of the Wadden Sea and the presence of species and species groups (computed after data from Smit, 1977 added with unpublished data on the occurrence of diving ducks and gulls from Swennen).

Table 63. Numbers of the main wader species in the Danish, German and Dutch Wadden Sea during three counts (after Prater, 1974, 1976c).

	January 1975	April 1975	September 1973
Oystercatcher *(Haematopus ostralegus)*	297,400	88,300	474,700
Ringed plover *(Charadrius hiaticula)*	4	1,150	17,500
Grey plover *(Pluvialis squatarola)*	4,000	6,500	33,600
Turnstone *(Arenaria interpres)*	1,700	3,400	3,760
Curlew *(Numenius arquata)*	84,300	45,400	155,500
Black-tailed godwit *(Limosa limosa)*	0	1,100	100
Bar-tailed godwit *(Limosa lapponica)*	21,500	104,900	78,100
Redshank *(Tringa totanus)*	15,600	18,000	55,400
Spotted redshank *(Tringa erythropus)*	3	1,640	5,500
Greenshank *(Tringa nebularia)*	9	390	9,500
Knot *(Calidris canutus)*	38,500	69,300	312,900
Dunlin *(Calidris alpina)*	220,100	458,000	728,800
Sanderling *(Calidris alba)*	1,950	1,800	6,200
Ruff *(Philomachus pugnax)*	150	350	3,400
Avocet *(Recurvirostra avosetta)*	2,500	6,340	45,300
Total, including other species	716,000	847,000	1,972,000

days are spent in the Dutch area, about 140 million of these being spent by waders (computed after Smit, 1977). In autumn especially Dunlin, Oystercatcher, Knot and Curlew are numerous, in winter Eider, Oystercatcher and Dunlin, and in spring Dunlin, Knot and Bar-tailed godwit (Rooth, 1966; Spaans, 1967; Smit, 1977; Boere & Zegers, 1974, 1975,1977; Swennen, in litt.).

5.2 THE WADDEN SEA COMPARED TO ESTUARIES ELSEWHERE

In Table 64 numbers of estuarine birds present in the Wadden Sea as a whole as well as in its four parts are compared to numbers present in other important estuarine areas. It appears that the Wadden Sea as a whole shows by far the highest values for the regular peak occurrences. Also for all separate parts numbers are high as compared to other estuaries.

Although the Wadden Sea may hold very large numbers of birds during autumn and spring, relatively low numbers winter in the area. For most of the species estuaries and coastal areas elsewhere in W Europe and NW Africa are more important as wintering grounds (Table 65, fig. 196). Among those areas especially the Banc d'Arguin in Mauritania takes a prominent position (Table 66).

The table lists, for each estuary/coast, the average number of waders in January and the numbers of individual wader and wildfowl species. Species are the row-labels of the printed table (with the "1% flyway population" threshold beside each); the estuaries/coasts form the columns. It is transcribed below with estuaries as rows and species as columns.

No.	Estuary/coast	Country	Average number of waders in January	Avocet (*Recurvirostra avosetta*)	Curlew sandpiper (*Calidris ferruginea*)	Dunlin (*Calidris alpina*)	Knot (*Calidris canutus*)	Redshank (*Tringa totanus*)	Bar-tailed godwit (*Limosa lapponica*)	Curlew (*Numenius arquata*)	Turnstone (*Arenaria interpres*)	Ringed plover (*Charadrius hiaticula*)	Grey plover (*Pluvialis squatarola*)	Oystercatcher (*Haematopus ostralegus*)	Barnacle goose (*Branta leucopsis*)	Brent goose (*Branta b. bernicla*)	Shelduck (*Tadorna tadorna*)	Red-breasted merganser (*Mergus serrator*)	Goosander (*Mergus merganser*)	Eider (*Somateria mollissima*)	Pintail (*Anas acuta*)	Wigeon (*Anas penelope*)	Teal (*Anas crecca*)	Mallard (*Anas platyrhychos*)
	1% flyway population			250	?	15000	7500	2500	3000	1600	250	450	450	5600	470	1100	1250	750	400	2000	500	4000	1500	10000
1–4	Wadden Sea	DK,D,NL	580000	45000	1500	728000	310000	55000	105000	155000	3800	17500	34000	475000	50000	120000	80000	12000	10000	250000	15000	250000	50000	100000
1	Wadden Sea	NL	400000	16000	700	205000	105000	35000	70000	80000	5000	6000	17000	210000	20000	54000	50000	12000	9500	160000	3500	120000	25000	26000
2	Wadden Sea, Schleswig-Holstein	D	60000	5000	1000	300000	300000	16000	60000	40000	2000	10000	13000	80000	20000	50000	60000		500	40000	10000	70000	20000	50000
3	Wadden Sea	DK	20000	7700		350000		10000	56000	5600	300	700	2900	65000		13000	24000	2000	3400	74000	7000	54000	11000	38000
4	Wadden Sea, Niedersachsen	D	100000	10000		+		+										1600	600	+	19000	+	14500	15000
5	Delta	NL	185000	8300		81000	23000	9200	7000	22000	2200	6600	5200	100000	40000	9500	11400				12300	62000	14500	80000
6	Banc d'Arguin	Mau	>750000–2200000	10000	173000	818000	366000	100000	543000	14000	17000	98000	23000	40000										
7	Solway Firth	GB	105000			20000	20000	5000	5000					9000			7700						6900	
8	Aiguillon	F	106000			50000								38000										
9	Morecambe Bay	GB	179000			45000	35000	9000	6000	6000			2000	42000		6000	13900							
10	Wash	GB	132000			70000	76000																	
11	Severn	GB	70000			50000						4000	500			10000	'2700				1500	62000	20000	4000
12	Dee	GB	73000			30000		5000	5000	4000	300	20000	3000								2000	10000		
13	Merja Zerga	N	50000	1900		20000							2000								6000	21000	20000	
14	Morbihan	F	46000			30000										13000	12800				4000	35000	5200	
15	Thames/Medway	GB	>44000												2500	30000					1000	15000		
16	England S	GB	>68000			40000										8000	9400					8000		
17	Denmark W	DK	?	+		+			+				4000											
18			138000																					
19	Baie Mt.St.Michel	F	>56000			75000	35000	7000	5000			1500										9600	4600	
20	Rio Tejo	P	50000																				4100	
21	Shannon	Ir	40000			25000						1500	2000	2000							1900			
22	Aveiro	P	30000			20000																		
23	Mersey	GB	18000														2300			105000	9900		8400	
24	Aalborg Bugt	DK	<10000														2900			175000				
25	Denmark Central	DK	<10000																					
26	Ribble	GB	63000																			10000		
27	Humber	GB	41000			30000	20000									2000	2200					12000		
28	Puerto Cansado	M	22000			20000	20000																	
29	Gironde/Les Landes	F	20000			+	2000																	
30	Blackwater/Dengie	GB	17000			30000																12000		
31	Firth of Forth	GB	52000			20000														49000		9500		
32	Oléron/Vendée	F	49000																					
33	Lindisfarne	GB	30000														1100			30000		12000		
34	Cromarty/Moray Firth	GB	270000			>20000	2000	6000												27000		9500		
35	Burry Inlet	GB	200000																					
36	Clyde	GB	16000																					
37	Faro	P	15000																					
38	Teesmouth	GB	14000										1500				4400							
39	Denmark S	DK	<100000																	100000				
40	Rio de Vigo	E	<100000																			5300		
41	Ireland N	Ir,GB	>36000																					
42	Baie de Bourgneuf	F	19000																					
43	Duddon	GB	19000																					
44	Conway Bay	GB	16000																					
45	Baie St. Brieuc	F	16000																					
46	Exe	GB	14000																					
47	Ireland S	Ir	>13000			>20000	2000																	
48	Sado	P	12000																					
49	Marismas	E	<100000																					
50	Albufera	E	<100000																					
51	Ebro	E	<100000																					
52	Camargue	F	<100000																					

For explanation see page 285.

Table 64. Regular peak occurrences in autumn, winter or spring in
estuaries and along coasts in W Europe and W Africa and average
numbers of waders wintering there. For site locations see fig. 193.
+ indicates that more than 1% of the flyway population is regularly
present, but exact numbers are not known. After Prater, 1974, 1976a,
1976c; Atkinson-Willes, 1975, 1976; Maheo, 1976; Ferns, 1977; Timmerman et al., 1976; Saeijs & Baptist, 1977; Piersma et al., 1980 and
data computed from Smit, 1977; Busche, 1980; Meltofte, 1980; reports
Vadefuglegruppen Dansk Ornithologisk Forening and unpublished data
from the Institut für Vogelforschung, Wilhelmshaven. Data for Niedersachsen have partly been estimated.

5.3 THE SIGNIFICANCE OF THE WADDEN SEA IN THE ANNUAL LIFE CYCLE OF THE BIRDS

Egg-laying, breeding and taking care of the young are much energy-con-
suming processes in the annual life cycle of birds. Shortly after this
period, migration towards the wintering quarters starts for many of
the species breeding in Arctic or Subarctic areas. These birds
generally are low in weight when they arrive at their first stop after
leaving the breeding areas. This applies for many of the birds
arriving in an estuarine area like the Wadden Sea. Moreover many species start moulting soon after they arrive in such an area. Boere et
al. showed that weights of such birds generally do not increase during the moulting period, but are stable due to a constant low fat con-

Table 65. Mean number in thousands of waders during January counts
in 1974-1978 along European coasts. Numbers of Purple sandpiper and
Turnstone are larger as both species occur also along rocky coasts
i.e. in Norway and Iceland which have not been incorporated in the
counts. + denotes that less than 50 birds are present, - that no
birds were present. After Prater (1978).

	Denmark without Wadden Sea	Wadden Sea	Delta	Belgium	Great-Britain	Ireland	France	Spain	Portugal	Total
Oystercatcher (*Haematopus ostralegus*)	+	304.5	100.0	0.6	200.0	15.3	30.2	2.0	0.2	652.8
Avocet (*Recurvirostra avosetta*)	-	2.1	0.4	+	0.1	-	9.6	0.5	10.4	23.1
Ringed plover (*Charadrius hiaticula*)	+	+	0.3	0.1	7.9	0.6	1.1	3.3	1.8	15.1
Kentish plover (*Charadrius alexandrinus*)	-	-	-	+	+	-	0.2	0.2	0.7	1.1
Grey plover (*Pluvialis squatarola*)	+	4.2	4.8	0.1	13.5	0.6	10.6	3.1	4.0	41.0
Knot (*Calidris canutus*)	+	45.8	20.0	-	189.5	17.1	7.3	10.0	1.0	290.7
Sanderling (*Calidris alba*)	-	2.7	2.5	0.7	10.0	0.4	0.5	+	0.3	17.1
Little stint (*Calidris minuta*)	-	-	-	+	+	+	0.1	0.2	0.5	0.8
Purple sandpiper (*Calidris maritima*)	-	0.2	0.2	0.3	18.5	+	1.3	-	-	20.5
Dunlin (*Calidris alba*)	2.3	210.3	81.0	1.1	561.4	51.3	322.6	14.0	47.7	1291.7
Ruff (*Philomachus pugnax*)	-	0.1	0.2	0.4	1.0	+	0.4	+	+	2.2
Black-tailed godwit (*Limosa limosa*)	-	+	-	-	4.5	2.5	10.3	6.4	11.4	35.1
Bar-tailed godwit (*Limosa lapponica*)	-	24.6	4.1	+	40.0	15.5	3.1	0.2	3.4	90.9
Whimbrel (*Numenius phaeopus*)	-	+	-	-	+	+	+	0.1	+	0.1
Curlew (*Numenius arquata*)	+	81.2	8.0	0.7	62.0	11.1	14.6	2.4	1.4	181.4
Spotted redshank (*Tringa erythropus*)	-	+	0.1	+	0.1	+	+	+	0.2	0.5
Redshank (*Tringa totanus*)	0.1	16.5	3.2	0.2	77.1	6.0	2.5	0.8	4.4	110.8
Greenshank (*Tringa nebularia*)	-	+	+	-	0.3	0.3	+	+	+	0.7
Green sandpiper (*Tringa ochropus*)	-	+	-	-	+	+	+	+	+	0.2
Common sandpiper (*Tringa hypoleucos*)	-	-	-	-	+	+	+	+	+	0.2
Turnstone (*Arenaria interpres*)	-	2.4	1.5	0.6	10.8	0.5	0.5	-	0.1	16.4
Wader spec.	-	-	-	-	-	-	+	0.1	0.1	0.2
	2.4	694.6	226.2	4.8	1096.7	121.2	414.9	43.3	87.6	3802.6

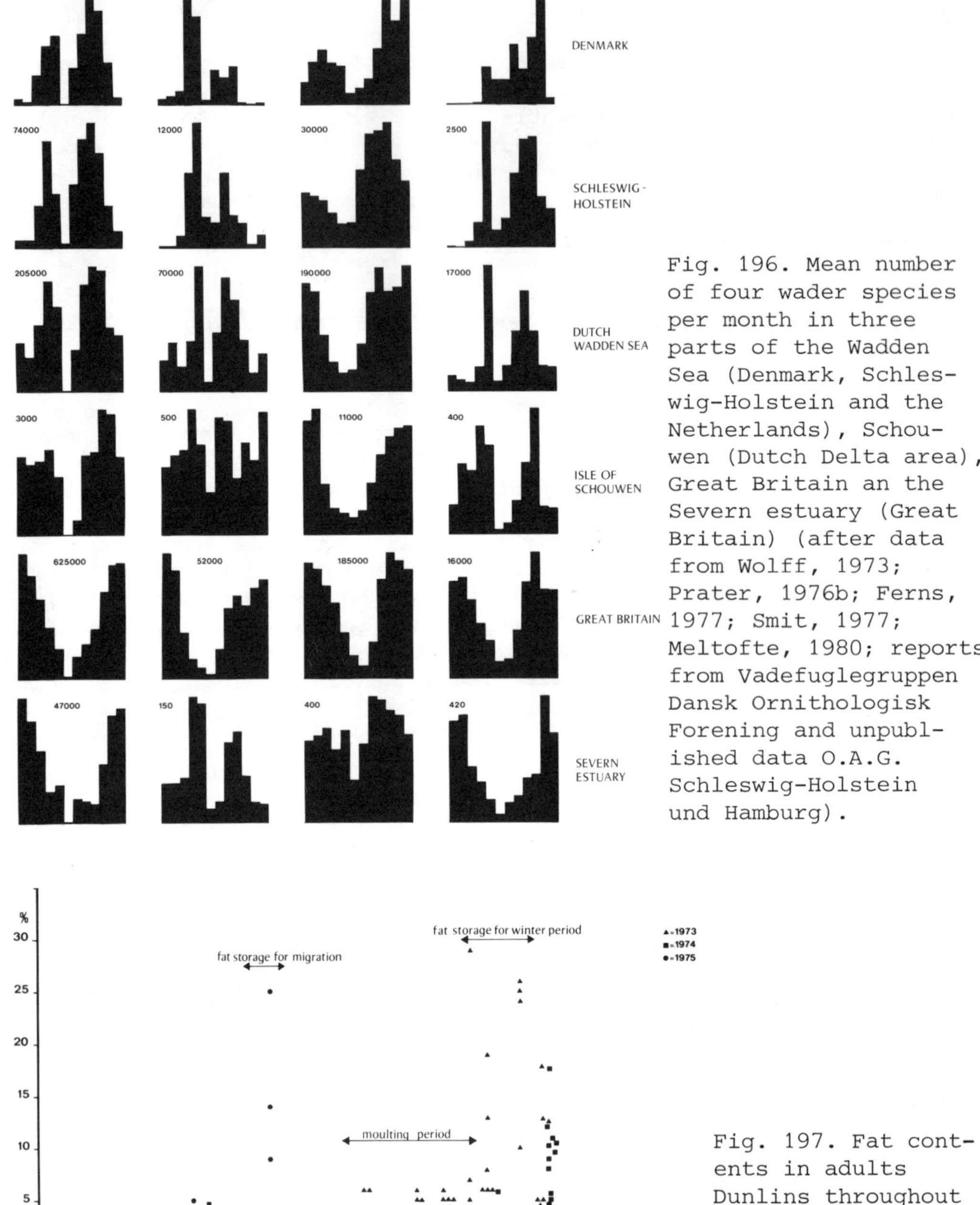

Fig. 196. Mean number of four wader species per month in three parts of the Wadden Sea (Denmark, Schleswig-Holstein and the Netherlands), Schouwen (Dutch Delta area), Great Britain an the Severn estuary (Great Britain) (after data from Wolff, 1973; Prater, 1976b; Ferns, 1977; Smit, 1977; Meltofte, 1980; reports from Vadefuglegruppen Dansk Ornithologisk Forening and unpublished data O.A.G. Schleswig-Holstein und Hamburg).

Fig. 197. Fat contents in adults Dunlins throughout the year. Fat content is expressed as a percentage of total body weight. After data collected on the island of Vlieland (Dutch Wadden Sea) from 1973-1975 (Boere, Pons & Drenth, unpublished).

Table 66. Results of the wader counts at the Banc d'Arguin (Mauritania) in January-March 1980 (Piersma et al., 1980) compared with earlier estimates of the wintering population by Prater (1976a) and Trotignon (1979).

	Prater (1976a)	Trotignon (1979)	Piersma et al. (1980)
Oystercatcher *(Haematopus ostralegus)*	3,000	6,800	9,000
Ringed Plover *(Charadrius hiaticula)*	13,000	136,500	98,000
Kentish plover *(Charadrius alexandrinus)*	3,000	6,500	17,000
Grey plover *(Pluvialis squatarola)*	3,500	14,200	23,000
Knot *(Calidris canutus)*	130,000	334,000	366,000
Sanderling *(Calidris alba)*	13,000	6,600	34,000
Little stint *(Calidris minuta)*	5,000	1,000	43,000
Curlew sandpiper *(Calidris ferruginea)*	38,000	129,000	173,000
Dunlin *(Calidris alpina)*	180,000	705,000	818,000
Bar-tailed godwit *(Limosa lapponica)*	210,000	538,000	543,000
Whimbrel *(Numenius phaeopus)*	3,500	-	16,000
Curlew *(Numenius arquata)*	2,500	21,400	14,000
Redshank *(Tringa totanus)*	100,000	31,100	70,000
Greenshank *(Tringa nebularia)*	800	850	1,450
Turnstone *(Arenaria interpres)*	10,000	6,000	17,000
	715,300	1,937,000	2,244,000

tent (fig. 197). The reason may be that also moult is a process consuming much energy. Moreover a low body weight is also a good adaptation to a reduced flight capacity because of missing flight feathers. Ducks and geese are not able to fly at all for some time when moulting primaries. In waders flight capacity is reduced because of the temporary loss of some of the primaries. Moulting places therefore have to be situated in places where enough food is available at short distances from quiet roosting places. Ducks and geese must be able to cover these distances by swimming or walking. All of these birds therefore select moulting areas that can offer sufficient food and rest, and where they can be safe from predators and disturbance as well. The Wadden Sea satisfies these conditions and is therefore a very important moulting area for many species.

By the end of the moulting period the birds build up some energy resources again and gain weight (Boere, 1976; chapter 3 of this report; fig. 197). Some species undergo a complete moult in the Wadden Sea, others arrest moult to finish it in staging areas further south or in the wintering areas. In spring when birds return from the wintering areas they have to build up energy resources for a flight (often a non-stop flight) back towards the breeding areas and for the breeding and egg-laying itself. For Brent geese has been demonstrated that the possibility of building up sufficient energy reserves in staging areas, like the Wadden Sea, in spring determines to a large extent the breeding success (Ebbinge et al., 1981). Of course this is of vital importance for the survival of this species. It is likely that this holds for other species as well.

References

Atkinson-Willes, G.L., 1973. Effectifs et distribution des Canard marins dans le Nord-Quest de l'Europe, janvier 1967-1973. Aves 12: p. 254-284.

Atkinson-Willes, G.L., 1976. The numerical distribution of ducks, swans and coots as a guide in assessing the importance of wetlands in midwinter. In: M. Smart (ed.), Proc. Int. Conf. on the Conserv. of Wetlands and Waterfowl, Heiligenhaven, 1974. IWRB Slimbridge, p. 199-254.

Boere, G.C., 1976. The significance of the Dutch Waddenzee in the annual life cycle of arctic, subarctic and boreal waders. Ardea 64: p. 210-291.

Boere, G.C. & P.M. Zegers, 1974. Wadvogeltelling in het Nederlandse Waddengebied in juli 1972. Limosa 47: p. 23-28.

Boere, G.C. & P.M. Zegers, 1975. Wadvogeltellingen in het Nederlandse Waddengebied in april en september 1973. Limosa 48: p. 74-81.

Boere, G.C. & P.M. Zegers, 1977. Watervogeltellingen in het Nederlandse Waddenzeegebied in 1974 en 1975. Watervogels 2: p. 161-173.

Braae, L. & H. Meltofte, 1975. Rapport fra vadefuglegruppen, 1975. Report Dansk Ornithologisk Forening: 16 pp.

Busche, G., 1980. Vogelbestände des Wattenmeeres von Schleswig-Holstein. Kilda, Greven (in press).

Drenckhahn, D., R. Heldt jun. & R. Heldt sen., 1971. Die Bedeutung der Nordseeküste Schleswig-Holsteins für einige eurasische Wat- und Wasservögel mit besonderer Berücksichtigung des Nordfriesischen Wattenmeeres. Natur und Landschaft 46: p. 338-346.

Ebbinge, B., A.K.M. St. Joseph, P. Prokosch & P. Zegers, 1981. The impact of spring weight on subsequent breeding success in the Dark-bellied Brentgoose (Branta bernicla). Wildfowl.

Ferns, P.N., 1977. Wading birds of the Severn estuary. Nature Conservancy Council: 114 pp.

Grosskopf, G., 1968. Die Vögel der Insel Wangerooge. Abhandl. Vogelk. 5. Institut für Vogelforschung, Wilhelmshaven: 293 pp.

Heldt, R., 1968. Uebersommernde Limikolen an der Westküste Schleswig-Holsteins. Corax 2: p. 108-130.

Joensen, A.H., 1974. Waterfowl populations in Denmark 1965-1973. A survey of the non-breeding populations of ducks, swans and coot and their shooting utilization. Dan. Rev. Game Biol. 9 (1): p. 1-206.

Maheo, R., 1976. The Brent geese of France, with special reference to the Golfe du Morbihan. Wildfowl 27: p. 55-62.

Meltofte, H., 1977. Rapport fra Vadefuglegruppen 1976. Report Dansk Ornithologisk Forening: 38 pp.

Meltofte, H., 1978. Rapport fra Vadefuglegruppen 1977. Report Dansk Ornithologisk Forening: 21 pp.

Meltofte, H., 1979. Rapport fra Vadefuglegruppen 1978. Report Dansk Ornithologisk Forening: 5 pp.

Meltofte, H., 1980. Fugle i Vadehavet. Vadefugletaellingen i Vadehavet 1974-1978. Miljøministeriet, Fredningsstyrelsen, København: 50 pp.

Meltofte, H., J. Sørensen & K. Hansen, 1974. Rapport fra Vadefugle-
    gruppen 1974. Report Dansk Ornithologisk Forening: 12 pp.
Meltofte, H. & S. Rønnest, 1975. Vadehavet 1974-1975. Report Dansk
    Ornithologisk Forening: 24 pp.
Panzer, W. & H. Rauhe, 1978. Die Vogelwelt an Elb- und Wesermündung.
    Heimatbund Männer vom Morgenstern, Bremerhaven: 336.
Piersma, T., M. Engelmoer, W. Altenburg & R. Mes, 1980. A wader-expe-
    dition to Mauritania. Wader Study Group Bull. 29: p. 14.
Prater, A.J., 1974. Wader research; coastal wader counts. IWRB Bull.
    37: p. 102-104.
Prater, A.J., 1976a. The distribution of coastal waders in Europe and
    North Africa. In: M. Smart (ed.). Proc. Int. Conf. on the Conser-
    vation of Wetlands and Waterfowl, Heiligenhafen, 1974. IWRB, Slim-
    bridge: p. 255-271.
Prater, A.J., 1976b. Birds of Estuaries Enquiry 1973-74. British Trust
    for Ornithol./Royal Soc. Protection of Birds/Wildfowl Trust: 47 pp.
Prater, A.J., 1976c. Wader Research Group. IWRB Bull. 41/42: p. 60-62.
Prater, A.J., 1978. Paper presented at 24th IWRB meeting Carthago,
    Tunesia.
Rooth, J., 1966. Vogeltelling in het hele Nederlandse Waddengebied in
    augustus 1963. Limosa 39: p. 175-181.
Rønnest, S., 1974. Vadefugleoptaellingen den 1. Sept. 1973 i Vadehavet.
    Report Dansk Ornithologisk Forening: 19 pp.
Saeijs, H.L.F. & H.J.M. Baptist, 1977. Watervogels in de veranderende
    delta van Zuidwest-Nederland. Limosa 50: p. 98-113.
Schlenker, R., 1968. Ueber das Wintervorkommen von Limikolen an der
    Westküste Schleswig-Holsteins. Corax 2: p. 108-130.
Smit, C.J., 1977. On the occurrence of 32 bird species in the Danish,
    German and Dutch Wadden Sea. Unpublished report Intern. Wadden Sea
    Working Group, 3 parts: 445 pp.
Spaans, A.L., 1967. Wadvogeltelling in het gehele Nederlandse Wadden-
    gebied in december 1966. Limosa 40: p. 206-215.
Temme, M., 1967. Vogelfreistätte Scharhörn. Jordsand Mitteilungen 3
    (1-4): p. 5-165.
Thelle, T. & B. Netterstrøm, 1971. Vadefugleoptaellingen i Vadehavet
    i juli og august 1969. Dansk Orn. Foren. Tidsskr. 65: p. 164-172.
Timmerman, A., M.F. Mörzer Bruyns & J. Phillipona, 1976. Survey of
    the winter distribution of palaearctic geese in Europe, Western
    Asia and North Africa. Limosa 49: p. 230-292.
Trotignon, J., 1979. Recensement hivernal des oiseaux aquatiques et
    des rapaces sur le Banc d'Arguin (Hiver 1978/1979). In: Comptes
    rendus d'activités scientifiques October 1977-Février 1979, Parc
    National du Banc d'Arguin, Mauritanie: p. 20-40.
Wolff, W.J., 1973. Resultaat van 5 jaar Steltlopertellingen op Schou-
    wen. Limosa 46: p. 21-41.

6 PRODUCTION OF BIOMASS BY INVERTEBRATES AND CONSUMPTION BY BIRDS IN THE DUTCH WADDEN SEA AREA
C.J. Smit

6.1 INTRODUCTION

In the past years some comparisons have been made between the production of macrobenthic invertebrates and fishes and the consumption by birds visiting the Dutch Wadden Sea (Hulscher, 1975; Swennen, 1976). Both authors found that a rather high percentage of the available biomass of the bottom fauna was eaten by birds. Their calculations however, were based on a rather small number of bird counts. Meanwhile more recent and extensive information became available from bird counts carried out all over the Dutch Wadden Sea (Smit, 1977). From these data and additional information from Swennen (in litt.) mean numbers of birds present per month in the area have been computed.

The availability of this information as well as the presence of more recent data on biomass of macrobenthic invertebrates (Beukema, 1976; Beukema et al., 1978) justify a new comparison between production and consumption. Because information from the German and Danish part of the Wadden Sea is insufficiently available this comparison has to be restricted to the Dutch part.

6.2 METHODS

The species selected for the calculations presented here are all carnivorous, at least in the Wadden Sea, and include all waders, gulls, terns and diving ducks (Table 67). Herbivorous species (dabbling ducks, geese) are not included though some are known to consume macrobenthic invertebrates at least during a part of the year. Information on macrobenthic consumption of dabbling ducks in the Wadden Sea however is very incomplete and an estimation of this consumption would be too speculative. It was taken into account that Herring gulls take food from refuse dumps, especially in winter. Some species occurring locally in small numbers have been neglected.

To determine the consumption of the birds present in the Wadden Sea the following equation was used:

$$C = N \times B \times 0.2 \times 5$$

where C = total consumption by a certain species per year in grams ash-free dry weight (g adw. year^{-1})

N = the number of bird-days of a species per year (bird.days)

B = basal metabolic rate (BMR) of a bird of a certain species (Kcal. $\text{day}^{-1}.\text{bird}^{-1}$)

0.2 = a factor to convert kcal into g adw (Winberg, 1971; Cummins & Wuycheck, 1971)

5 = a multiplication factor for converting BMR into the birds consumption (Hulscher, 1975; Wolff et al., 1976).

B = has been computed from body weights by using the equation

$$\log B = \log 78.3 + 0.723 \log W$$

where W = body weight in kg (Lasiewski & Dawson, 1967).

Body weights were taken from literature (Swennen, 1975; Wolff et al., 1976).

For all species which are foraging on the tidal flats a mean bill length (s) per month was computed using the equation:

$$s = \frac{(s_a . n_a) + (s_b . n_b) + \dots\dots\dots\dots\dots (s_x . n_x)}{n_a + n_b + \dots\dots\dots\dots\dots n_x}$$

in which s_a = bill length of species a in mm
n_a = mean number per month of species a

Bill lengths have been taken from Haftorn (1971); Bauer & Glutz (1968) and Glutz et al. (1975, 1977).

6.3. RESULTS

6.3.1 Total consumption by birds

Data on the average daily consumption per bird species is presented in Table 67 and fig. 198. It appears that about 75% of the total consumption is taken by only five species: Eider 28%, Oystercatcher 19%, Curlew 11%, Herring gull 8% and Shelduck 7%. All other species con-

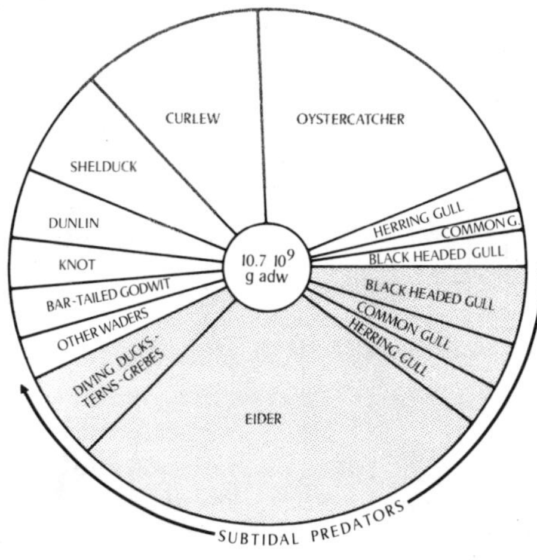

Fig. 198. Estimate of the average daily consumption by carnivorous birds in the Dutch part of the Wadden Sea and the distribution of consumption over species and species groups. Dark shaded: birds foraging in the subtidal part of the area (computed after Smit, 1977 added with unpublished information on the occurrence of diving ducks ans gulls from Swennen).

Table 67. Mean numbers, Basal Metabolic Rate and mean consumption of carnivorous birds in the Wadden Sea. 1) = partly foraging on tidal flats. 2) = data from Swennen (1975). 3) = computed from Swennen (1976).

	Mean number.day^{-1}	Mean annual weight (kg)	BMR (Kcal.day^{-1}.bird^{-1})	Mean consumption.day^{-1} (gram adw.10^6)
Species depending on subtidal area				
Great crested grebe (*Podiceps cristatus*)[2]	1000	1.102	84	0.08
Scaup (*Aythya marila*) [2]	4500	0.900	73	0.29
Goldeneye (*Bucephala clangula*)[2]	1800	0.872	71	0.13
Long-tailed duck (*Clangula hyemalis*)[2]	250	0.700	61	0.02
Velvet scoter (*Melanitta fusca*)[2]	250	1.400	100	0.02
Common scoter (*Melanitta nigra*)[2]	8000	1.062	82	0.65
Eider (*Somateria mollissima*)[3]	62000	2.118	135	8.33
Red-breasted merganser (*Mergus serrator*)[2]	3900	1.084	83	0.32
Goosander (*Mergus merganser*)[2]	2300	1.624	111	0.04
Herring gull (*Larus argentatus*)[1]	15000	0.991	78	1.00
Black-backed gull (*Larus marinus*)[2]	1500	1.702	115	0.17
Common tern (*Sterna hirundo*)	1300	0.135	18	0.02
Arctic tern (*Sterna paradisea*)[2]	55	0.110	16	0.01
Little tern (*Sterna albifrons*)	150	0.075	12	0-0.01
Sandwich tern (*Sterna sandvicensis*)	900	0.250	29	0.02
Species depending on tidal flats				
Oystercatcher (*Haematopus ostralagus*)	120000	0.497	47	5.62
Grey plover (*Pluvialis squatarola*)	5600	0.223	27	0.15
Ringed plover (*Charadrius hiaticula*)	1100	0.070	11	0.01
Kentish plover (*Charadrius alexandrinus*)	150	0.047	9	0-0.01
Turnstone (*Arenaria interpres*)	2000	0.106	16	0.03
Curlew (*Numenius arquata*)	44000	0.921	74	3.26
Bar-tailed godwit (*Limosa lapponica*)	30000	0.248	29	0.85
Redshank (*Tringa totanus*)	10000	0.128	18	0.18
Spotted redshank (*Tringa erythropus*)	700	0.149	20	0.01
Greenshank (*Tringa nebularia*)	1100	0.180	23	0.03
Knot (*Calidris canutus*)	50000	0.128	18	0.88
Dunlin (*Calidris alpina*)	120000	0.057	10	1.16
Curlew sandpiper (*Calidris ferruginea*)	150	0.070	11	0-0.01
Avocet (*Recurvirostra avosetta*)	5800	0.340	36	0.21
Shelduck (*Tadorna tadorna*)	21000	1.237	91	1.95
Common gull (*Larus canus*)	34000	0.369	38	1.29
Black-headed gull (*Larus ridibundus*)	50000	0.241	28	1.40
Herring gull (*Larus argentatus*)	15000	0.991	78	1.00

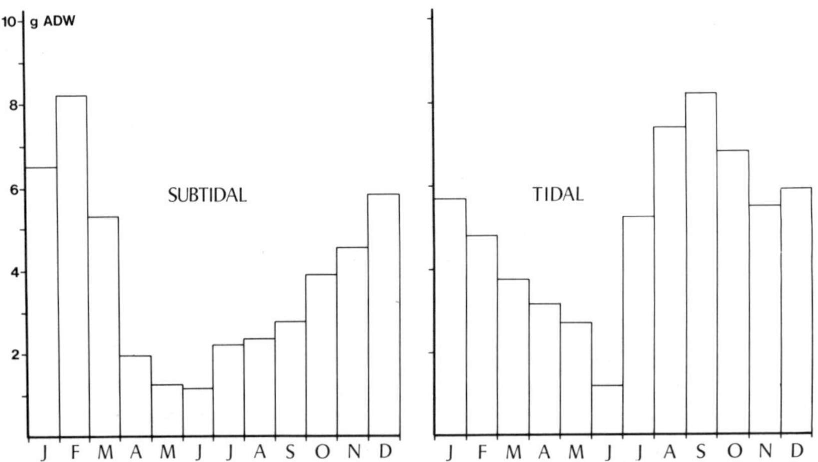

Fig. 199. Estimated average consumption per month by carnivorous birds on the tidal flats and in subtidal areas in the Dutch part of the Wadden Sea (data in g ash-free dry weight.10^8), based on the assumptions that food requirements do not change throughout the year and that Eiders forage entirely in the subtidal areas.

Table 68. The estimated consumption in g adw per year of carnivorous bird species in the Dutch Wadden Sea.

	Consumption in g adw. year^{-1}.10^8	% of total consumption	Consumption in g adw.m^{-2}.year^{-1}
Subtidal area			
Eider (*Somateria mollissima*)	3.03	28%	2.16
Diving ducks, grebes	6.02	6%	0.44
Gulls	11.3	11%	0.81
Terns	0.2	0.2%	0.01
Total	48.0	45%	3.43
Tidal flats			
Waders	45.6	42%	3.80
Shelduck (*Tadorna tadorna*)	7.1	7%	0.59
Gulls	6.7	6%	0.56
Total	59.4	55%	4.95
Total intertidal and subtidal	107.4		4.13

sume less than 5%. Consumption of all species in the Dutch Wadden Sea appears to amount to $10.7 \cdot 10^9$ g adw.year^{-1}. It is assumed that waders and Shelduck, and a part of the Herring-, Common- and Black-headed gulls forage on the tidal flats. It is also assumed that diving ducks, terns, Black-backed gulls and a part of the Herring-, Common- and Black-headed gulls forage in the subtidal part of the area. For the tidal flats and subtidal areas consumption can be estimated at 4.9 and 3.4 g adw.m^{-2}.year^{-1}, respectively, based on 120,000 ha of tidal flats and 140,000 ha of subtidal area (Table 68; fig. 199).

Probably however consumption on the tidal flats of the Dutch Wadden Sea is higher. The figure of 3.4 g adw.m^{-2}.year^{-1} for the subtidal part is based on the assumption that the Eider, being the most important consumer, does not forage on the tidal flats, which is not true. To what extent food is taken from the flats however is unknown. The figure of 4.9 g adw.m^{-2}.year^{-1} therefore must be regarded as a minimum, and consequently the figure of 3.4 g adw.m^{-2}.year^{-1} as a overestimate. The figure for the overall consumption by birds in the Wadden Sea (4.1 g adw.m^{-2}.year^{-1}) is of the same magnitude as earlier estimates by Hulscher and Swennen. Hulscher (1975) arrived at 56.10^9 kcal.year^{-1} for the entire Dutch Wadden Sea, yielding an average figure of about 4.3 g adw.m^{-2}.year^{-1}. Swennen (1976) found 3.7 g adw.m^{-2}.year^{-1}.

6.3.2 Consumption per month

As has been shown in chapter 5 the maximum number of birds in the Dutch part of the Wadden Sea occurs in August and September. From October till February a rather constant number is present. Fig. 200 shows that the amount of consumption by birds differs considerably between tidal and subtidal areas. Overall consumption reaches a peak level in September, decreases somewhat in the next two months but in January and February reaches levels of the same magnitude as those found in September, about $3.7 \cdot 10^7$ g adw per day. The most important reason for the increase in winter is the increasing number of Eiders arriving as autumn proceeds, reaching a peak in February. As it is a heavy bird occurring in large numbers Eiders have a relatively large impact on the total consumption.

Large and numerous waders like Oystercatcher and Curlew, important consumers in winter, leave the Dutch Wadden Sea already by the end of winter and in early spring. As a result of this, consumption decreases considerably in March, April and May. An influx of large numbers of Dunlins and Bar-tailed godwits does not counteract this trend drastically. Consumption on the tidal flats therefore appears to peak from August to October and to decrease considerably in the months following January. Consumption in the subtidal areas increases gradually from July till February, to drop in the months following. Once again it should be noted that this picture might be obscured by Eiders foraging on the flats during high tide.

For the 12 most important consumers assumptions about prey species taken per month, can be made on basis of literature data. These assumptions are listed in Table 69. This is a more or less discutable approach since prey species may vary considerably from place to place, from month to month and from individual to individual. These changes cannot be indicated sufficiently. Nevertheless the assumptions give a rough indication of the food spectrum of the birds present in the Dutch Wadden Sea throughout the year (fig. 200). The yearly consumption of all species comprises:

Molluscs	$7.31 \cdot 10^9$ g adw	= 69%
Crustaceans	$1.18 \cdot 10^9$ g adw	= 11%
Fishes	$1.16 \cdot 10^9$ g adw	= 11%
Polychaetes	$1.03 \cdot 10^9$ g adw	= 10%

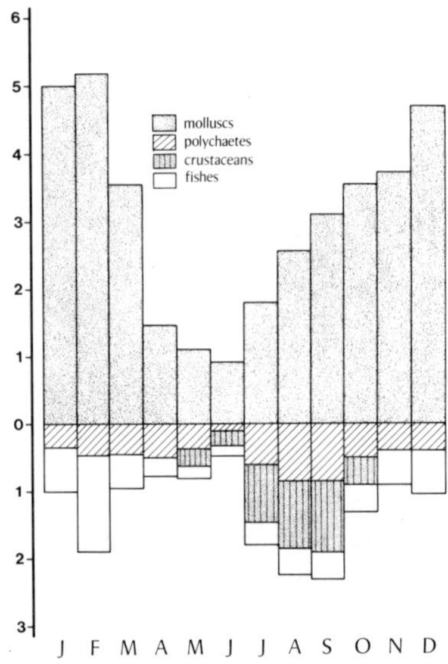

Fig. 200. Estimated average consumption of molluscs, polychaetes, crustaceans and fishes in g ash-free dry weight ($\times 10^7$) per day by carnivorous birds in the Dutch part of the Wadden Sea.

Molluscs appear to be the favourite prey dominating especially in
winter, late autumn and early spring. In summer and early autumn
crustaceans and polychaetes are important as well.

6.3.3 Bill lengths and weights

Mean bill lengths of birds foraging on the tidal flats increase from
August-February (fig. 201) and decrease from March-May. The data for
June and the first part of July have been omitted because of the very
low numbers of birds in the Wadden Sea area in that period. The mean
length for the winter period (December-February) is 65.3 mm, for spring
(March-May) 54.1 mm and for autumn (September-November) 59.7 mm.

Mean weights per season of the birds foraging on the tidal flats
of the Wadden Sea show the same trend (fig. 202). In winter (December-
February) mean weight amounts to 389 g, in spring (March-May) to 246
g and in autumn (September-November) to 328 g. In the coldest months
of the year the average wader and gull in the Dutch Wadden Sea is rather heavy. It also carries a somewhat longer bill as compared to the
other seasons, especially spring.

Species foraging in the subtidal areas have not been included in
bill length calculations. The number of macrobenthos eating species
is limited, Eider, Common scoter, Scaup and Goldeneye being the most
numerous. Clear differences in bill lengths per month do not occur
and might not be expected because of the dominance of the Eider.

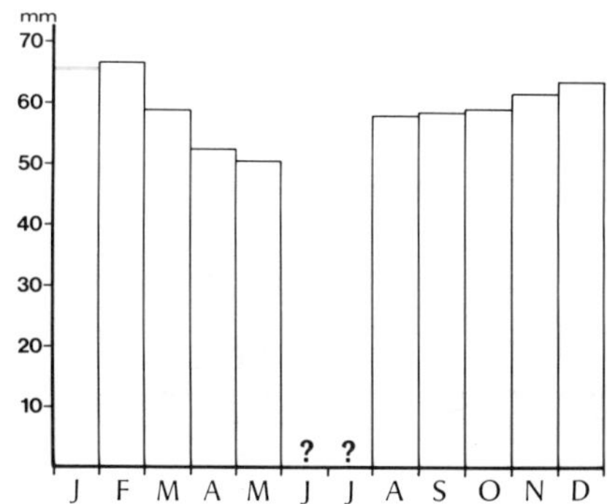

Fig. 201. Mean length of bills
in mm of carnivorous birds
foraging on the tidal flats of
the Dutch part of the Wadden
Sea. Data for June and July
were not taken into account
(for sources see text).

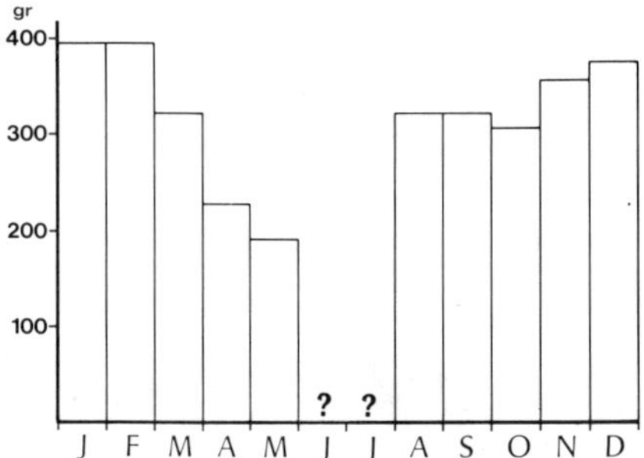

Fig. 202. Mean weight in grams
of carnivorous birds present
on the tidal flats in the Dutch
part of the Wadden Sea. Data
for June and July were not
taken into account (for sources
see text).

6.3.4 Production of macrobenthic invertebrates

The average amount of biomass present on the tidal flats of the Dutch Wadden Sea has been estimated in 1971/1972 at 26.6 g adw.m^{-2} (Beukema, 1976). Biomass is dominated by 10 species together comprising 95%. The total biomass consists of molluscs (66%), polychaetes (31%) and crustaceans (3%). In the subtidal part the amount of biomass amounts to about 10-15 g adw.m^{-2} (Beukema, 1976). Samples collected in 1977 showed that biomass decreased somewhat as compared to 1971/1972, especially because of a decrease of large Clams *(Mya arenaria)*. Some other species however took benefit of a series of mild winters and increased *(Lanice, Nephtys, Angulus)* (Beukema et al., 1978). The amount of biomass on the tidal flats also shows seasonal fluctuations (fig. 203). However, average biomass is not a very exact indication of what is available to birds and other predators. What should be known

Table 69. Assumptions about prey items consumed by birds in the Dutch Wadden Sea. M = molluscs, P = polychaetes, C = crustaceans, F = fishes.

	Spring	Summer	Autumn	Winter
Eider *(Somateria mollissima)*	M	M	M	M
Oystercatcher *(Haematopus ostralegus)*	M	M	M	M
Curlew *(Numenius arquata)*	MP	CP	CP	MP
Herring gull *(Larus argentatus)*	MF	CMF	CMF	MF
Shelduck *(Tadorna tadorna)*	M	M	M	M
Dunlin *(Calidris alpina)*	PM	PM	PM	PM
Knot *(Calidris canutus)*	M	M	M	M
Bar-tailed godwit *(Limosa lapponica)*	PM	PM	PM	PM
Common gull *(Larus canus)*	MPF	MFPC	MFPC	MPFC
Black-headed gull *(Larus ridibundus)*	MPF	MFPC	MFPC	MPFC
Common scoter *(Melanitta nigra)*	M	M	M	M
Red-breasted merganser *(Mergus serrator)*	F	F	F	F
Remaining species	MPF	MFPC	MFPC	MPF

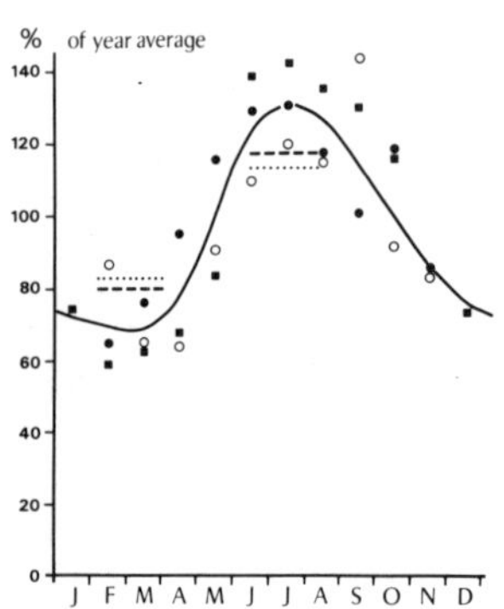

Fig. 203. Relative amounts of biomass at three sampling stations, expressed as mean percentages of year-average of biomass for each month at each station. The solid line was fitted by eye to the points referring to shallow-living animals. Deep-living animals have been indicated separately: Arenicola (dotted line) and Mya (dashed line), for 2 seasons only instead of 12 months. After Beukema (1974).

is the rate of production of biomass. For example, species with a high reproduction rate and a short life span are generally small and rather unimportant with respect to their biomass at any moment, but because of their high rate of production of biomass they may be relatively important food sources to birds. Unfortunately until now only very few data have been published on the production of biomass in the Dutch Wadden Sea. Beukema (1974, 1976, 1980) assumes that the annual production by benthic animals in the Dutch Wadden Sea is about 35 g adw.m^{-2}. His estimate for the tidal flats is about 45 g adw.m^{-2}.year^{-1} and for the subtidal areas about 25 g adw.m^{-2}.year^{-1}. However, part of this production is not available to birds, either because the animals are too small (meiofauna) or because they are too big or live too deep (adult *Arenicola, Mya)*, or because they die in large numbers at the same time due to other factors than predation (frost). Also the production of liberated gametes should be subtracted from the figures mentioned above. The result is that probably not more than 15-20 g adw.m^{-2}.year^{-1} is available to the birds and the other large predators. The figure for the tidal flats will be somewhat higher, that for the subtidal areas lower.

6.4 DISCUSSION

6.4.1 Relation between production and consumption of biomass

In chapter 6.3.1 the total consumption by birds has been estimated at 4.9 g adw.m^{-2}.year^{-1} for the tidal flats, 3.4 g for the subtidal parts and 4.1 g for the whole Dutch Wadden Sea.

Zijlstra (1979) estimates consumption by fishes in the German Wadden Sea at 2.8-4.6 g adw.m^{-2}.year^{-1} on the flats and 2.75-5.5 g adw. m^{-2}.year^{-1} in the subtidal part. Kuipers (1977) found that Plaice *(Pleuronectes platessa)* consumed about 5 g adw.m^{-2}.year^{-1} on the flats of the Balgzand. This species is consuming in all months of the year but consumption has a maximum in summer (De Vlas, pers. comm.). The composition of the food of this species, the most important among the fishes, shows large differences from that taken by birds. Kuipers (1977) found 13% molluscs, 15% crustaceans and 72% polychaetes. It may be assumed that fishes take about 5 g adw.m^{-2}. year^{-1} of the benthic fauna in the Dutch Wadden Sea (Beukema, 1980).

Beukema (1981) also gives estimates of the predation by invertebrates on the benthic fauna and on the effects of fisheries. The total annual predation pressure on the benthic fauna is estimated by him at 4 (birds) + 5 (fishes) + 3 (invertebrate predators) + 1 à 2 (fisheries) = 13 à 14 g adw.m^{-2}.year^{-1}.

In chapter 6.3.4 the amount available to birds and other large predators was estimated at 15-20 g adw.m^{-2}.year^{-1}. Hence it appears that the annual food requirements of large predators approach the annual food supply. Consequently food may be in short supply, especially in periods when the benthic biomass is at its lowest level (compare also Beukema, 1981), since bird consumption (fig. 199) and biomass present (fig. 203; Beukema, 1976) show a considerable discrepancy.

These calculations make clear that food might be limiting the numbers of birds in the Wadden Sea, but they do not prove this. Such proof can only be obtained by studies of the population dynamics of the birds themselves in relation to food supply.

6.4.2. Availability of prey organisms

Birds foraging on the flats depend on the periods of low tide when the flats emerge. In the calculations made in chapter 6.3.2 the size of these flats was estimated at 120,000 ha. A part of this area however has only a very short emersion time and therefore in part of the area foraging is only possible in a small period. Great differences appear to exist between the Western and the Eastern part of the Dutch Wadden Sea area. From data presented by Boere (1973) it can be computed that in the Western part two hours after low tide only 47% of the flats is still dry, after 4 hours only 4%. In the Eastern part these figures are 92% and 41%, respectively. The availability of preys thus will be higher in this Eastern part, if they would be distributed evenly over the tidal flats. However, emersion time of the flats not only determines how long birds can forage but is also a key factor in the feeding possibilities of the benthic fauna. Wolff et al. (1976) found in the Grevelingen estuary that macrobenthic biomass increased as submersion time increased. Zwarts (1978) had comparable results in the sedimentation fields along the mainland coast of Friesland and Groningen. Wolff et al. (1976) conclude that above half-tide level benthic biomass is the prime factor governing bird predation, but that at lower levels the emersion time of the flats (= availability of preys for birds) becomes important.

When after a high tide the tidal flats emerge many waders and gulls follow the edge of the tide when feeding. Vader (1964) showed that some time after emersion the infauna species show vertical downward movements into the sediment. The reverse occurs when the tide comes in. From the behaviour of the birds can be concluded that for many birds this behaviour of a prey species is a very important factor in determining its availability.

In his presentation of biomass data Beukema (1974) already made a seperation between shallow-living *(Cerastoderma, Macoma, Pectinaria)* and deep-living species *(Mya, Arenicola, Lanice)*. Specimens of this last group may be buried so deep (20 cm or more) that they are always unavailable to birds. Many invertebrate species appear to bury deeper into the sediment in winter and therefore become unavailable to birds only during a part of the year. Reading & McGrorty (1978) for example showed that in June about 90% of *Macoma balthica* was within reach of Knots while in December the amount was only 5%.

Macoma is not the only species that is living deeper in the sediment in winter. This phenomenon is also well known for the polychaetes, *Arenicola* and *Nereis* (Muus, 1967). For several molluscs, however, siphon length limits the depth to which the animals can bury into the sediment.

Not only the vertical movements of some of the prey species have a negative effect on the amount of food available to birds. In late autumn some of the prey species leave the tidal flats of Wadden Sea and migrate to the deeper gullies and the North Sea *(Crangon, Carcinus)*.

It is unclear what impact migration of prey species, a decrease in the amount of food present in the bottom (fig. 203) and downward movements of macrobenthic invertebrates in winter have on both numbers and species composition of birds in the Wadden Sea. Fact is that several species do not or hardly winter in the area. Fig. 201 and 202 show that especially shorter billed and relatively small species become less dominant in winter as compared to autumn and spring. It could very well be possible that a decrease in the amount of shallow living prey animals cause poor feeding conditions, especially for these birds. Goss-Custard et al. (1977a) showed that waders in the Wash may have problems in winter too. Especially small birds (Dunlins, Knots, Redshanks) used a much higher percentage of the available time for foraging. Small as well as large birds meet the problem that the activity of prey organisms is decreased in winter. Prey therefore is more difficult to detect. Finally prey organisms of a certain size and species contain less energy in winter than in summer (Beukema & De Bruin, 1977).

The preceding chapters have shown that birds meet many problems in obtaining the amount of food they need. It is not astonishing then that in all estuaries birds appear to have a highly developed niche partitioning (chapter 4; Davidson, 1971; Baker & Baker, 1973; Eadie et al., 1979; Goss-Custard et al., 1977b). How inter- and intraspecific interactions between birds work and how the structure of the complex food web between birds and prey animals is, however has not been studied sufficiently, neither in the Wadden Sea nor elsewhere. Many questions therefore still cannot be answered, like:
- which factor(s) actually limit the numbers of a certain bird species in the Wadden Sea;
- what is the reason that especially so many small waders leave the Wadden Sea to winter in the British Isles, France, Southern Europe or Africa;
- what is the reason for the large differences in bird numbers (see chapters 3 and 5) between the Danish, German and Dutch part of the Wadden Sea?

A profound study of the carrying capacity and the foodweb relations, the availability of preys and the energy use of the birds throughout the year is necessary for answering such questions.

References

Ankney, C.D., 1979. Quantifying interspecific variation in foraging behaviour of syntopic Anas (Anatidae). Can. J. Zool. 57: p. 412-415.

Baker, M.C. & A.E.M. Baker, 1973. Niche relationships among six species of shorebirds on their wintering and breeding ranges. Ecol. Monogr. 43: p. 193-212.

Bauer, K.M. & U.N. Glutz von Blotzheim, 1968. Handbuch der Vögel Mitteleuropas, Vol. 2. Akademische Verlagsgesellschaft, Frankfurt/Main: 535 pp.

Beukema, J.J., 1974. Seasonal changes in the biomass of the macrobenthos of a tidal flat area in the Dutch Wadden Sea. Neth. J. Sea Research 8: p. 94-107.

Beukema, J.J., 1976. Biomass and species richness of the macro-benthic animals living on the tidal flats of the Dutch Wadden Sea. Neth. J. Sea Research 10: p. 223-235.

Beukema, J.J., 1981. The role of the larger invertebrates in the Wadden Sea ecosystem. In: N. Dankers, H. Kühl & W.J. Wolff (eds.). Invertebrates of the Wadden Sea. Balkema, Rotterdam.

Beukema, J.J. & W. de Bruin, 1977. Seasonal changes in dry weight and chemical composition of the soft parts of the tellinid bivalve Macoma balthica in the Dutch Wadden Sea. Neth. J. Sea Res. 11: p. 42-55.

Beukema, J.J., W. de Bruin & J.J.M. Jansen, 1978. Biomass and species richness of the macrobenthic animals living on the tidal flats of the Dutch Wadden Sea: long term changes during a period with mild winters. Neth. J. Sea Res. 12: p. 58-77.

Boere, G.C., 1973. Aantallen, aantalsfluctuaties en verspreiding van 13 soorten steltlopers in het Nederlandse Waddengebied. Een eerste kwantitatief onderzoek. Unpubl. report Waddenzeecommissie: 89 pp.

Cummins, K.W. & J.C. Wuycheck, 1971. Caloric equivalents for investigations in ecological energetics. Mitt. int. Verein. theor. angew. Limnol. 18: p. 1-158.

Davidson, P.E., 1971. Some foods taken by waders in Morecambe Bay, Lancashire. Bird Study 18: p. 177-186.

Eadie, J.M., T.D. Nudds & C.D. Ankney, 1979. Quantifying interspecific variation in foraging behaviour of syntopic Anas (Anatidae). Can. J. Zool. 57: p. 412-415.

Goss-Custard, J.D., R.A. Jenyon, R.E. Jones, P.E. Newbery & R.L.B. Williams, 1977a. The ecology of the Wash. II. Seasonal variation in the feeding conditions of wading birds (Charadrii). J. Appl. Ecol. 14: p. 701-719.

Goss-Custard, J.D., D.G. Kay & R.M. Blindell, 1977b. The density of migratory and overwintering Redshanks, Tringa totanus (L.) and Curlew, Numenius arquata (L.) in relation to the density of their prey in south-east England. Est. Coastal Mar. Sc. 5: p. 497-510.

Glutz von Blotzheim, U.N., K.M. Bauer & E. Bezzel, 1975. Handbuch der Vögel Mitteleuropas, Vol. 6. Akademische Verlagsgesellschaft, Wiesbaden: 840 pp.

Glutz von Blotzheim, U.N., K.M. Bauer & E. Bezzel, 1977. Handbuch der Vögel Mitteleuropas, Vol. 7. Akademische Verlagsgesellschaft, Wiesbaden: 893 pp.

Haftorn, S., 1971. Norges fugler. Universitetsforlaget, Oslo; 862 pp.

Hulscher, J.B., 1975. Het wad, een overvloedig of schaars gedekte tafel voor vogels? In: C. Swennen, P.A.W.J. de Wilde & J. Haeck (eds.)

Symposium Waddenonderzoek. Med. Werkgroep Waddengebied 1: p. 57-82.

Kuipers, B.R., 1977. On the ecology of juvenile plaice on a tidal flat in the Wadden Sea. Neth. J. Sea Research 11: p. 56-91.

Lasiewski, R.C. & W.R. Dawson, 1967. A re-examination of the relation between standard metabolic rate and body weight in birds. Condor 69: p. 13-23.

Muus, B.J., 1967. The fauna of Danish estuaries and lagoons, Distribution and ecology of dominating species in the shallow reaches of the mesohaline zone. Medd. Danmarks Fisk og Havunders. NS 5(1): p. 1-34.

Reading, C.J. & S. McGrorty, 1978. Seasonal variation in the burying depth of Macoma balthica (L.) and its accessibility to wading birds. Est. Coastal Mar. Sc. 6: p. 135-144.

Smit, C.J., 1977. On the occurrence of 32 birdspecies in the Danish, German and Dutch Wadden Sea. Unpubl. report Intern. Wadden Sea Working Group, 3 parts: 474 pp.

Swennen, C., 1975. Aspecten van voedselproductie in Waddenzee en aangrenzende zeegebieden in relatie met de vogelrijkdom. Vogeljaar 21: p. 141-156.

Swennen, C., 1976. Wadden Seas are rare, hospitable and productive. In: M. Smart (ed.). Proceedings Int. Conf. on the Conservation of Wetlands and Waterfowl, Heiligenhafen, 1974. IWRB, Slimbridge: p. 184-198.

Vader, W.J.M., 1964. A preliminary investigation into the reaction of the infauna of tidal flats to tidal fluctuations in water level. Neth. J. Sea Research 2: p. 189-222.

Winberg, G.G., 1971. Symbols, units and conversion factors in studies of fresh water productivity. I.B.P. London, 24 pp.

Wolff, W.J., A.M.M. v. Haperen, A.J.J. Sandee, H.J.M. Baptist & H.L.F. Saeijs, 1976. The trophic role of birds in the Grevelingen estuary, the Netherlands, as compared to their role in the saline lake Grevelingen. Proc. 10th Europ. Symp. Mar. Biol., Ostend, Belgium, 1975; Vol. 2: p. 673-689.

Zijlstra, J.J., 1979. Quantitative aspects of the role of fishes in Wadden Sea food chains. In: N. Dankers, W.J. Wolff, and J.J. Zijlstra (eds.). Fishes and fisheries of the Wadden Sea. Final report of the section "Fishes and fisheries" of the Wadden Sea Working Group; p. 124-132.

Zwarts, L., 1978. De biologische waardering van de landaanwinningswerken; een eerste benadering. Jaarverslag Rijksdienst IJsselmeerpolders 1976: p. 96-101.

7 THREATS TO THE BIRDS OF THE WADDEN SEA
F. Goethe, C.J. Smit & W.J. Wolff

7.1 GENERAL CONSIDERATIONS

Many factors influence the birds of the Wadden Sea. Natural as well as anthropogenic factors both may have adverse as well as positive influences. Since it may be assumed that the regular Wadden Sea birds are adapted to their natural environment, it follows that they must be able to withstand adversely acting natural factors, e.g. high floods in the breeding season or food shortage in severe winters. Indeed birds possess many morphological, physiological, ecological and behavioural mechanisms to compensate for the negative effects of natural environmental factors.

It seems reasonable to assume that birds may use the same mechanisms to compensate for the negative effects of anthropogenic environmental factors, such as aircrafts causing disturbance. In some cases this compensation will be successful, in other ones the negative influence results in a negative effect on the birds. For example the disturbance caused by a single, temporary visitor to birds feeding on the tidal flats probably may be compensated for by a slightly longer feeding period. Severe disturbance by longer periods of military training in a moulting area of Shelduck, however, possibly cannot be compensated for and the birds then will loose weight and eventually may die or show a bad reproduction next year. In this case we can speak of a threat to the Shelduck, but it will be clear from the examples given that not all human influences are threats to the birds.

We propose as a criterion for which factors are a threat to the birds of the Wadden Sea, the presence of a potential negative effect on the size of the bird populations breeding in or visiting otherwise the Wadden Sea area. Any factor potentially causing a decline of these populations will be considered a threat. Of course cumulative or even synergistic effects have to be considered as well. By this definition of a threat we are able to differentiate between human influences, to investigate them quantitatively and to apply the results of this research in the management of the Wadden Sea area. We strongly recommend an intensification of this kind of research.

7.2 THREATS TO THE ENTIRE WADDEN SEA ECOSYSTEM

Previous chapters have shown that the Wadden Sea is of vital importance for millions of birds. Although large areas have already been
designated as nature reserves, especially in Germany, this does not
prevent that the birds of the Wadden Sea are exposed to numerous
threats. In this chapter these threats are enumerated and discussed.
Some threats only concern the bird populations, others the entire
Wadden Sea ecosystem.

Threats potentially influencing the entire nature and landscape
of the Wadden Sea will be discussed in more detail in Report 11 of
the Wadden Sea Working Group (Dankelman et al., 1981). The major ones
will be considered here shortly, especially with regard to birdlife.

Reclamation for agriculture and other embankments change parts of
the Wadden Sea into completely different types of landscape. Usually
this means the end of the estuarine ecosystem and consequently the
end of the conditions of life essential for all estuarine birds. Plans
exist for reclamation of many thousands of hectares of tidal land-
scapes in all countries bordering the Wadden Sea. A single example
may demonstrate the impact on bird life. The planned embankment of
the Nordstrander Bucht (3430 ha) would mean the loss of 25% of the
feeding areas of the Brent goose in Schleswig-Holstein (Prokosch,
1977).

Embankments for large deep-sea harbours will have similar effects.
Sikkema (1976) observed that the construction of the Ems harbour, the
Netherlands, caused the disappearance of 20,000-30,000 birds.

Pollution is another problem threatening the entire Wadden Sea
ecosystem. Its significance has already been discussed in Report 8
of the Wadden Sea Working Group (Essink & Wolff, 1978). One group of
pollutants justifies a closer consideration, viz. those compounds
which show concentration and biomagnification in the food chains. Since
many Wadden Sea birds are final links of estuarine food chains it is
evident that many species run a high risk to be exposed to hazardous
concentrations of these compounds. The dramatic decline of Sandwich
tern in the Dutch Wadden Sea around 1965 due to the insecticides telodrin and dieldrin is a classic example of this danger (Koeman et
al., 1967; Koeman, 1971). Although the telodrin-dieldrin problem has
been solved, chlorinated hydrocarbons remain suspect. Reijnders (1980)
makes probable that polychlorinated biphenyls (PCB's) have a negative
effect on the reproduction of Harbour seals *(Phoca vitulina)* and shows
also that Wadden Sea birds contain amounts of PCB's only little lower
than those in seals. Some heavy metals also deserve attention, e.g.
mercury and cadmium.

Introduction of exotic species may pose a threat to the Wadden Sea
ecosystem because it may cause important shifts in the species composition of the biocenosis. One of the most important introductions
until now is that of the cord-grass *(Spartina anglica)* which changed
the vegetation of the salt marshes considerably. Effects on bird life
have never been analysed properly, however.

Fisheries have also been mentioned as a threat to the birds of the

Wadden Sea. Several effects occur. The fisheries for shrimp *(Crangon crangon)* and flatfish in the tidal channels and off the barrier islands take potential food for birds, but on the other hand the discards form actual food for mainly gulls, thus increasing the population size of e.g. Herring gulls *(Larus argentatus)* (Spaans, 1971). The fisheries for small mussels *(Mytilus edulis)*, cockles *(Cerastoderma edule)* and lugworms *(Arenicola marina)* on the tidal flats of the Wadden Sea also take potential bird food, but the effects on birds have not yet been sufficiently investigated. The fishery for mullets *(Mugil spec.)* in the Dutch Wadden Sea with large set nets on the tidal flats clearly has negative effects on birds. Relatively large numbers of Eiders *(Somateria mollissima)* of the Dutch breeding population have been found drowned in such nets (Swennen, pers. comm.).

7.3 THREATS MAINLY TO BIRDS

7.3.1 Breeding birds

The breeding birds of the Wadden Sea area are adapted to a highly dynamic environment. High floods sometimes destroy all eggs and young in a breeding colony but on the other hand such floods maintain the right condition of the breeding places. The erosive action of wind and water also may destroy breeding places, but elsewhere these natural forces will create new possibilities. Over long periods such natural changes of the habitat may certainly not be held responsible for the decline of any bird population.

Locally the management of breeding places may have effects. Many bird species require on their breeding sites types of vegetation which have to be maintained by grazing, either by cattle or sheep. Overgrazing as well as too little grazing may cause vegetation changes detrimental to breeding birds. Wintering geese also prefer grazed salt marshes as feeding areas.

A requirement for breeding colonies of coastal birds is the absence of ground predators. Especially Rats *(Rattus norvegicus)*, but also Stoat *(Mustela putorius)*, Fox *(Vulpes vulpes)* and Hedgehog *(Erinaceus europaeus)* may destroy large numbers of nests and even catch the adult breeding birds. As described in Report 10 of the Wadden Sea Working Group (Smit & Wolff, 1981) these predatory species have been introduced to several Wadden Sea islands, either through the construction of dams connecting islands to the mainland, or mostly, accidentally, by boat transport. For example, the temporary disappearance of the breeding colony of terns on Scharhörn has been attributed to the introduction of Rats (Temme, 1967).

Gulls, especially Herring gulls, often take young specimens of other bird species. For this reason Herring gulls have been killed and their eggs have been destroyed for many years, in Germany as well as in the Netherlands. In the latter country this practice has been stopped in 1967, in the former it is still in use. Nevertheless there are no signs that the populations of other bird species breeding in the Netherlands are declining since the Herring gulls were protected. On the contrary several species, e.g. Shelduck, Eider, Sandwich tern, Common tern, and Little tern (Swennen, 1976; chapters 3.5, 3.31, 3.32, 3.33) showed an increase of their breeding population in the period since 1967. This led to the conclusion that Herring gulls, although they predate on juveniles of other species, normally do not influence the population size of other birds (Spaans, 1971; Rooth, pers. comm.). In some small reserves however (i.e. Jordsand, Memmert, Mellum, Lütje Hörn) Herring gulls may take over breeding places from terns due to competition for space in the colonies (Goethe, 1961, 1962, 1976; Jepsen, 1975).

Finally, disturbance has to be mentioned as a threat to breeding birds. Most breeding colonies of estuarine birds in the Wadden Sea region nowadays are situated in nature reserves. If in these reserves wardens are appointed, they usually are able to keep out unwanted visitors. Without wardens and in breeding colonies outside the nature reserves tourists and other people may disturb breeding birds. The decline of Ringed and Kentish plover (chapters 3.15 and 3.16) probably may be attributed to this cause.

Another cause of disturbance are low-flying airplanes. Small (sport) planes, helicopters and jet fighters all may cause panic-like reactions in colonies of breeding birds. Bell (1972) even attributed a mass hatching-failure in a Sooty tern colony in Florida, USA, to physical damage to eggs by sonic booms caused by low-flying jet fighters. Similar observations have not yet been made in the Wadden Sea area.

7.3.2 Non-breeding birds

Hunting is one of the most drastic human influences on the birds of the Wadden Sea. Table 70 shows the open seasons for the 32 most important species of Wadden Sea birds (compare chapter 3).

On all state property in the Dutch part of the Wadden Sea, constituting over 95% of the total area, shooting is not allowed. Only on some privately owned salt marshes along the Dollard embayment and

Table 70. Open hunting seasons in 1979 of the 32 most important Wadden Sea species (compare chapter 3). Species not recorded are protected in all Wadden Sea countries. After Wiese (1979).

	Netherlands	German Federal Republic				Denmark
		Niedersachsen	Bremen	Hamburg	Schleswig-Holstein	
Brent goose (*Branta bernicla*)	–	–	1/11-15/1	1/11-15/1	1/11-15/1	–
Mallard (*Anas platyrhynchos*)	24/7-31/1	1/9-15/1	1/9-15/1	1/9-15/1	1/9-15/1	16/8-31/12
Wigeon (*Anas penelope*)	1/9-31/1	1/10-15/1	1/10-15/1	1/10-15/1	1/10-15/1	16/8-31/12
Teal (*Anas crecca*)	18/8-31/1	1/10-15/1	1/10-15/1	1/10-15/1	1/10-15/1	16/8-31/12
Pintail (*Anas acuta*)	1/9-31/1	1/10-15/1	1/10-15/1	1/10-15/1	1/10-15/1	16/8-31/12
Eider (*Somateria mollissima*)	–	–	–	–	–	1/10-29/2
Red-breasted merganser (*Mergus serrator*)	–	1/10-15/1	1/10-15/1	1/10-15/1	1/10-15/1	1/10-29/2
Goosander (*Mergus merganser*)	–	1/10-15/1	1/10-15/1	1/10-15/1	1/10-15/1	1/10-29/2
Oystercatcher (*Haematopus ostralegus*)	–	–	–	–	–	1/8-31/12
Grey plover (*Pluvialis squatarola*)	–	–	–	–	–	1/8-31/12
Bar-tailed godwit (*Limosa lapponica*)	–	–	–	–	–	1/8-31/12
Curlew (*Numenius arquata*)	–	–	–	–	–	1/8-31/12
Redshank (*Tringa totanus*)	–	–	–	–	–	1/8-31/12
Black-headed gull (*Larus ridibundus*)	–	16/7-1/5	16/7-1/5	16/7-1/5	16/7-1/5	16/8-29/2
Common gull (*Larus canus*)	–	16/8-1/5	16/8-1/5	16/8-1/5	16/8-1/5	16/8-29/2
Herring gull (*Larus argentatus*)	–	16/8-1/5	16/8-1/5	16/8-1/5	16/8-1/5	16/8-30/4

along the coast of Groningen hunting still occurs (Projectbureau Waddenzee, 1979).

In the German part no shooting occurs in several large nature reserves, but in the other areas hunting is fairly intensive, especially on the salt marshes. This is not only detrimental for the species hunted, but also for all other species, which are severely disturbed at their high-tide roosts.

In the Danish Wadden Sea hunting is very intensive. Fig. 204 shows the maximum number of hunters counted per census area in 1978, fig. 205 the number of shooting shelters in this area.

The Brent goose population in NW Europe only could increase after in Denmark the season for this species was closed in 1972 (compare chapter 3.4), so it may be assumed that large-scale hunting influences the population size of this Wadden Sea bird. For many other species this is probably not the case, but even for these species and for several completely protected species hunting forms a major cause of disturbance, which may influence the condition and thereby the population size of bird species.

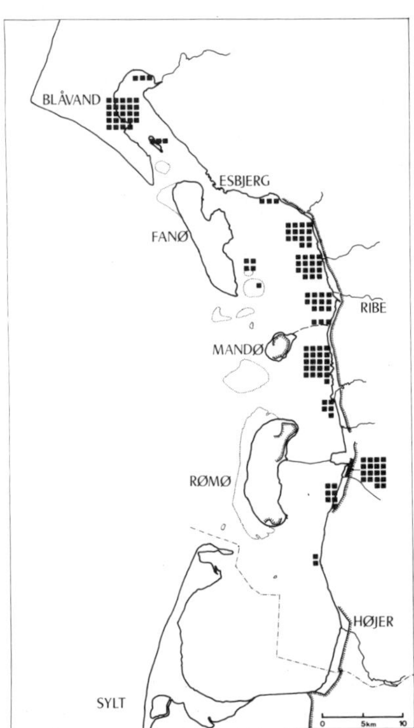

Fig. 204. The distribution of sport hunters in the Danish part of the Wadden Sea during high tide. The highest number recorded per census area during 1978 is plotted. After Meltofte (1980).

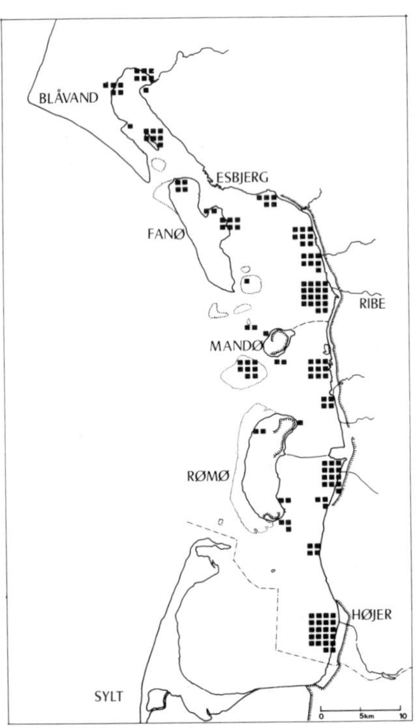

Fig. 205. The distribution of shooting shelters along the coast of the Danish Wadden Sea. After Meltofte (1980).

Duck-decoys are only allowed to operate in the Netherlands. Presently 11 duck-decoys in the Wadden Sea area still have a commercial licence and an estimated 30,000 ducks (nearly exclusively *Anas* species) are caught annually (Projectbureau Waddenzee, 1979).

Disturbance of non-breeding Wadden Sea birds is caused by people visiting the tidal flats and high-tide roosts. A single person may cause the disappearance of tens of thousands of birds from a high-tide roost. People visiting the tidal flats cause the disappearance of nearly all birds in an area of several hectares around them. In all parts of the Wadden Sea the development of recreation facilities, especially along the mainland coast, brought about that disturbance increased considerably in summer. Major problems are the increasing number of small boats and wind-surfers in the area, with which people may land on the tidal flats and the increasing number of groups of people walking on the flats. In the Danish and Dutch part of the area these problems seem to be not as large yet as in the German part. The effects of disturbance on the size of the bird populations are still very difficult to quantify and further study of this problem is urgently required. The same applies to disturbance by low-flying airplanes and military operations. These have been observed to disturb birds, but often do not. It should be investigated under what conditions disturbance occurs and what are the ultimate effects.

A major threat to the birds of the Wadden Sea is oil pollution. For example, a major oil spill near the Elbe estuary in August could exterminate a very large part of the NW European Shelduck population. Further details are to be found in Report 8 of the Wadden Sea Working Group (Essink & Wolff, 1978).

References

Bell, W.B., 1972. Animal response to sonic booms. J. Acoust. Soc. America 51: p. 758-765.

Dankelman, I., M.F. Mörzer Bruyns, C.J. Smit & W.J. Wolff (eds.), 1981. Nature management and physical planning in the Wadden Sea area. In prep.

Essink, K. & W.J. Wolff (eds.), 1978. Pollution of the Wadden Sea area. Report 8 of the Wadden Sea Working Group. Balkema, Rotterdam: 61 pp.

Goethe, F., 1961. Das Naturschutzgebiet Mellum als Grossreservat für nordische Strand- und Wasservögel ausserhalb der Brut. Bericht No. 1 Deutsche Sektion Internat. Rat Vogelschutz: p. 32-45.

Goethe, F., 1962. Das Seevogelschutzgebiet Lütje Hörn. Festschrift Vogelschutzwarte Hessen, Rheinland-Pfalz, Saarland: p. 67-76.

Goethe, F., 1976. Zur Lenkung der Möwenbestände an der Nordseeküste. Forum Umwelt Hygiene 2: p. 28-31.

Jepsen, P.U., 1975. Vadehavet vildtreservat med øen Jordsand. Danske vildtundersøgelser 24: 80 pp.

Koeman, J.H., 1971. Het voorkomen en de toxicologische betekenis van enkele chloorkoolwaterstoffen aan de Nederlandse kust in de periode van 1965 tot 1970. Thesis, University of Utrecht: 136 pp.

Koeman, J.H., A.A.G. Oskamp, J. Veen, E. Brouwer, J. Rooth, P. Zwart, E. v.d. Broek & H. van Genderen, 1967. Insecticides as a factor in the mortality of the Sandwich tern (Sterna sandvicensis). A preliminary communication. Meded. Rijksfac. Landbouwwet. Gent 32: p. 841-854.

Meltofte, H., 1980. Fugle i Vadehavet. Vadefugletaellinger i Vadehavet 1974-1978. Fredningsstyrelsen, Miljøministeriet, København: 50 pp.

Projectbureau Waddenzee, 1979. De interprovinciale struktuurschets voor het Waddenzeegebied. I. Inventarisatierapport. Leeuwarden: 110 pp.

Prokosch, P., 1977. Plans for the diking-in of feeding habitat of Branta b. bernicla in Schleswig-Holstein, Federal Republic of Germany. Paper presented at the ICGWB/IWRB Technical meeting on Western palaearctic migratory bird management, Paris, 1977.

Reijnders, P.J.H., 1980. On the causes of the decrease in the Harbour seal (Phoca vitulina) population in the Dutch Wadden Sea. Thesis, Agricultural University of Wageningen: 119 pp.

Sikkema, C.P., 1976. Wadvogeltellingen bij de Eemsmond. Gevolgen van de aanleg en uitbreiding van de Eemshaven voor de vogels van de Waddenzee en daarbuiten. Unpublished report: 17 pp.

Smit, C.J. & W.J. Wolff, 1981. Terrestrial animals of the Wadden Sea area. In prep.

Spaans, A.L., 1971. On the feeding ecology of the Herring gull Larus argentatus Pont. in the northern part of the Netherlands. Ardea 59: p. 73-188.

Swennen, C., 1976. Populatie-structuur en voedsel van de Eidereend Somateria m. mollissima in de Nederlandse Waddenzee. Ardea 64: p. 311-371.

Temme, M., 1967. Vogelfreistätte Scharhörn. Jordsand Mitteilungen 3 (1-4): p. 5-165.

Wiese, M., 1979. DJV-Handbuch. Deutschen Jagdschutz-Verband. Hoffmann, Mainz: 478 pp.